T0289624

Advances in Food and Agriculture Engineering

Advances in Food and Agriculture Engineering

Edited by Brandon Corbyn

New York

Published by Syrawood Publishing House,
750 Third Avenue, 9th Floor,
New York, NY 10017, USA
www.syrawoodpublishinghouse.com

Advances in Food and Agriculture Engineering
Edited by Brandon Corbyn

© 2023 Syrawood Publishing House

International Standard Book Number: 978-1-64740-418-5 (Hardback)

This book contains information obtained from authentic and highly regarded sources. Copyright for all individual chapters remain with the respective authors as indicated. All chapters are published with permission under the Creative Commons Attribution License or equivalent. A wide variety of references are listed. Permission and sources are indicated; for detailed attributions, please refer to the permissions page and list of contributors. Reasonable efforts have been made to publish reliable data and information, but the authors, editors and publisher cannot assume any responsibility for the validity of all materials or the consequences of their use.

Trademark Notice: Registered trademark of products or corporate names are used only for explanation and identification without intent to infringe.

Cataloging-in-publication Data

Advances in food and agriculture engineering / edited by Brandon Corbyn.
p. cm.
Includes bibliographical references and index.
ISBN 978-1-64740-418-5
1. Food industry and trade. 2. Agricultural engineering. 3. Food science. I. Corbyn, Brandon.
TP370 .A38 2023
664--dc23

TABLE OF CONTENTS

PREFACE

Food engineering refers to a broad field of food technology that includes several disciplines like agriculture, chemistry, food science, engineering and microbiology. It involves the use of engineering principles for the storage, processing, and distribution and delivery of food materials. Agricultural engineering is the field of engineering that deals with the design, development and improvement of agricultural machinery and equipment. It combines agricultural science, and engineering expertise and knowledge to create technologies and processes that increase the production and productivity of agriculture by making better use of natural resources and conserving them for future use. The agrifood industry encompasses traditional and innovative food production technologies. These include vertical farming and precision farming. This book unravels the recent studies in the field of food and agriculture engineering. It consists of contributions made by international experts. As this field is emerging at a rapid pace, the contents of this book will help the readers understand the modern concepts and applications of the subject.

After months of intensive research and writing, this book is the end result of all who devoted their time and efforts in the initiation and progress of this book. It will surely be a source of reference in enhancing the required knowledge of the new developments in the area. During the course of developing this book, certain measures such as accuracy, authenticity and research focused analytical studies were given preference in order to produce a comprehensive book in the area of study.

This book would not have been possible without the efforts of the authors and the publisher. I extend my sincere thanks to them. Secondly, I express my gratitude to my family and well-wishers. And most importantly, I thank my students for constantly expressing their willingness and curiosity in enhancing their knowledge in the field, which encourages me to take up further research projects for the advancement of the area.

Editor

1

Impact of Resistant Maltodextrin Addition on the Physico-Chemical Properties in Pasteurised Orange Juice

Elías Arilla [1], Marta Igual [1,*], Javier Martínez-Monzó [1], Pilar Codoñer-Franch [2,3] and Purificación García-Segovia [1]

1 Food Investigation and Innovation Group, Food Technology Department, Universitat Politècnica de València, Camino de Vera s/n, 46022 Valencia, Spain; elarco@upv.es (E.A.); xmartine@tal.upv.es (J.M.-M.); pugarse@tal.upv.es (P.G.-S.)
2 Department of Pediatrics Obstetrics and Gynecology, University of Valencia, Avenida de Blasco Ibáñez, No. 15, 46010 Valencia, Spain; pilar.codoner@uv.es
3 Department of Pediatrics, University Hospital Dr. Peset, Foundation for the Promotion of Health and Biomedical Research in the Valencian Region (FISABIO), Avenida Gaspar Aguilar, No. 90, 46017 Valencia, Spain
* Correspondence: marigra@upvnet.upv.es

Abstract: Resistant maltodextrin (RMD) is a water-soluble fibre that can be fermented in the colon and exert prebiotic effects. Therefore, its addition to food and beverage products could be beneficial from both technological and nutritional viewpoints. However, to date, most studies have focused on the stability of the prebiotic fibre rather than its impact in the original food matrices. Therefore, this work aimed to evaluate the addition of RMD on the physico-chemical properties of pasteurised orange juice (with and without pulp). °Brix, pH, acidity, particle size distribution, density, turbidity, rheology, and colour were measured in orange juices with increasing RMD concentrations (2.5, 5, and 7.5%). Control samples without RMD were also prepared. RMD added soluble solids to the orange juice, affecting the °Brix, density, turbidity, and rheology. Slight colour differences were observed, and lower citric acid content was achieved because of orange juice replacement with RMD. Differences in particle size distribution were exclusively because of pulp content. Further studies are needed to elucidate if potential consumers will appreciate such physico-chemical changes in organoleptic terms.

Keywords: prebiotic; resistant maltodextrin; orange juice; physico-chemical properties

---

1. Introduction

Adding new food components to formulate novel nutritious and safe food products has become a method to improve the quality of human diets. One major food component that has tremendous interest from scientists, companies, and consumers is prebiotics fibres. Prebiotics are defined as "selectively fermented ingredients that allow specific changes, both in the composition and/or activity in the gastrointestinal microflora that confers benefits upon host wellbeing and health" [1]. The most studied and accepted prebiotic are non-digestible carbohydrates that include inulin, fructo-oligosaccharides, galacto-oligosaccharides, lactulose, and human milk oligosaccharides [2]. However, other food components, such as resistant maltodextrin (RMD), could exert functional effects too and, therefore, attract considerable interest. RMD is a water-soluble and fermentable fibre produced by the heat treatment of corn starch, indigestible in the small intestine but fermentable in the colon, resulting in enhanced short-chain fatty acid production [3]. Figure 1 shows the glycosidic linkages and molecular structure of RMD, which comprises a small ratio of saccharides with a degree of polymerisation (DP) 1–9 and many polysaccharides with a DP 10 or more. RMD contains a random distribution of 1–2 and

1–3 linkages, which are formed during the dextrinization process, and 1–4 and 1–6 linkages naturally found in starch [4]. In terms of food technology and food processing, RMD is a very user-friendly fibre because of its low viscosity and its high stability to heat and acid conditions. In addition, it is tasteless and flavourless, so it could be easily added to a wide range of food products.

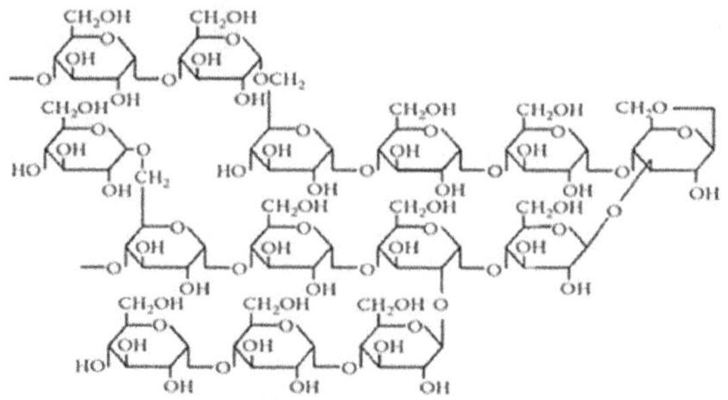

Figure 1. RMD chemical structure model.

In past studies, RMD has shown its potential prebiotic effect. In a double blind, randomised controlled crossover study, the daily intake of 25 g of RMD for 3 weeks (followed by a 2-week washout) increased faecal bifidobacteria counts and stool wet weight [5], suggesting health benefits. However, using RMD in food products is not limited to its potential prebiotic effect. For instance, in another double blind, randomised controlled crossover study, RMD demonstrated short-term decreased hunger and increased satiety hormones when ingested with a meal [6]. In addition, according to a systematic review of randomised placebo-controlled trials, the functional effect of RMD seems to be more effective in liquid foods rather than solid foods [7], so its use in beverages could be beneficial. These named studies suggest that RMD could have an adequate functional effect in a liquid matrix.

These functional effects can be linked to the maintenance of intestinal homeostasis because the gastrointestinal microflora plays a key role in the overall health status, affecting important human bodily functions such as metabolic, trophic, and protective [8]. Several factors have been shown to influence gastrointestinal microflora composition, being the environmental factors associated with diet and lifestyle the most predominant ones to shape it [9].

To provide a better diet, the food and beverage industry works to develop novel products that meet consumers' requirements and population wellbeing. Therefore, the beverage industry, and especially the juice industry, is probably one of the most dynamic and innovative sectors. In accordance with the growing consumer inclination toward healthier food products [10], the beverage industry is increasing the number of healthier ingredients in their juices, for example, by developing functional beverages through adding prebiotics [11].

In terms of flavour, orange juice is the most demanded [11], as it is the most consumed fruit juice worldwide. Aspects like the fruit origin, fruit cultivar, maturity, juice processing conditions, packaging, and storage conditions affect the physico-chemical properties of the juice. Moreover, orange juice, one of the most representative citrus juices, contains many nutritive and biologically active micronutrients, besides its natural sugar content. The most significant micronutrients for orange juice are potassium, copper, folate, vitamin C, flavonoids (mostly hesperidin), and dietary fibres [12]. From the viewpoint of juice processing, one of the primary considerations is the technology applied to assure the microbial stabilisation. Although alternative processing treatments have been developed (for example, high-pressure processing technologies or pulsed electric fields), thermal pasteurisation is still the most cost-effective method to reduce microbial populations and enzyme activity [13].

Because of their technological characteristics, food components perceived as prebiotics have shown not only an upgrade in the nutritious quality of the food product but also an improvement in the quality regarding sensory properties, texture, and physico-chemical properties [14]. Thus, based on previous evidence, adding RDM could be beneficial from nutritional and technological viewpoints, to give a novel potentially prebiotic pasteurised orange juice. However, it is necessary to specifically evaluate how each functional food component affects the matrix in which it is added. In addition, most of the studies that have been previously completed in this field focused on the stability of the prebiotic fibre in fruit juice processing and storage conditions or in the functionality of the finished beverages [15–18]. These are undoubtedly of great importance, but few studies show how the prebiotic fibres added affect the physico-chemical properties of the finished beverage. In addition, most of the studies that have been done were focused on the effect of prebiotic addition on stability, storage conditions, or functionality of the prebiotic fibre in the finished beverages [15–18]. Only a few studies show how the addition of prebiotic fibres impact the physico-chemical properties of the finished beverage. Therefore, this study aimed to evaluate the addition of RMD on the physico-chemical properties of pasteurised orange juice (with and without pulp). How the RMD affects °Brix, pH, acidity, particle size distribution, density, turbidity, rheology, and colour were elucidated to help develop novel functional products that could be accepted by a consumer.

2. Materials and Methods

2.1. *Raw Materials*

This study was conducted with freshly squeezed orange juice supplied by Refresco Iberia S.A.U. (Valencia, Spain). All oranges were from Spanish origin. RMD (Fibersol-2) added to the juice was purchased from ADM/Matsutani, LLC (Decatur, IL, USA). Frozen pasteurised orange pulp was provided by a local fruit processing company (Zumos Valencianos del Mediterráneo, Valencia, Spain).

2.2. *Sample Preparation and Pasteurisation*

A total of eight samples of orange juice were prepared. Four were orange juice with pulp (OJP) and the other four were orange juice without pulp (OJWP). Fresh orange juice was directly collected from the industrial squeezed lines. Orange pulp (2.5%) was added to the OJP samples, and increasing RMD concentrations (2.5, 5, and 7.5%) were added into both OJP and OJWP samples. Both orange pulp and RMD concentrations were applied in weight/weight percentage. Pulp content was homogenised using a stirrer (LH Overhead Stirrer, VELP Scientifica, Usmate, Italy), by applying 200 rpm for 5 min. Increasing RMD concentrations were mixed and stirred (200 rpm, 15 min) into both OJP and OJWP samples. Thus, for a finished beverage portion of 200 g, either 5, 10, or 15 g of RMD would be ingested, enough to display functional effects according to other studies [5–7]. Control samples without RMD addition (OJP0 and OJWP0) were also prepared, and they complied with the European Fruit Juice Association (AIJN) orange juice guidelines [19], so no adulteration or deviation occurred during the juice extraction. Finally, all samples were pasteurised (Fruchtsaftdispenser, Mabo Steuerungselemente GmbH, Eppingen, Germany) at 85 °C for 10 s, and were hot filled into 250 mL polyethylene terephthalate (PET) bottles. All bottles were immersed in a cold-water bath (<10 °C) for 30 min to cool down their temperature after the heat treatment.

2.3. *Physico-Chemical Determinations*

2.3.1. °Brix, Acidity, and pH

Measurement of total soluble solids (°Brix) was conducted using refractometry (Abbemat 200, Anton Paar, Graz, Austria). Acidity, expressed as grams of citric acid per 100 mL of orange juice (gCA/100 mL), was determined using a DL53 acid titrator (Mettler Toledo, Greifensee, Switzerland). Determination of pH was made using a Basic 20 pH meter (Crison, Alella, Spain). All determinations were performed in triplicate in accordance with AOAC guidelines [20].

2.3.2. Particle Size Distribution

Juice particle size distribution was determined by applying the laser diffraction method and Mie theory, following the ISO13320 regulation [21], by using a particle size analyser (Malvern Instruments

Ltd., Mastersizer 2000, Malvern, UK) equipped with a wet sample dispersion unit (Malvern Instruments Ltd., Hydro 2000 MU, Malvern, UK). Laser diffraction reports the volume of material of a given size, since the light energy reported by the detector system is proportional to the volume of material present. The Mie theory requires the information on both the sample and the dispersant optical property. For orange juice, the particle refraction and absorption were 1.52 and 0.1, respectively, and the water refraction index was 1.33. The sample was dispersed in distilled water and pumped through the optical cell under moderate stirring (1800 rpm) at 20 °C. The volume (%) against particle size (in μm) was obtained and the size distribution was characterised by the volume mean diameter (*D* (4,3)). The standard percentile *d* (0.1) or size of particle below which 10% of the sample lies and *d* (0.9) or size of particle below which 90% of the sample lies were also considered for juice characterisation.

2.3.3. Rheological Measurements

Juice flow behaviour was analysed using a controlled shear stress rheometer coupled to a thermostatic bath (Thermo Electron Co., Haake RheoStress 1, Waltham, MA, USA) with coaxial cylinders (Z34 DIN) using sensor system set at 20 °C following the Igual et al., [22] methodology. A relax time of 900 s was selected for the sample before running the test. Shear rate, (γ; s^{-1}), was increased from 0 to 150 s^{-1} in 20 step (fixed duration for each step 30 s) and shear stress σ (Pa), was recorded.

2.3.4. Density

Density was determined by using a pycnometer (50 mL) and distilled water at 25 °C as a reference.

2.3.5. Turbidity

Orange juice was centrifuged at 3000 rpm for 10 min. The turbidity of the upper layer solution was determined using a spectrophotometer UV-VIS (Thermo Scientific, Helios Zeta UV-Vis, Loughborough, LE, UK) at 600 nm, as described by Chandler and Robertson [23]. Sample transmittance (T) was obtained in relation to distilled water, and the turbidity (Tb) was calculated using Equation (1) [24].

$$\mathrm{Tb}\ (\%) = 100 - \mathrm{T} \tag{1}$$

2.3.6. Colour Measurement

Colour values were obtained from the reflection spectrum. Samples colour was measured using a colorimeter (CM-700d, Konica Minolta, Tokyo, Japan) with a standard illuminant D65 and a visual angle of 10°. Results were obtained in terms of L* (brightness: L* = 0 (black), L* = 100 (white)), a* (−a* = greenness, +a* = redness), and b* (−b* = blueness, +b* = yellowness), according to the CIELab system [25]. Chroma, C*ab (saturation) and hue angle, h*ab were also calculated, using equations 2 and 3, respectively. The colour difference was calculated regarding the control samples in each case, in all OJP and OJWP samples, to evaluate the RMD addition effect.

$$C^*_{ab} = ((a^{*2} + b^{*2}))^{1/2} \tag{2}$$

$$h^*_{ab} = \arctan\left(\frac{b^*}{a^*}\right) \tag{3}$$

2.4. Statistical Analysis

Analysis of variance (ANOVA) was applied with a confidence level of 95% ($p < 0.05$), to evaluate the differences among samples. Furthermore, a correlation analysis among studied properties of juices and RMD concentration was made, with a 95% significance level. Statgraphics (Centurion XVII Software, version 17.2.04, Statgraphics Technologies, Inc. The Plains, VA, USA) was used.

3. Results and Discussions

°Brix, acidity, and pH were evaluated as the basic control parameters, as is the general protocol in the juice industry (Table 1). Increasing concentrations of RMD implied a significant increase in total soluble solids in both OJP and OJWP samples ($p < 0.05$). This makes sense since, as explained, RMD is a water-soluble fibre. OJWP samples showed slightly higher values of °Brix ($p < 0.05$), mainly because a small percentage of orange juice was replaced in OJP samples by orange pulp, which is an insoluble fibre. Ghavidel et al. [17] also reported than an increase in fibre content (fructo-oligosaccharides, FOS) produced an increase in the total soluble solids in an orange juice-based sugar-added beverage. In addition, the substitution of sugar by other prebiotic fibres, namely oligofructose and inulin, did not change the °Brix range of papaya nectar [26]. Because of this soluble solid addition to the matrix and its light sweetness taste, such fibres have been proposed as sugar replacers, among other food technology applications [27,28]. This could be beneficial in acidic food products, such as orange juice, to help balance its sensory profile without adding sugar but adding functional ingredients.

Table 1. Mean values (and standard deviations) of °Brix, pH, and acidity of pasteurised orange juice.

Sample	°Brix	Acidity (g CA/100 mL)	pH
OJP0	11.36 (0.04) ^{a}	0.773 (0.003) ^{h}	3.36 (0.04) ^{a}
OJP2.5	13.58 (0.03) ^{c}	0.756 (0.003) ^{g}	3.46 (0.08) ^{b}
OJP5	15.83 (0.08) ^{e}	0.733 (0.002) ^{f}	3.50 (0.07) ^{bc}
OJP7.5	17.98 (0.04) ^{g}	0.7133 (0.0006) ^{e}	3.44 (0.04) ^{b}
OJWP0	11.52 (0.05) ^{b}	0.686 (0.003) ^{d}	3.55 (0.03) ^{cd}
OJWP2.5	13.75 (0.03) ^{d}	0.671 (0.002) ^{c}	3.59 (0.03) ^{d}
OJWP5	15.96 (0.04) ^{f}	0.656 (0.002) ^{b}	3.54 (0.03) ^{cd}
OJWP7.5	18.18 (0.08) ^{h}	0.6380 (0.0005) ^{a}	3.57 (0.03) ^{cd}

The same letter in superscript within column indicates homogeneous groups established by ANOVA ($p < 0.05$). OJP, orange juice with pulp; OJWP, orange juice without pulp.

However, higher RMD concentrations significantly decreased acidity values in both OJP and OJWP samples ($p < 0.05$). This is because RMD addition helped to reduce the quantity of raw orange juice, and therefore citric acid. Moreover, OJP samples had significantly higher acidity values than OJWP ($p < 0.05$). In terms of pH, the differences were small but significant ($p < 0.05$), as OJP samples presented lower pH values than OJWP. Therefore, orange pulp addition showed an impact on the acidity and pH values ($p < 0.05$), whereas RMD addition decreased citric acid content by replacing orange juice content ($p < 0.05$). The acidity of the orange pulp could be higher than the acidity from the orange juice, thus leading to an increase in acidity and a decrease in pH in OJP samples. RMD addition had less impact on the pH.

°Brix, pH, and acidity values of the control samples are hard to compare with those in other studies, since the oranges used for the juice extraction were all from Spanish origin. Citric acid values reported from Mexican orange juices were lower [29]. However, °Brix and pH were almost the same as those reported from Cortés et al. [30], who used the Navel cultivar from Spain. This enhances the importance of the raw material origin and the complexity to properly compare the physico-chemical properties in fruit-derived products.

Juice density is also an important quality control parameter in the juice industry [31]. Figure 2a shows that density values were not affected by pulp addition ($p > 0.05$) and that they increased as RMD concentration increased ($p < 0.05$). This could be explained because RMD was dissolved completely in the orange juice. Moreover, it is widely known that, in fruit juices, the soluble solids quantity affects density values, while insoluble solids, such as cloud and pulp, contribute little to density measurements [31]. The relationship between density and soluble solids has been extensively studied and regression models have been developed, like the one obtained by Ramos and Ibarz [32].

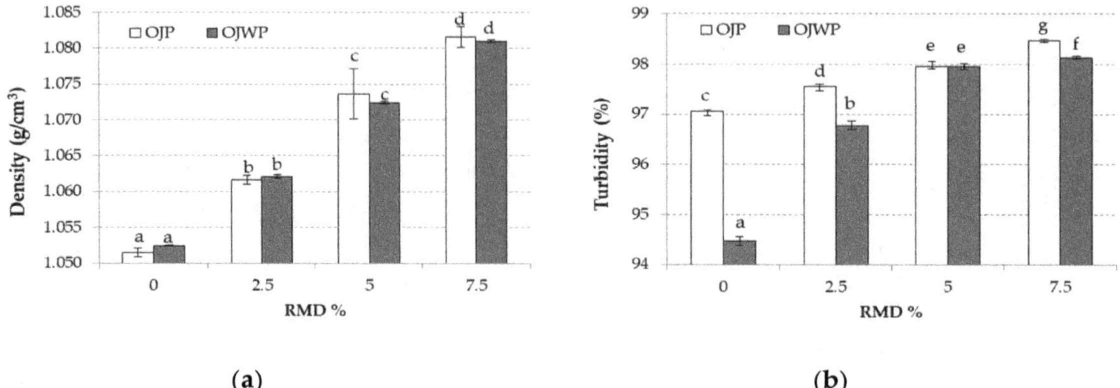

Figure 2. (**a**) Mean values (and standard deviations) of density of pasteurised orange juice; (**b**) Mean values (and standard deviations) of turbidity of pasteurised orange juice. Letters indicate homogeneous groups established by the ANOVA ($p < 0.05$) for each parameter analysed. OJP, orange juice with pulp; OJWP, orange juice without pulp.

Turbidity is represented in Figure 2b. Turbidity provides a measure of the concentration of disperse particles in a solution by measuring its light-scattering properties [33]. Therefore, as RMD was completely dissolved, higher RMD concentrations led to higher turbidity values ($p < 0.05$). However, in contrast to the density measurements, pulp content played a role in the turbidity values ($p < 0.05$), clouding OJP samples. This can be checked by comparing turbidity values of the control samples (OJP0 and OJWP0). However, the clouding effect because of pulp content was limited as turbidity difference between OJP and OJWP samples decreased as RMD concentration raised. For example, no turbidity difference was found in both 5% RMD-added samples.

Volume particle size distribution is represented for the OJP samples (Figure 3a) and OJWP samples (Figure 3b). Both OJP and OJWP samples showed a similar trend, as it can be observed in Figure 3. Table 2 compiles the mean values (and standard deviations) of volume mean diameter D (4,3) and the standard percentiles d (0.1), d (0.5), and d (0.9). Particles size of OJP samples presented significantly greater volume mean diameter than OJWP samples ($p < 0.05$). RMD addition did not have an impact in the volume mean diameter ($p > 0.05$), as it was dissolved in the orange juice. Therefore, the difference between OJP and OJWP samples regarding volume mean diameter were exclusively because of pulp content ($p < 0.05$). This phenomenon is demonstrated by comparing the standard percentiles. As the number of analysed particles grows, the particle size distribution becomes homogenised; this is seen in Table 2 by comparing d (0.1) values to d (0.9) values. In the first percentile d (0.1), greater differences in particle size distributions were found between all orange juice samples ($p < 0.05$). However, by increasing the number of particles analysed (d (0.9)), the same relationship is obtained as with the volume mean diameter. Thus, OJP obtained a greater quantity of particles of larger size ($p < 0.05$). The particle size distribution obtained for these orange juice samples differed from the one performed by Stinco et al. [34], who, on average, reported larger particle size for the industrially squeezed orange juices (both fresh and pasteurised). Achieving smaller particles could be beneficial as the food matrix is one of the key factors related to the release of bioactive compounds, such as carotenoids [34].

Regarding rheology, all orange juices showed a non-Newtonian, non-time dependent, pseudoplastic behaviour as observed in Figure 4. The obtained flow curves were well-fitted ($R^2 \geq 0.99$) to the Ostwald de Waele model (Equation (4)), where k is the consistency index ($Pa \cdot s^n$) and n is the flow index (Table 3). This mathematical relationship is useful because of its simplicity [35].

$$\sigma = k \, \dot{\gamma}^{\,n} \tag{4}$$

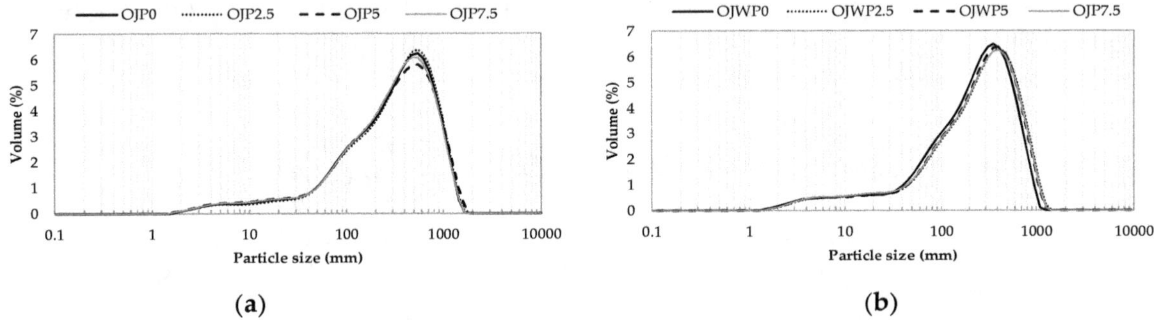

(a) (b)

Figure 3. (**a**) Volume particle size distributions (representative curves) of pasteurised orange juice with pulp; (**b**) Volume particle size distributions (representative curves) of pasteurised orange juice without pulp. OJP, orange juice with pulp; OJWP, orange juice without pulp.

Table 2. Mean values (and standard deviations) of volume mean diameter (µm) *D* (4,3), standard percentiles (µm) *d* (0.1), *d* (0.5), and *d* (0.9) of pasteurised orange juice.

Sample	*D* (4,3)	*d* (0.1)	*d* (0.5)	*d* (0.9)
OJP0	368 (27) ^{b}	44 (4) ^{d}	305 (25) ^{cd}	791 (55) ^{b}
OJP2.5	382 (33) ^{b}	49 (6) ^{e}	321 (30) ^{d}	814 (71) ^{b}
OJP5	374 (39) ^{b}	38 (5) ^{c}	299 (30) ^{b}	825 (92) ^{b}
OJP7.5	361 (27) ^{b}	40 (4) ^{c}	296 (22) ^{b}	783 (61) ^{b}
OJWP0	300 (14) ^{a}	33 (3) ^{b}	253 (12) ^{a}	637 (33) ^{a}
OJWP2.5	297 (20) ^{a}	28 (3) ^{a}	248 (16) ^{a}	638 (44) ^{a}
OJWP5	288 (12) ^{a}	30 (2) ^{ab}	241 (10) ^{a}	615 (28) ^{a}
OJWP7.5	289 (15) ^{a}	27 (3) ^{a}	240 (13) ^{a}	623 (32) ^{a}

The same letter in superscript within column indicates homogeneous groups established by ANOVA ($p < 0.05$). OJP, orange juice with pulp; OJWP, orange juice without pulp.

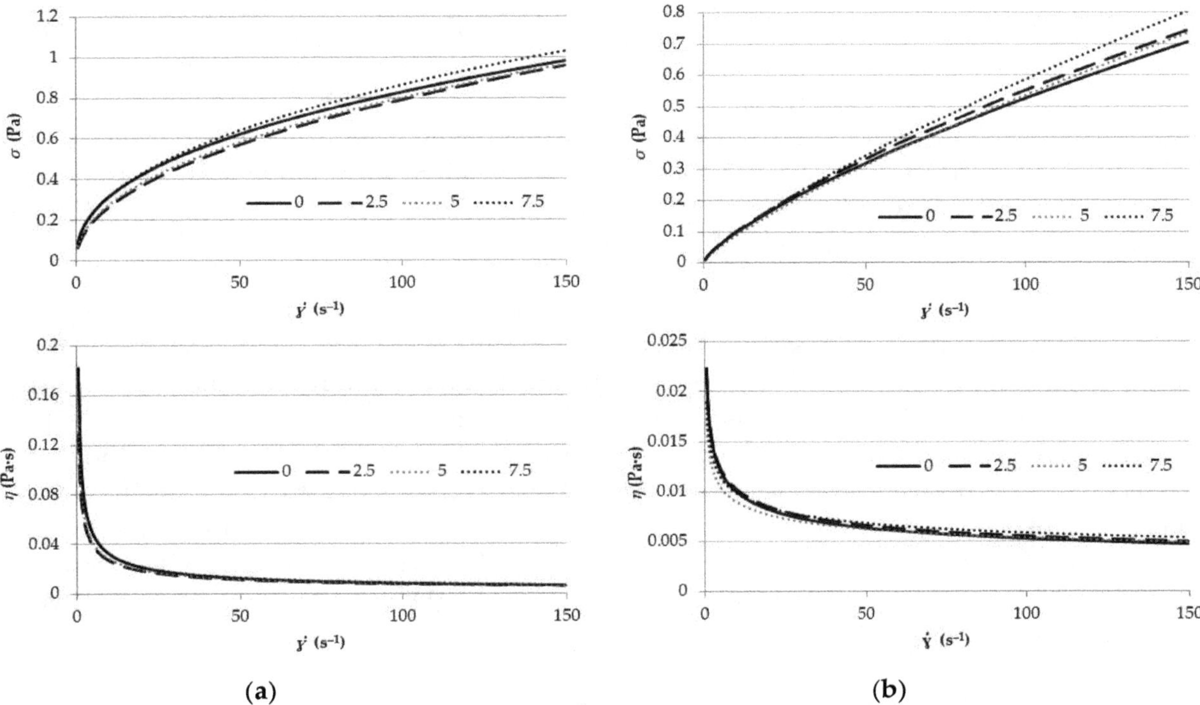

(a) (b)

Figure 4. (**a**) Flow behaviour (σ-γ) and viscosity profiles (η-γ) of pasteurised orange juice with pulp (OJP); (**b**) Flow behaviour (σ-γ) and viscosity profiles (η-γ) of pasteurised orange juice without pulp (OJWP).

Table 3. Mean values (and standard deviations) of consistency index (k), flow index (n) of pasteurised orange juice.

Sample	k (Pa·s^n)	n
OJP0	0.121 (0.009) ^{c}	0.4276 (0.0102) ^{a}
OJP2.5	0.0891 (0.0009) ^{b}	0.47 (0.02) ^{c}
OJP5	0.096 (0.005) ^{b}	0.461 (0.012) ^{bc}
OJP7.5	0.117 (0.015) ^{c}	0.43 (0.02) ^{ab}
OJWP0	0.0184 (0.0002) ^{a}	0.728 (0.002) ^{d}
OJWP2.5	0.0186 (0.0002) ^{a}	0.73675 (0.00106) ^{d}
OJWP5	0.01477 (0.00009) ^{a}	0.7800 (0.0012) ^{e}
OJWP7.5	0.01618 (0.00006) ^{a}	0.7799 (0.0013) ^{e}

The same letter in superscript within column indicates homogeneous groups established by ANOVA ($p < 0.05$). OJP, orange juice with pulp; OJWP, orange juice without pulp.

OJP samples showed significantly higher k values than OJWP samples ($p < 0.05$), meaning that pulp addition could slightly increase the viscosity of the orange juice. This can be observed in Figure 4 comparing viscosity profiles of OJP and OJWP. Rega et al. [36] also demonstrated that viscosity increased as pulp content increased. This could be explained because suspended solids increases the apparent viscosity of the fruit juice [37], probably because of the pectin content of orange pulp [38]. RMD addition did not produce a significant change in k of OJWP samples ($p > 0.05$). Moreover, RMD addition did not produce a clear trend in the k of both OJP and OJWP samples. Despite soluble solids supposedly increasing viscosity [37], in this study, RMD addition did not show differences in the viscosity profiles. However, OJWP samples marked significantly higher n values than OJP samples ($p < 0.05$), thus, its flow behaviour was slightly more Newtonian than OPJ samples. This suggests that OJP exhibits a more pseudoplastic behaviour than OJWP (Figure 4). RMD addition had also a significant effect on the n behaviour in OJWP, as n values in samples with 5 and 7.5% RMD increased ($p < 0.05$). However, orange pulp addition seems to have a stronger impact on the rheology of orange juice samples.

A statistical correlation was conducted to explain the relationship between the studied parameters in the orange juice samples. There were significant correlations among particle size parameters and rheological values according to Pearson coefficients ($p < 0.05$). As was observed in other citric fruit juices [22], k showed a positive correlation with D (4,3) (0.9238), d (0.1) (0.8512), d (0.5) (0.9068), and d (0.9) (0.9263); n showed a negative correlation with D (4,3) (−0.9609), d (0.1) (−0.8932), d (0.5) (−0.9467), and d (0.9) (−0.9609). Therefore, the higher the particle size in juices, the higher the consistency index in juices but the lower the flow index. Moreover, k and n were correlated with pH and acidity showing −0.7875 and 0.8766, and 0.7628 and −0.9168 Pearson coefficients, respectively.

Finally, colour is one of the most important parameters for consumers, as it is related to quality perception [30]; colour changes are shown in Figure 5. It was shown that higher RMD concentrations significantly decreased L* values ($p < 0.05$) (Figure 5a), thus that all orange juices turned slightly darker, as RMD concentration was raised. This could be explained as higher RMD concentrations helped to reduce orange juice content. Orange pulp showed no significant effect on the L* values ($p > 0.05$). L* presented significant correlations with density (0.9493) and turbidity (0.7467) ($p < 0.05$). Igual et al. [22] also found a high correlation between L* and turbidity in grapefruit juices. In addition, OJP0 and OJP2.5 had the lowest a* values ($p < 0.05$) (Figure 5b), which implies that they were greener than the rest. This could probably be explained because RMD addition led to a protective effect on the carotenoids content [39], therefore leading to a protective effect in the reddish tones during the pasteurisation process. For the yellow-blue content (b*, Figure 5c), RMD addition showed no significant effect ($p > 0.05$), while orange pulp content had a significant effect ($p < 0.05$) on b* values. Thus, in the OJWP samples, an increasing addition of RMD seemed to lower the yellowish tones, as OJWP5 and OJWP7.5 had lower b* values. However, few differences exist in the OJP because of RMD addition. This suggests that orange pulp content could have a protective effect on the natural yellow/orange tones of orange juice. C* values (Figure 5d) resulting from Equation (2) were almost the same as b*, as a* values were all close to 0. Regarding h*, RMD addition on OJWP samples did not play a significant role ($p > 0.05$). However, h* values (Figure 5e) were significantly increased ($p < 0.05$) as RMD concentration was higher in OJWP samples.

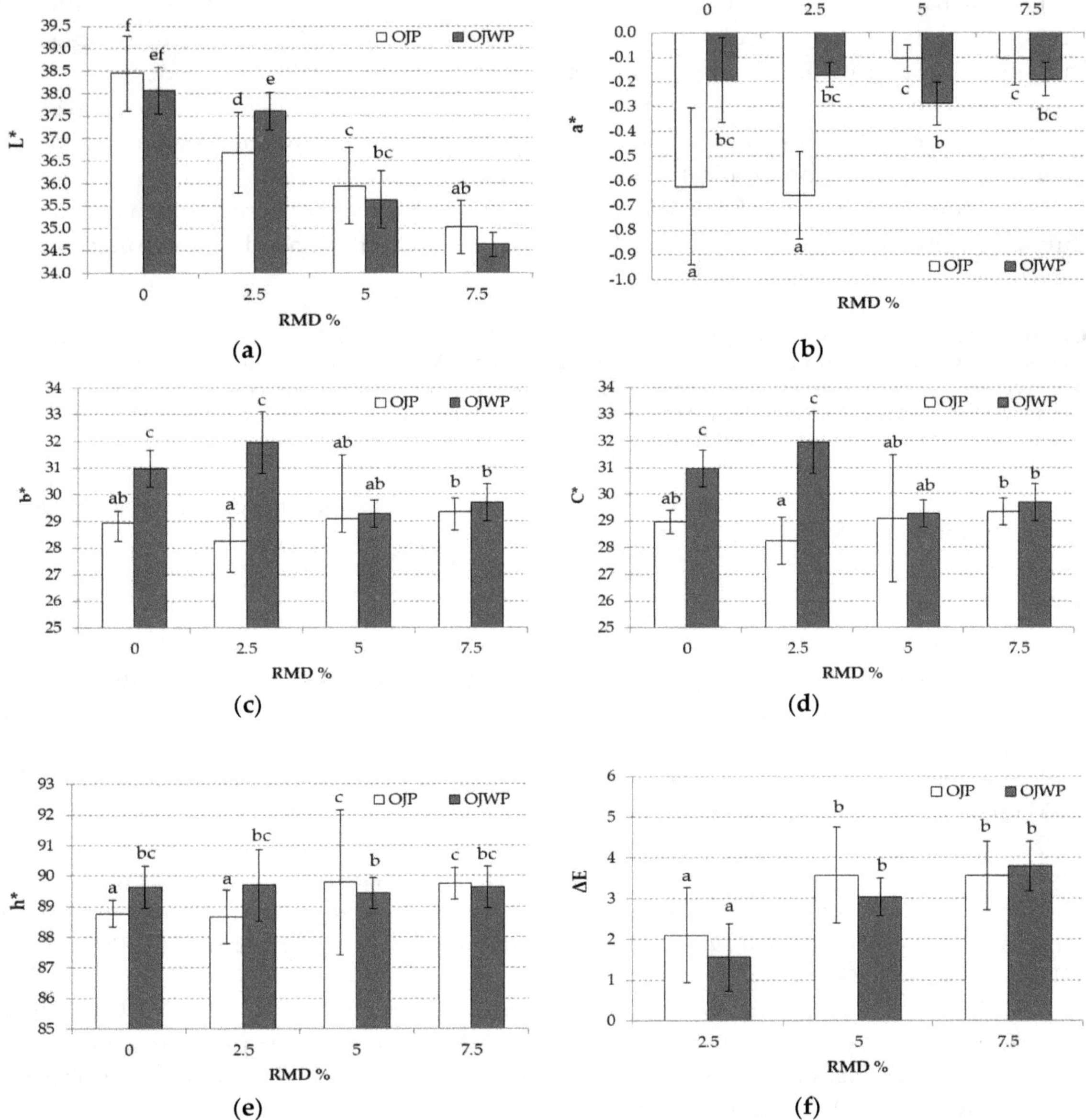

Figure 5. Mean values (and standard deviations) of colour coordinates L* (**a**), a* (**b**), b* (**c**), C* (**d**), and h* (**e**) and total colour differences (ΔE, (**f**)) of pasteurised orange juice. Letters indicate homogeneous groups established by the ANOVA ($p < 0.05$) for each parameter analysed. OJP, orange juice with pulp; OJWP, orange juice without pulp.

RMD addition in OJP and OJWP samples had almost the same result in terms of total colour difference (ΔE) regarding control samples (Figure 5f), thus, that pulp content did not interfere in colour differences ($p > 0.05$). Total colour differences between juices with RMD 2.5% and the control samples were lower than three units. Therefore, they are not perceptible to human eye, which only distinguishes colour differences if ΔE is larger than three [40]. Increasing RMD addition implied small but significant ($p < 0.05$) colour differences, but they were limited because samples with 5% and 7.5% RMD had almost no colour difference between them.

4. Conclusions

In this study, physico-chemical differences were found due to adding RMD to orange juice with and without pulp. Samples with 7.5% added RMD presented a greater impact on the physico-chemical properties in both OJP and OJWP samples. With RMD, a soluble-water fibre, its addition to orange juice increased the total soluble solids content, which raised Brix, density, and turbidity values, the last more evident in OJWP samples. Citric acid content was lowered because of the orange juice replacement by RMD, and small but significant changes were observed in terms of pH. Differences in particle size distribution were exclusively because of pulp content. Orange pulp content, and not RMD addition, appears to have an impact on the orange juice rheology. Slight colour differences were found; however, only higher RMD concentration would be perceptible by the human eye.

This study demonstrates that RMD addition in a wide range of concentrations is feasible from a food technology viewpoint. However, the optimal dose of RMD will depend on the functional effect to be achieved.

Author Contributions: Conceptualisation, E.A., M.I., J.M.-M., P.C.-F. and P.G.-S.; methodology, E.A. and M.I.; formal analysis, E.A.; investigation, E.A. and M.I.; resources, J.M.-M., and P.G.-S.; data curation, E.A., M.I., P.C.-F. and P.G.-S.; writing—original draft preparation and editing, E.A.; writing—review, M.I., P.C.-F., P.G.-S. and J.M.-M.; supervision, M.I., P.C.-F. and P.G.-S.; project administration, J.M.-M.; funding acquisition, P.G.-S. and J.M.-M. All authors have read and agreed to the published version of the manuscript.

Acknowledgments: We would like to thank Refresco Iberia S.A.U. for its support and for providing the squeezed orange juice to perform this study.

References

1. Gibson, G.R.; Probert, H.M.; Van Loo, J.; Rastall, R.A.; Roberfroid, M.B. Dietary modulation of the human colonic microbiota: Updating the concept of prebiotics. *Nutr. Res. Rev.* **2004**, *17*, 259–275. [CrossRef] [PubMed]
2. Corzo, N.; Alonso, J.L.; Azpiroz, F.; Calvo, M.A.; Cirici, M.; Leis, R.; Lombó, F.; Mateos-Aparicio, I.; Plou, F.J.; Ruas-Madiedo, P.; et al. Prebiotics: Concept, properties and beneficial effects. *Nutr. Hosp.* **2015**, *31*, 99–118. [CrossRef] [PubMed]
3. Lockyer, S.; Nugent, A.P. Health effects of resistant starch. *Nutr. Bull.* **2017**, *42*, 10–41. [CrossRef]
4. Hashizume, C.; Okuma, K. Fibersol®-2 resistant maltodextrin: Functional dietary fiber ingredient. In *Fiber Ingredients—Foods Applications and Health Benefits*, 1st ed.; Cho, S.S., Samuel, P., Eds.; CRC Press: Boca Raton, FL, USA, 2009; pp. 61–78.
5. Burns, A.M.; Solch, R.J.; Dennis-Wall, J.C.; Ukhanova, M.; Nieves Jr, C.; Mai, V.; Christman, M.C.; Gordon, D.T.; Langkamp-Henken, B. In healthy adults, resistant maltodextrin produces a greater change in fecal bifidobacteria counts and increases stool wet weight: A double-blind, randomised, controlled crossover study. *Nutr. Res.* **2018**, *60*, 33–42. [CrossRef]
6. Ye, Z.; Arumugam, V.; Haugabrooks, E.; Williamson, P.; Hendrich, S. Soluble dietary fiber (Fibersol-2) decreased hunger and increased satiety hormones in humans when ingested with a meal. *Nutr. Res.* **2015**, *35*, 393–400. [CrossRef]
7. Livesey, G.; Tagami, H. Interventions to lower the glycemic response to carbohydrate foods with a low-viscosity fiber (resistant maltodextrin): Meta-analysis of randomized controlled trials. *Am. J. Clin. Nutr.* **2009**, *89*, 114–125. [CrossRef]
8. Guarner, F.; Malagelada, J.R. Gut flora in health and disease. *Lancet* **2003**, *361*, 512–519. [CrossRef]
9. Rothschild, D.; Weissbrod, O.; Barkan, E.; Kurilshikov, A.; Korem, T.; Zeevi, D.; Costea, P.I.; Godneva, A.; Kalka, I.N.; Bar, N.; et al. Environment dominates over host genetics in shaping human gut microbiota. *Nature* **2018**, *555*, 210–215. [CrossRef]
10. Saba, A.; Sinesio, F.; Moneta, E.; Dinnella, C.; Laureati, M.; Torri, L.; Peparaio, E.; Saggia Civitelli, E.; Endrizzi, I.; Gasperi, F.; et al. Measuring consumers attitudes towards health and taste and their association with food-related life-styles and preferences. *Food Qual. Prefer.* **2019**, *73*, 25–37. [CrossRef]

11. Priyadarshini, A.; Priyadarshini, A. Market dimensions of the fruit juice industry. In *Fruit Juices: Extraction, Composition, Quality and Analysis*, 1st ed.; Rajauria, G., Tiwari, B.K., Eds.; Elsevier: London, UK, 2018; pp. 15–32. [CrossRef]
12. Ivanova, N.N.; Khomich, L.M.; Perova, I.B. Orange juice nutritional profile. *Voprosy Pitaniia* **2017**, *86*, 103–113. [CrossRef]
13. Perez-Cacho, P.R.; Rouseff, R. Processing and storage effects on orange juice aroma: A review. *J. Agric. Food Chem.* **2008**, *56*, 9785–9796. [CrossRef] [PubMed]
14. De Paulo Farias, D.; de Araújo, F.F.; Neri-Numa, I.A.; Pastore, G.M. Prebiotics: Trends in food, health and technological applications. *Trends Food Sci. Technol.* **2019**, *93*, 23–35. [CrossRef]
15. Yousaf, M.S.; Yusof, S.; Abdul Manap, M.Y.B.; Abd-Aziz, S. Storage stability of clarified banana juice fortified with inulin and oligofructose. *J. Food Process. Preserv.* **2010**, *34*, 599–610. [CrossRef]
16. Davim, S.; Andrade, S.; Oliveira, S.; Pina, A.; Barroca, M.J.; Guiné, R.P.F. Development of fruit jams and juices enriched with fructooligosaccharides. *J. Fruit Sci.* **2015**, *15*, 100–116. [CrossRef]
17. Ghavidel, R.A.; Karimi, M.; Davoodi, M.; Jahanbani, R.; Asl, A.F.A. Effect of fructooligosaccharide fortification on quality characteristic of some fruit juice beverages (apple & orange juice). *Int. J. Farm. Allied Sci.* **2014**, *3*, 141–146.
18. Renuka, B.; Kulkarni, S.G.; Vijayanand, P.; Prapulla, S.G. Fructooligosaccharide fortification of selected fruit juice beverages: Effect on the quality characteristics. *LWT Food Sci. Technol.* **2009**, *42*, 1031–1033. [CrossRef]
19. AIJN-European Fruit Juice Association. Orange Juice Guideline. Available online: https://aijn.eu/en/publications/aijn-papers-guidelines/juice-quality (accessed on 13 October 2020).
20. Latimer, G.W.; AOAC International. *Official Methods of Analysis of AOAC International*, 19th ed.; AOAC International: Gaithersburg, MD, USA, 2012; Volume 2.
21. Spanish Association for Standardization and Certification (AENOR). *Particle Size Analysis-Laser Diffraction Methods*; AENOR: Madrid, Spain, 2009.
22. Igual, M.; Contreras, C.; Camacho, M.M.; Martínez-Navarrete, N. Effect of thermal treatment and storage conditions on the physical and sensory properties of grapefruit juice. *Food Bioprocess Technol.* **2014**, *7*, 191–203. [CrossRef]
23. Chandler, B.U.; Robertson, G.L. Effect of pectic enzymes on cloud stability and soluble limonin concentration in stored orange juice. *J. Sci. Food Agric.* **1983**, *34*, 599–611. [CrossRef]
24. Matsui, K.N.; Gut, J.A.W.; de Oliveira, P.V.; Tadini, C.C. Inactivation kinetics of polyphenol oxidase and peroxidase in green coconut water by microwave processing. *J. Food Eng.* **2008**, *88*, 169–176. [CrossRef]
25. Commission Internationale de l'Eclairage (CIE). *Colorimetry*, 2nd ed.; Commission Internationale de l'Eclairage: Vienna, Austria, 1986.
26. Braga, H.F.; Conti-Silva, A.C. Papaya nectar formulated with prebiotics: Chemical characterization and sensory acceptability. *LWT Food Sci. Technol.* **2015**, *62*, 854–860. [CrossRef]
27. Priya, B.N. A role of prebiotics in food and health: A review. *J. Crit. Rev.* **2020**, *7*, 782–785. [CrossRef]
28. Wang, Y. Prebiotics: Present and future in food science and technology. *Food Res. Int.* **2009**, *42*, 8–12. [CrossRef]
29. Farnworth, E.R.; Lagace, M.; Couture, R.; Yaylayan, V.; Stewart, B. Thermal processing, storage conditions, and the composition and physical properties of orange juice. *Food Res. Int.* **2001**, *34*, 25–30. [CrossRef]
30. Cortés, C.; Esteve, M.J.; Frígola, A. Color of orange juice treated by high intensity pulsed electric fields during refrigerated storage and comparison with pasteurized juice. *Food Control* **2008**, *19*, 151–158. [CrossRef]
31. Kimball, D.A. *Citrus Processing: Quality Control and Technology*; Springer Science & Business Media: New York, NY, USA, 2012.
32. Ramos, A.M.; Ibarz, A. Density of juice and fruit puree as a function of soluble solids content and temperature. *J. Food Eng.* **1998**, *3*, 57–63. [CrossRef]
33. Vaillant, F.; Pérez, A.M.; Acosta, O.; Dornier, M. Turbidity of pulpy fruit juice: A key factor for predicting cross-flow microfiltration performance. *J. Membr. Sci.* **2008**, *325*, 404–412. [CrossRef]
34. Stinco, C.M.; Fernández-Vázquez, R.; Escudero-Gilete, M.L.; Heredia, F.J.; Meléndez-Martínez, A.J.; Vicario, I.M. Effect of orange juice's processing on the color, particle size, and bioaccessibility of carotenoids. *J. Agric. Food Chem.* **2012**, *60*, 1447–1455. [CrossRef]

35. Steffe, J.F. *Rheological Methods in Food Process Engineering*, 2nd ed.; Freeman Press: East Lansing, MI, USA, 1996.
36. Rega, B.; Fournier, N.; Nicklaus, S.; Guichard, E. Role of pulp in flavor release and sensory perception in orange juice. *J. Agric. Food Chem.* **2004**, *52*, 4204–4212. [CrossRef]
37. Hernandez, E.; Chen, C.S.; Johnson, J.; Carter, R.D. Viscosity changes in orange juice after ultrafiltration and evaporation. *J. Food Eng.* **1995**, *25*, 387–396. [CrossRef]
38. Schalow, S.; Baloufaud, M.; Cottancin, T.; Fischer, J.; Drusch, S. Orange pulp and peel fibres: Pectin-rich by-products from citrus processing for water binding and gelling in foods. *Eur. Food Res. Technol.* **2018**, *244*, 235–244. [CrossRef]
39. Arilla, E.; Martínez-Monzó, J.; García-Segovia, P.; Igual, M.A. Effect of resistant maltodextrin on bioactive compounds of orange pasteurized juice. In Proceedings of the 1st International Electronic Conference on Food Science and Functional Foods, Online, 10–25 November 2020. [CrossRef]
40. Bodart, M.; de Peñaranda, R.; Deneyer, A.; Flamant, G. Photometry and colorimetry characterisation of materials in daylighting evaluation tools. *Build. Environ.* **2008**, *43*, 2046–2058. [CrossRef]

2

Modelling Future Agricultural Mechanisation of Major Crops in China: An Assessment of Energy Demand, Land use and Emissions

Iván García Kerdan [1,2,3,*], Sara Giarola [4], Ellis Skinner [1], Marin Tuleu [1] and Adam Hawkes [1]

1. Department of Chemical Engineering, Imperial College London, London SW7 2AZ, UK; ellis.skinner14@imperial.ac.uk (E.S.); marin.tuleu14@imperial.ac.uk (M.T.); a.hawkes@imperial.ac.uk (A.H.)
2. Department of the Built Environment, School of Design, University of Greenwich, London SE10 9LS, UK
3. Instituto de Ingeniería, Universidad Nacional Autónoma de México, Mexico City 04510, Mexico
4. Department of Earth Science & Engineering, Imperial College London, London SW7 2AZ, UK; s.giarola10@imperial.ac.uk

* Correspondence: i.garcia-kerdan@imperial.ac.uk

Abstract: Agricultural direct energy use is responsible for about 1–2% of global emissions and is the major emitting sector for methane (2.9 $GtCO_2eq\ y^{-1}$) and nitrous oxide (2.3 $GtCO_2eq\ y^{-1}$). In the last century, farm mechanisation has brought higher productivity levels and lower land demands at the expense of an increase in fossil energy and agrochemicals use. The expected increase in certain food and bioenergy crops and the uncertain mitigation options available for non-CO_2 emissions make of vital importance the assessment of the use of energy and the related emissions attributable to this sector. The aim of this paper is to present a simulation framework able to forecast energy demand, technological diffusion, required investment and land use change of specific agricultural crops. MUSE-Ag & LU, a novel energy systems-oriented agricultural and land use model, has been used for this purpose. As case study, four main crops (maize, soybean, wheat and rice) have been modelled in mainland China. Besides conventional direct energy use, the model considers inputs such as fertiliser and labour demand. Outputs suggest that the modernisation of agricultural processes in China could have the capacity to reduce by 2050 on-farm emissions intensity from 0.024 to 0.016 $GtCO_2eq\ PJ_{crop}^{-1}$ (−35.6%), requiring a necessary total investment of approximately 319.4 billion 2017$US.

Keywords: energy; agriculture; modelling; mechanisation; land use; China

---

1. Introduction

1.1. Emissions and Energy Use in Agriculture

Agricultural lands, due to their sheer proportion of overall land use and rigorous management, have a significant environmental impact on the earth's carbon and nitrogen cycles. Houghton and Nassikas [1] estimated that since 1850, over 145 ± 16 GtC ($\mu \pm 1\sigma$) from biomass and soils has been emitted worldwide as a consequence of land-use and land-use changes. Deforestation due to crop and pastureland expansion accounted for 77% of the total emissions. Additionally, the sector is responsible for 2.9 $GtCO_2eq\ y^{-1}$ emissions of methane (CH_4) due to manure management and enteric fermentation, and around 2.3 $GtCO_2eq\ y^{-1}$ of nitrous oxide (N_2O) due to widespread use of fertiliser, increasing rice cultivation, manure inputs and soil degradation [2,3]. As a whole, the agriculture, forestry and land use sector is responsible for around 10–12 $GtCO_2eq\ y^{-1}$, which represents 25% of net anthropogenic greenhouse gas (GHG) emissions [4]. These estimates have an inherent high uncertainty and might vary considerably from study to study depending on the selected system boundaries.

Agricultural and land use emissions are especially critical for developing countries, where it is estimated that to cover the future food demand by 2050, productivity will have to increase to at least the double [5]. According to the Food and Agriculture Organization [6], current average daily energy food intake is about 11.6 MJ day^{-1} (2770 kcal day^{-1}). Projections suggest that by 2050, due to an increase in population income, average daily intake could reach about 12.8 MJ day^{-1} (3050 kcal day^{-1}) and 12.4 MJ day^{-1} (2970 kcal day^{-1}) per person for developed and developing economies, respectively. Considering an increase in global population, it is estimated that the global food demand will increase by 60%–110% [5,7–9]. Moreover, it is expected that livestock production will increase driven by meat-based demand in developing countries. Thornton [10] expects that by 2050 food demand for livestock products will double in South Asia and sub-Saharan Africa, followed by South American and former-Soviet Union countries, while developed OECD countries will experience minimal change. In the case of China, He et al. [11] projects that with current diet patterns, by 2032, demand for land and water could increase around 14%; however, if recommended nutrition is followed, it is expected that China's existing land would be insufficient to meet future demand, especially if appropriate measures and technological change is in place. If current trends of food production are kept (by intensification in developed economies and land clearing in developing economies), about 1 Gha of land would need to be cleared by 2050 for agricultural purposes only [7]. This represents emissions of about 3 $GtCO_2eq\ y^{-1}$ and 250 Mt y^{-1} of Nitrogen (N) from land management only.

Depending on the country, agricultural direct energy demand is between 2–8% [6,12] of the country's demand. If the entire food supply chain is considered, this means including inputs for food processing, manufacturing and transport, this figure could increase to up to 30% [13]. Globally, in 2014 the sector was responsible for the consumption of 8.2 EJ or 0.46 $GtCO_2eq\ y^{-1}$. The main energy sources were electricity and diesel, representing three quarters of the whole consumption (Figure 1). In the past, several authors have investigated the impact of energy-related greenhouse gas (GHG) emissions in the sector [14–20]. Robaina-Alves and Moutinho [21] and Li et al. [22] found that two main cost-efficient paths exist to reduce sectoral emissions: (i) increase use of improved machinery (efficiency) and farming practices and (ii) promotion of production and utilisation of renewable energy sources.

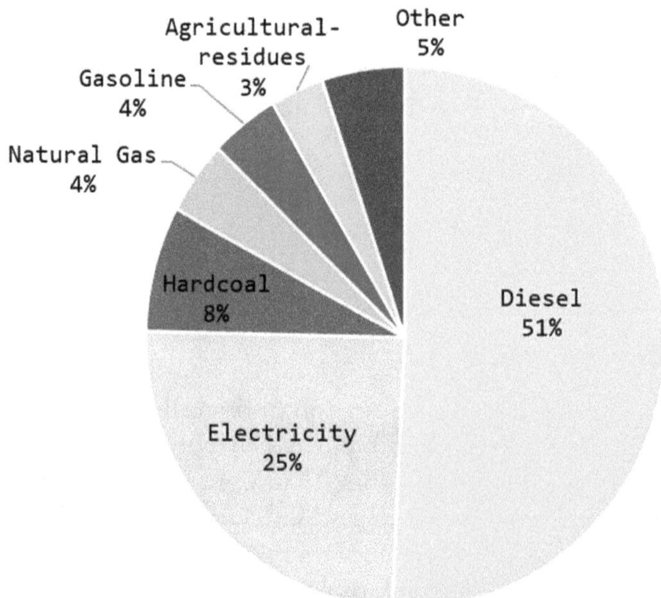

Figure 1. Global fuel share in the agricultural sector in 2014. (Electricity production share- Coal: 41%, natural gas: 22%, nuclear: 11%, hydro: 16%, other renewable (solar, wind, biomass): 6%, oil at 4%) [6].

Energy Inputs in Agricultural Crops

The agricultural sector is being and will be even more affected by climate change: even in a 2 °C world, the chance of major heatwaves will go to 49% with risk for maize heat damage equal to 18%, compared to 5% chance in 1981–2010 [23]. Global and regional impacts of climate change at different levels of global temperature increase). However, it is still missing a framework encompassing energy, environmental, and economic aspects in a single tool supporting effectively policy decision-making in the field. In this respect, the lack of comprehensive datasets is worsened by high heterogeneity of the sector.

On the energy use accounting side, although national energy balances publish the amount of energy demand and related emissions in the agriculture sector, the lack of disaggregated data by crop type is the main problem for a reliable understanding of the sector's dynamics. Researchers have tried to disaggregate the impacts on GHG emissions due to individual crops. For example, Woods et al. [24] described the importance of a precise analysis of agricultural products, especially if emissions are to be modelled. Camargo et al. [25] used the 'Farm Analysis Tool' to compare direct energy use and emissions for different crops in the north-east of the USA. Similarly, Pimentel [26] analysed the required direct and indirect energy inputs for twelve different crops in developed and developing nations.

As each crop is characterised by specific farming practices and machinery, the lack of crop disaggregation in the statistics generate a huge variability also in the labour inputs. For example, employment from the agricultural sector could make up anywhere within 50% to 90% of the population in developing economies [27], represented mainly by small independent farming operations with variation due to factors such as weather, technology diffusion and other geographical factors [28,29]. This variability reflects differences between developed and developing countries. For example, it has been estimated that the average input of human labour in the USA is around 11.4 h ha^{-1}, while in developing countries such as India and Indonesia are about 634 h ha^{-1} [26].

Overall, the lack of comprehensive databases limits the modelling capabilities and in turn limits agricultural energy policy design.

1.2. *Agricultural Mechanisation*

In the last century, mechanisation in developed economies has dramatically reshaped the agricultural landscape [30]. Farm mechanisation have also brought higher yields and lower demand for land at the expense of higher inputs in the form of fossil-based energy and fertilisers. As the sector is very sensitive to social impacts and energy prices, rising prices usually force producers to seek new technological advances or methods to stay competitive, especially for large farms. The sector is very sensitive to environmental and social impacts as well as energy prices. Population and demand increase, diet change behaviour, impacts of climatic variability and weather extremes and rising energy prices usually force producers to seek new technological advances or methods to stay competitive [31]. Effects are more relevant for large farms as they have a more flexible structure which enables them to respond to market dynamics faster [32].

Economic growth and the commercialisation of agricultural systems is leading to further mechanisation of agricultural systems in developing countries in Asia and Latin America [33]. Over recent decades and mainly due to the introduction of new technologies, yields for most crops have been increasing between 1–2% y^{-1} [6]. Nevertheless, it is likely that these rates will be reduced for the foreseeable future [5]. The FAO [6] forecasts that crop yields will continue to grow at an average rate of 0.8% y^{-1}; this value is half compared to previous decades (1.7% y^{-1}). In fact, the appropriate mechanisation planning and the use of improved and efficient machinery and farming practices have the potential to enhance crop productivity while reducing energy intensity, land use and related emissions [22,34]. For instance, the European Commission [35] has developed a series of 28 agri-environmental indicators. One of those, number 12 refers to intensification/extensification. This indicator not only showcases the current farming intensity and agricultural productivity of each country members, but also provides an index that could be used for planning for sustainable agricultural practices in the region as well as to show the potential future socioenvironmental impacts of the sector.

Many authors have tried to characterise a mechanisation index to define the level of adoption of agricultural technologies. Nevertheless, the definition of a mechanisation index remains highly uncertain due to the simplification of the agricultural production process and the impact of other biophysical factors such as weather, water availability and soil characteristics. Conforti and Giampietro [18] defined a mechanisation index based on output-input ratio of global crop production, which was then later used by Ozkan et al. [36] for Turkey. Later, Singh [37] proposed a mechanisation index based on the ratio of cost of use of machinery to the total cost of use of human labour, draught animals and machinery.

Mechanisation has mainly concerned land preparation [38,39], tractorisation [40] and harvesting practices (crop rotation management [41]), but also the integration of reforestation practices [42–45] and which have led to energy consumption reduction. Finally, for the case of China, for example, Yang et al. [46] investigated the rapid mechanisation diffusion in China in in the previous decade; they show a five-fold increase in horsepower stock for medium and large tractors, and 25-fold for small tractors. Considering farms of around 133 ha of size, the study shown that the implementation of mechanisation and tractorisation has allowed farms to reduce harvesting time to only 2–3 h per farm.

1.3. Agricultural-Based ESM and Relevance for Policy Makers

To properly implement any energy-related technological measure in the sector, a rigorous analysis must be undertaken by simulation models to estimate the projected value of technical advances that would aid in the implementation of informed decisions by key actors in the sector. Generally, the agriculture sector has been neglected in energy systems modelling (ESM) due its lack of adequate direct and indirect energy use and emissions data that could aid to better characterise production processes in the sector. The increased role of bioenergy in the last years has been the main reason that a handful of ESMs started to consider the agricultural sector in its modelling approaches [47]. The sector's interactions with other energy vectors as well with ecosystems makes it important for the inclusion in climate change mitigation studies and policies. For example, important dynamics exist such as the indirect energy input from fertilisers and other agrochemicals as well as the amount of total emissions related to the entire food, forestry products and bioenergy production processes [48].

Walker [15] described a method for modelling integrated energy systems in agriculture. The modelling algorithm helped identify the different classes of an agricultural process (including the material's transformation, transport and storage). Each class is also associated with a cost, which can be monetary, labour, energy or land. Uri [49] provided a model to address fuel substitution between alternative types of energy focusing in the US agricultural sector. The author found the responsiveness of the sector to fuel price change. Later, he provided similar studies focusing on sectoral price responsiveness on diesel, gasoline, LPG and natural gas [16,17]. More recently, some authors have linked energy system models and detailed agricultural modelling to provide answer to specific questions [50–53].

Daioglu et al. [54] used the IMAGE, a spatial model with the capability of determining supply and demand of biomass and bioenergy products and its potential effect on the wider energy system by 2050. The model has been used to determine the future share of bioenergy in a 1.5 °C scenario. Results show that biomass could represent up to 20% (180 EJ/year) of the global energy consumption; nevertheless, for this to be achieved significant improvement in agricultural production and land zoning is needed. Similarly, Wu et al. [55] estimated bioenergy potential under carbon constrained scenarios and societal transformation measures. The authors based the analysis using a Computable General Model combined with a spatial Land Use and Environmental model. Authors found similar results for bioenergy production (up to 200 EJ/year by 2050) to the previous paper, with supply prices between USD 5/GJ and USD 10/GJ.

Other efforts where the modelling of the agricultural sector is essential is on the water-food-energy nexus research area. For instance, Li et al. [56] develop a stochastic model to optimise irrigated agriculture. This allows policy makers to develop programmes to optimise land, energy and water resources while maximising environmental and economic benefits of the agriculture sector. Elkadeem et al. [57] presented a techno-economic optimisation analysis for renewable energy integration aiming at agriculture and irrigation electrification. This study has the ability to present different technological systems structure that minimise environmental impact and maximise economic returns.

While the ESMs which integrate the agricultural sector, have a simplified description of the mechanisation in terms of emissions, energy demand and land use for specific crop types, the agriculture-oriented models include specific crops [58] but lacks of the integration with the rest of the energy system. To the best of the authors' knowledge, there is no ESM which integrates a technology-rich agricultural sector modelling with a dynamic modelling of the whole energy sector. This modelling approach guarantees a comprehensive assessment of the energy use, agrochemical demand, land use as well as the emissions associated to the agricultural sector.

1.4. Study's Aim

The aim of the paper is to assess the potential decarbonisation in the agricultural sector by proposing a comprehensive methodology in which mechanisation levels diffusion interplays with food and feed demand. The methodology is based on a modelling framework, the MUSE-Ag & LU model [59] which belongs to the MUSE modelling suite [60]. The proposed approach allows us to model the cross-dependencies between (i) socio-economic development, driving an expectedly high growth in demand of agricultural commodities, (ii) the energy demand linked to the technology development in agriculture, as represented by mechanisation levels, low-carbon fuels, and agrochemicals demand, and (iii) the sector emissions.

In addition, this tool has the capability to model the techno-economics as well as the environmental benefits of mechanisation on the harvest of specific crops. The main difference of MUSE-Ag & LU with other agricultural energy models is that energy technologies in the sector are represented by different levels of mechanisation; therefore, the model is capable to model and inform future mechanisation adoption in the sector. The model is part of a broader ESM which considers all the sectors in the economy, providing straightforward connections to model the dynamics between agriculture and the rest of energy vectors in the wider energy system. To demonstrate the model's capabilities, this study will focus on analysing the following: (i) future specific crop demand for maize, soybean, wheat and rice under different scenarios (such as low and high food and feed demand and population growth), (ii) modelling of mechanisation adoption, labour and necessary investment, and (iii) analysing energy demand, land use change and related emissions for the selected crops.

To demonstrate the applicability of the novel framework, China has been selected as case study mostly due to data availability from the FAO [6] and USDA [61]. Another reason for the selection has been due to their importance in global food production, the potential for modern agricultural mechanisation and the expected dietary increases for the population. According to the FAO [6], currently the average vegetable-based daily food production for China is 9.9 MJ day^{-1} (2382 kcal day^{-1}), and 58.4 and 37.1 g day^{-1} for protein and fat, respectively. However, if feed demand is also considered, these figures would increase by 803 kcal day^{-1} or 25%. As comparison, the USA food production stands at stands at 11.3 MJ day^{-1} (2697 kcal day^{-1}), with supplies of 39.8 and 93.9 g day^{-1} of protein and fat, respectively; if feed demand is considered, the total value would increase by 23%. Usually a low share in feed demand (>20%) can be related to ruminants-based production systems, while larger shares is due non-ruminants fed production systems. Additionally, Pradhan et al. [62] considers that for the case of China, if the expected diet changes projections take place, feed demand share could increase to 1203 kcal day^{-1}, representing 33% of the crop production share.

The paper is organized as follows. First the advancement compared to the MUSE-Ag & LU is described. Secondly, an overview of the modelled scenarios is presented. Next, the paper showcases the results, followed by discussions and conclusions.

2. Materials and Methods

2.1. *MUSE-Ag & LU Additions*

The general MUSE framework aims to simulates future decarbonisation scenarios in regional energy systems [63]. The Agriculture and Land Use (Ag & LU) module [59] is a model within the MUSE framework [63]. MUSE-Ag & LU, similar to other demand sectors in MUSE (buildings, transport and industry), contains investment decisions, as well as energy technologies representation with a particular land-use oriented modelling approach. Furthermore, the model determines the impact of the sector on global warming, quantifying emissions due to direct and indirect energy use as well as land use management and land use change. In doing so, the model applies a bottom-up approach and assesses the effects of technology innovation. The model produces a time series of fuel, agrochemicals and land demand to meet four general agricultural services: (a) crops, (b) animal-based food, (c) forestry products and (d) bioenergy [59]. The model accounts for CO_2 emissions linked with on-farm energy use as well as N_2O emissions (e.g., from fertiliser utilisation) and CH_4 emissions (e.g., from rice production or manure management). The Ag & LU module generic simulation framework and integration with MUSE is represented in Figure 2.

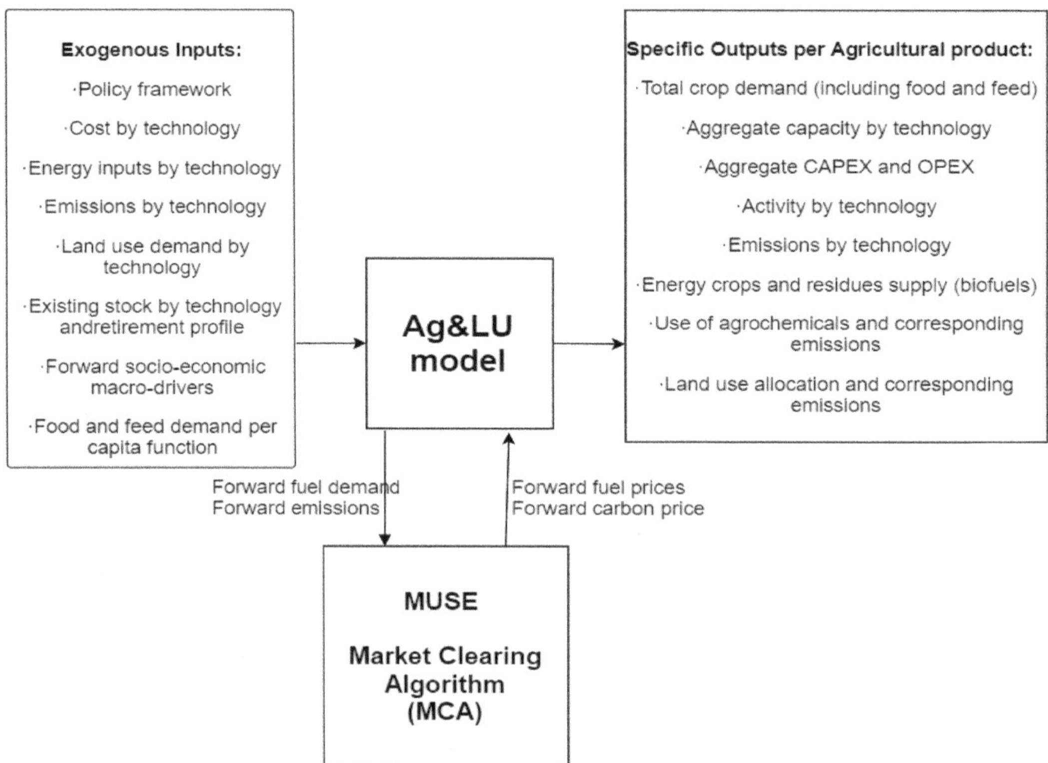

Figure 2. MUSE-Ag & LU framework and data exchange. Source: [59].

The model has been expanded to model specific agricultural crops. In this sense, demand projections are necessary. To cover this demand, new mechanisation technologies have been characterised using cluster analysis and optimisation techniques. It is to this mechanisation processes, that data on energy inputs, yields, fertiliser demand, costs and related emissions have to be defined.

The modelling approach proposed does not explicitly model regional soil, climate, and non-technological factors affecting farming practices, but uses crop yields as a proxy which reflect the abovementioned factors. Crop yields are based on datasets which reflect the variability of local soil, climate and farming practices at an aggregated level. This paper will focus on the land, energy, emissions, and cost implications related to four main crops. However, land implications include constraints which refer the use for animal food, forest-based products, and bioenergy.

The following subsections explain the main steps followed in the development of the model.

2.1.1. Crop Selection

Originally, MUSE-Ag & LU [59] only modelled four general agricultural services: (a) food crops, (b) meat-based products, (c) forestry products and (d) bioenergy. In ESMs this simplification is a common approach due to its low global share of direct energy consumption and related greenhouse gas emissions [59]. In this study, the expansion of food crops is suggested. Table 1 displays the top 6 crops by global production for China as well as for the USA for comparison purposes.

Table 1. Crop production in 2016 [6].

Crop	Global (Mt)	USA (Mt)	Relative Importance by Mt Produced USA (Rank)	China (Mt)	Relative Importance by Mt Produced China (Rank)
Sugar cane	1891	30	5th	123	5th
Maize	1060	385	1st	232	1st
Wheat	749	63	3rd	132	4th
Rice, paddy	741	10	8th	211	2nd
Potatoes	377	20	6th	99	6th
Soybeans	334	117	2nd	12	27th

Although, sugarcane is the most produced crop is also very sensitive to temperature, humidity, and other factors that vary largely by country. Thus, food has been disaggregated into four crops: maize, wheat, rice, and soybeans, mainly because their price is highly correlated with oil prices [64]. In China, where these crops represent about 33% of Chinese crop production [6], this would potentially have higher implications on food security.

2.1.2. Crop Demand Projections

MUSE uses exogenously given macrodrivers (population and GDP) [27] to project future demands of services. Based on historical data, future crop demands are derived using regression models. As demand services are specified in Joules in MUSE, data on total calorie production per crop collected from the FAO [6] has been converted into Joules through the raw calorie density per crop type (Table 2).

Table 2. Energy densities per selected crops.

Crop	kcal/kg	MJ/kg
Maize	3650	15.27
Soybean	4460	18.66
Wheat	3390	14.18
Rice	3570	14.93

Following, and similarly to van Ruijven et al. [65], linearised regression models have been used to identify that related economic activity (GDP per capita) to specific crops. In this case, four linear regression models were tested and implemented to forecast future crop demand:

$$Linear(L){:}\ C = a + b \times GDP_{pc} \quad (1)$$

$$Exponential(E){:}\ C = a \times e^{b\ GDPpc} \quad (2)$$

$$Semi\text{-}log(SL): C = a + b \times \ln(GDP_{pc}) \quad (3)$$

$$Log\text{-}log(LL): lnC = a + b \times \ln(GDP_{pc}) \quad (4)$$

In which a and b are constants estimated in the regression and serve as an input to the model. Both the R^2 value and the root mean square error (RMSE) on per capita consumption values for all models are reported. The R^2 values for the linear models are not all comparable, since some are for absolute consumption levels and others are for logarithmic values (ln(C)). Hence, the lowest RMSE model is selected for future predictions. The historical crop production per capita data, as well as the regressions tables highlighting the selected linear function for each crop model, are presented in Figure 3 and Table 3, respectively.

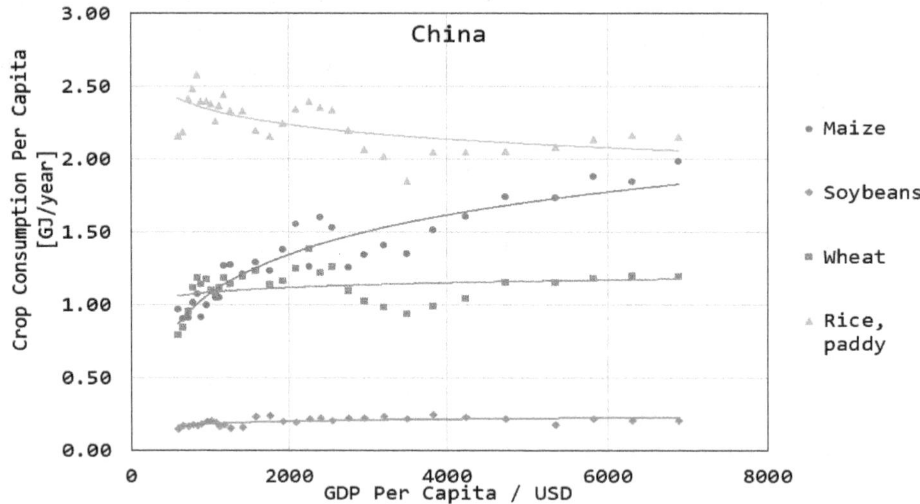

Figure 3. Crop production (including food and feed demands) per capita versus GDP per capita in China. Data: FAO [6].

Table 3. Correlations per crop for China.

Crop	Equation Form	R2	M	C
Maize	Linear	0.132	3.44×10^{-10}	3.08×10^{-7}
Maize	Exponential	0.891	9.77×10^{-5}	1.03×10^{-6}
Maize	Semi-Log	0.943	8.28×10^{-7}	-5.33×10^{-6}
Maize	Log-Log	0.928	4.98×10^{-1}	-1.75×10^{-1}
Soybeans	Linear	0.328	-9.53×10^{-12}	2.63×10^{-7}
Soybeans	Exponential	0.301	-4.36×10^{-5}	2.67×10^{-7}
Soybeans	Semi-Log	0.350	-4.92×10^{-8}	6.33×10^{-7}
Soybeans	Log-Log	0.324	-2.26×10^{-1}	-1.34×10^{-1}
Wheat	Linear	0.838	7.75×10^{-11}	7.14×10^{-7}
Wheat	Exponential	0.826	7.18×10^{-5}	7.66×10^{-7}
Wheat	Semi-Log	0.895	4.00×10^{-7}	-2.29×10^{-6}
Wheat	Log-Log	0.885	3.71×10^{-1}	-1.69×10^{-1}
Rice, paddy	Linear	0.718	6.98×10^{-11}	1.71×10^{-6}
Rice, paddy	Exponential	0.698	3.44×10^{-5}	1.74×10^{-6}
Rice, paddy	Semi-Log	0.763	3.59×10^{-7}	-9.89×10^{-7}
Rice, paddy	Log-Log	0.746	1.77×10^{-1}	-1.46×10^{-1}

2.1.3. Mechanisation Levels

Similarly to [34], the model considers three levels of mechanisation: traditional, transitional and modern, with distinct inputs for fuel use, costs and production efficiencies. A basic representation of the agricultural mechanisation concept applied in the model is shown in Figure 4.

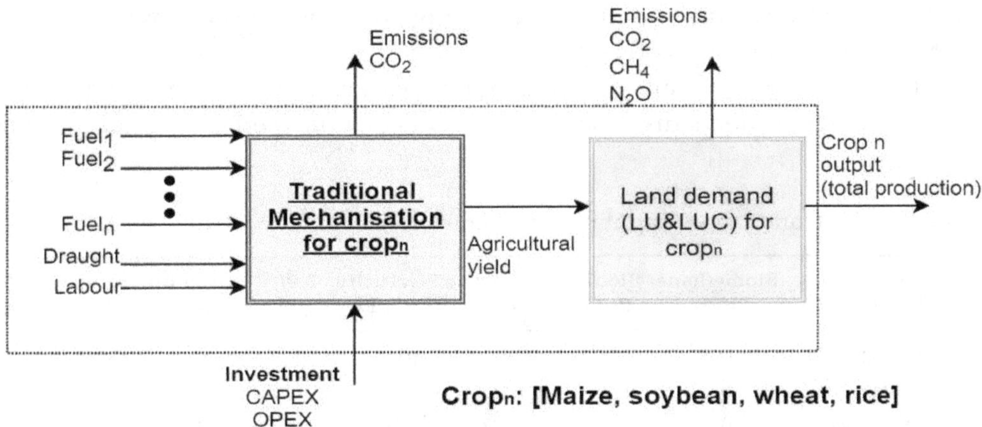

Figure 4. MUSE-Ag & LU mechanisation production process [59].

Traditional mechanisation considers both manual and some level of mechanised agriculture (e.g., tillers) with a minimum amount of direct and indirect energy. Transitional mechanisation presents mechanisation in some parts of the agricultural production chain. Tractors, tilling machines, mechanical heating/drying and irrigation using mainly electricity and other traditional fossil fuels are considered in this process. The use of fertilizers and agrochemicals is also common in this level.

Modern mechanisation, apart from a fully mechanised supply chain, it is considered technologically advanced equipment and other inputs in the form of irrigation water, fertilizers and pesticides. Additionally, an extra mechanisation level called "modern renewable" is represented. This level represents the data outliers or cutting-edge agricultural practices; therefore, it is assumed higher efficiency levels (input/output), a larger share of renewable energy and an optimised fertiliser utilisation.

Modern mechanisation levels (with high production rates) would come at much higher capital costs, representing how technical improvements are not attractive by default as farmers are al-ready operating in an area of decreasing marginal return. In MUSE, the demand for higher farm mechanisation emerges at the point when it becomes cost-effective for farmers to use it over other available options.

The identification of a mechanisation level and its characterisation requires the qualitative and quantitative assessment of a mechanisation index, and its impact on agricultural production (yield) and economic factors (cost of cultivation, deployment of mechanical power, agrochemical demand, animal power, etc.). The index should incorporate the significance and economic utility of using equipment with animate and electro-mechanical power for different farm operations in different crops. To generate the modelling technologies, a pre-processing approach is required and detailed in Appendix A. The characterisation of technologies has been obtained by applying the non-hierarchical method approach [18], based on cluster analysis of different countries at different levels of data observations from FAO [6].

To create technologies which adequately represent the heterogeneity of farming practices, mechanisation, and the relationship between climate, energy and agrochemical inputs and yields and mechanisation levels, cluster analysis has been conducted on national data. Among diverse economic variables, we have used GDP agricultural share (%GDPagr) aiming at providing a more robust insight into the agricultural technological development in a region. As described in [59] k-means clustering was used to reduce heterogeneity. For this, the following Hartigan–Wong minimization algorithm [66] was used:

$$W(C_k) = \sum_{x_i \in C_k} (X_i - \mu_k)^2 \tag{5}$$

where X_i is the actual cluster data, μ_k is the cluster k data mean value. To identify the optimal amount of clusters, the Average Silhouette Method [67] was used:

$$S_i = \frac{b_i - a_i}{\max(b_i, a_i)} \tag{6}$$

where b_i and a_i is the lowest average distance and average distance of data i with respect to the entire observations within one cluster, respectively.

Inputs for labour and draught are based in hours per crop output obtained from [37]. Table 4 presents the input values per mechanisation process for each analysed crop used in this study.

Table 4. Fuel, animal and labour input per mechanisation level for each analysed crop in China.

Crop	Mechanisation Level	Biomethane PJ/PJ	Biodiesel PJ/PJ	Diesel PJ/PJ	Electricity GWh/PJ	Gas PJ/PJ	Gasoline PJ/PJ	Coal PJ/PJ	Draught hrs/GJ	Labour hrs/GJ
Maize	*Traditional*	0	0	0	0	0	0	0.002	0.9	150.0
Maize	*Transitional*	0	0	0.023	0.097	0	0.001	0.015	0.1	16.3
Maize	*Modern Fossil*	0	0	0.027	0.110	0.010	0.007	0.066	0	0.5
Maize	*Modern Renewable*	0.004	0.007	0.012	0.120	0	0	0	0	0
Wheat	*Traditional*	0	0	0.105	0	0	0	0	1.1	123.5
Wheat	*Transitional*	0	0	0.198	0.366	0.003	0.005	0.129	0.1	16.3
Wheat	*Modern Fossil*	0	0	0.262	0.636	0.102	0.073	0.648	0	6.6
Wheat	*Modern Renewable*	0.031	0.062	0.104	0.727	0	0	0	0	0
Soybean	*Traditional*	0	0	0.061	0	0	0	0	1.7	237.8
Soybean	*Transitional*	0	0	0.063	0.103	0.001	0	0.041	0.4	55.1
Soybean	*Modern Fossil*	0	0	0.063	0.195	0.024	0.001	0.155	0	6.6
Soybean	*Modern Renewable*	0.010	0.020	0.033	0.278	0	0.020	0	0	0
Rice	*Traditional*	0	0	0	0	0	0	0	44.9	286.2
Rice	*Transitional*	0	0	0.133	1.272	0.002	0.003	0.087	20.6	131.6
Rice	*Modern Fossil*	0	0	0.266	2.300	0.104	0.074	0.657	0	9.2
Rice	*Modern Renewable*	0.030	0.060	0.100	2.600	0	0	0	0	0

2.1.4. Crop Yields

A yield needs to be assigned for each mechanisation level for each crop. The crop yield has been obtained from historical data. For instance, Figure 5 displays the historical maize yields for three different regions of interest demonstrating different levels of agricultural development. As presented in [59], the 'Least Developed Countries' (countries considered have a %GDP agriculture share of above 16%) can be found to have yields between 1000 and 1400 kg ha^{-1}; thus in this study, it has been assigned around 1200 kg ha^{-1} for traditional mechanisation for maize. On the other hand, the highest levels (represented by countries with a %GDP agricultural share of below 2%) present yields to around 8500 kg ha^{-1}, which has been considered the yield for modern mechanisation processes The same approach was applied to the four crops assessed in this work. Yields values for each crop and mechanisation level is shown in Table 5.

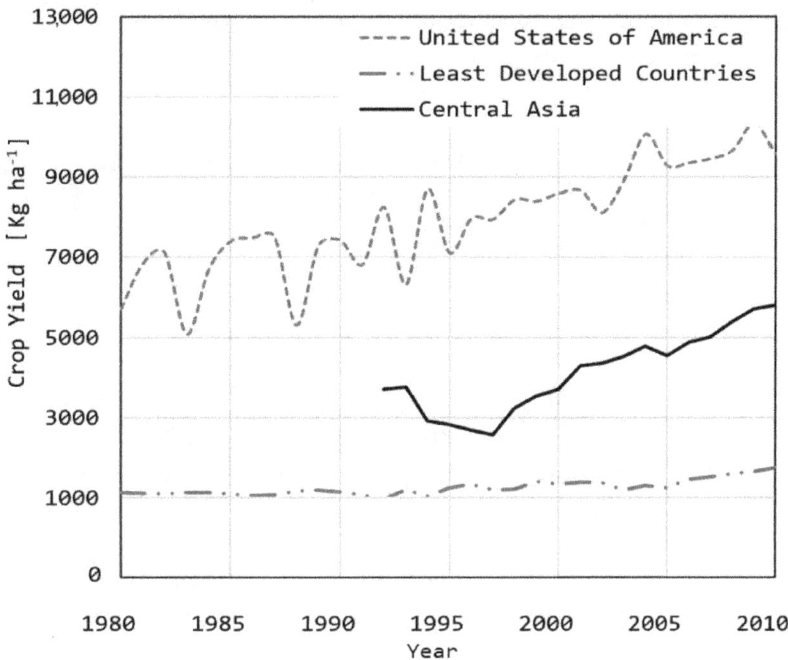

Figure 5. Maize Yields for Selected Countries. Data: FAO [6].

Table 5. Crop yields per technological level.

Technological Level	Maize Yield ($kg\ ha^{-1}$)	Soybean Yield ($kg\ ha^{-1}$)	Wheat Yield ($kg\ ha^{-1}$)	Rice Yield ($kg\ ha^{-1}$)
Traditional	1200	1000	1750	1750
Transitional	5000	1900	3000	3000
Modern	8500	2500	6000	7000
Modern Renewable	11,000	3500	9000	9000

2.1.5. Fertiliser Demand

Modern synthetic fertilisers are usually a combination of three main nutrients: nitrogen, phosphorous, and potassium (N, P_2O_5 and K_2O). Levels of Nitrogen-based fertiliser per ton of production were obtained from FAO data [6] and USDA [61]. Observing the trends of global fertiliser use over the past century, a steady increase in total N-based fertiliser use for all crop types can be observed. This increase is linear for countries in time periods determined as traditional to modern but then lowers when a country reaches a cutting-edge level (above modern mechanisation levels). For instance, maize and wheat production in the USA, have experienced constant production growth and fertiliser use reductions. As time passes and yields increase with modern technologies, fertiliser application rates could become even more efficient. As Calaby–Floody [68] suggest, the inclusion of biotechnology and nanotechnology in the form of smart fertilisation with controlled nutrient release, could dramatically improve fertiliser efficiency.

To represents nutrient input for each mechanisation level and crop type historical data [61] a cluster analysis (using fertiliser input against crop yield) have been used. A full tabulated summary of fertiliser use for each level can be found in Table 6. Data analysis shows that the progression from traditional to modern is almost linear, but further modernisation to modern renewable lowers the demand of fertiliser, as application is starting to be optimised [6].

Table 6. N-based fertiliser Inputs by Crops and Mechanisation levels.

Mechanisation Level	Maize	Soybean	Wheat	Rice
	(kg ha^{-1})			
Traditional	139	42	82	32
Transitional	215	51	92	105
Modern	270	57	106	181
Modern Ren.	245	45	93	155

2.1.6. Integrating CH_4 and N_2O Emissions

FAO [6] data show that more than 70% of N_2O emissions linked with farming are the result of synthetic and organic N fertilisers and crop residues. The sources of N_2O are extremely varied for crop growth. To account for N_2O emissions in the model, the following equation is used and highlights the different factors used to estimate direct N_2O emissions [69]:

$$N_2O_{tot,emm} = (F_{sn} + F_{on} + F_{cr} + F_{vol} + F_{leach}) \times EF_1 \tag{7}$$

where $N_2O_{tot,emm}$ is total emissions by crop, F_{sn} is the N input from fertiliser use, F_{on} is the organic N applied as fertiliser (e.g., animal manure, compost, sewage sludge, rendering waste), F_{cr} is the above and below-ground N content of each crop, F_{vol} is N volatilization, F_{leach} is N from leaching processes, and EF_1 is the emission factor per kg N fertiliser applied (assumed as 0.01 kg N_2O-N). These concern all nitrogen-related processes, such as synthetic and organic fertilisation and crop residues. The sum of both direct and indirect emissions from these sources is then reduced to units of kt PJ^{-1} of crop produced for each mechanisation level (which uses different fertilisation amounts) and entered as a model input. For example, for transitional maize production, the F_{sn} = 215 kg ha^{-1}; therefore, the associated emissions due to fertiliser input are considered at 2.15 kg N_2O ha^{-1}, which in turn converts to 0.027 kt N_2O PJ^{-1} of maize.

Similarly, methane (CH_4) emissions are considered due to their significance in the agricultural sector's contribution to global GHG releases. All modelled values for GHG per mechanisation level are illustrated in Table 7.

Table 7. Nitrous Oxide and Methane emissions per mechanisation for China.

	China (kt PJ^{-1})				
	N_2O Maize	N_2O Wheat	N_2O Soybean	N_2O Rice	CH_4 Rice
Traditional	0.065	0.007	0.021	0.012	4.725
Transitional	0.027	0.007	0.024	0.027	2.756
Modern	0.021	0.006	0.020	0.026	1.181
Modern Ren.	0.016	0.006	0.015	0.020	0.919

2.1.7. Techno-Economics of Mechanisation

MUSE-Ag & LU uses a bottom-up approach for technology characterisation, on which the net present value (NPV) is estimated to determine the investment in new technologies. In particular, new investments would favour technologies with a higher NPV.

For a detailed techno-economic characterisation of each mechanisation level, data on costs and return estimates from USDA [61] has been collected. The values are available in USD per planted acre. These values have been converted into USD per hectare using historic yield data [6] and then to USD per PJ y^{-1} to show unit price for installed capacity (also, USD values were scaled for inflation). The data has been plotted against the achieved yield (for instance, maize yields in Figure 6), and by locating the mechanisation level according to the yield, the capital, fixed and variable costs have been assigned. The obtained costs per mechanisation level for each crop are displayed in Table 8.

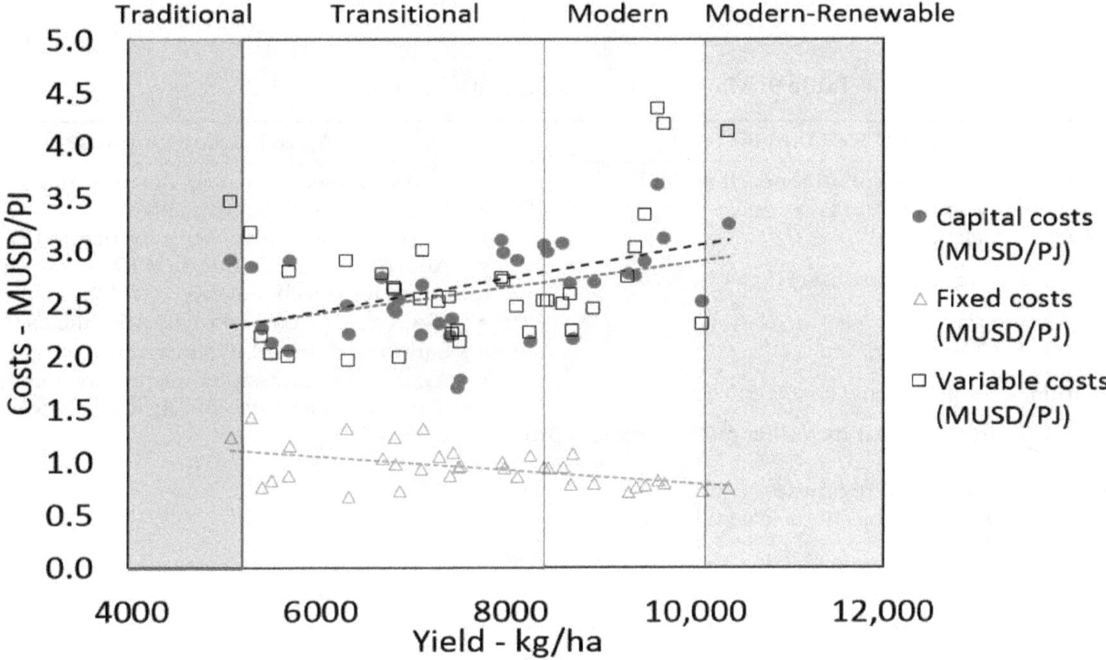

Figure 6. Costs per planted maize energy content, USD2010 PJ y^{-1}. Data: USDA [61].

Table 8. Techno-economic data for different mechanisation levels per type of crop.

Crop	Maximum Capacity Addition for China (PJ y^{-1})			
	Traditional	**Transitional**	**Modern**	**Modern Renew.**
Maize	30.0	90.0	200.0	40.0
Soybean	15.0	20.0	140.0	50.0
Wheat	-	0	3.0	1.0
Rice	4.0	30.0	200.0	80.0
	Capital Costs (MUSD/PJ y^{-1})			
	Traditional	**Transitional**	**Modern**	**Modern Renew.**
Maize	1.6	2.2	2.8	2.9
Soybean	2.8	6.3	6.6	6.9
Wheat	3.8	5.3	6.3	8
Rice	1.8	3.7	4.8	5.1
	Fixed Costs (MUSD/PJ y^{-1})			
	Traditional	**Transitional**	**Modern**	**Modern Renew.**
Maize	1.3	1.1	1	0.8
Soybean	1.2	2	2.1	1.8
Wheat	2.1	2.1	2	1.7
Rice	0.9	1.5	1.9	1.6
	Variable Costs (MUSD/PJ y^{-1})			
	Traditional	**Transitional**	**Modern**	**Modern Renew.**
Maize	1.4	2.2	3.1	2.5
Soybean	1.5	4.2	4.4	3.8
Wheat	1.3	2.7	3.6	3
Rice	4.8	5.4	5.8	4.9

2.1.8. Model's Inputs/Outputs

A summary of the main inputs used in the model as well as outputs provided by the model is shown in Table 9.

Table 9. Main inputs and outputs from the model.

Ag & LU Key Inputs	Ag & LU Key Outputs
Techno-economic characterisation for each agriculture crop • Input by energy source (PJ/PJ) • Conversion efficiency (%) • Agrochemical input (N fertiliser) (kt PJ^{-1}) • Yields (Mha PJ^{-1}) • Emissions ($KtCO_2eq$ PJ^{-1}) • Unit capital and operational cost (USD PJ^{-1}) Existing stock for the base year including their retirement pro le (PJ y^{-1}) Policy framework and fiscal regimes Macro-economic drivers' projections (e.g., GDPcap, population, urbanisation)	Agricultural mechanisation detail outputs • Fuel demand by source (PJ) • Agricultural commodity production (PJ) • Aggregated CAPEX and OPEX of new installed technologies (mechanisation) (USD) • Aggregated demand of agrochemicals (kt) • Land use demand by agricultural crop (Mha) • Aggregated Emissions due to direct energy use, fertilisers and Land use change ($KtCO_2eq$)

2.2. Calibration and Validation

Values on agriculture energy use, emissions, and land use have been calibrated using 2010 as a base year [6]. This year has been selected because of the existence and reliability of data for all the sectors as well as being a year without unusual political, economic or environmental circumstances. Additionally, the model has been validated by projecting values for 2010–2014 on food demand, fuel consumption, fertiliser demand, and emissions [6].

Due to lack of explicit data to account for the base-year installed capacities of the proposed mechanisation levels, a linear programming (LP) optimisation problem has been developed in the General Algebraic Modelling System [70] aiming at minimising the difference between actual and modelled sectoral emissions based on a set of constraints (share of mechanisation level and share of fuel per technology). To obtain modelled sectoral emissions, this pre-processing procedure considers the generated mechanisation levels by the LP, its fuel input share and the related emissions factors given by the [4]. By minimising the difference between modelled and actual emissions in the sector, the optimisation model is capable to calculate installed capacity of mechanisation levels per technology (t), crop (s) and region (r). The model mathematical formulation is presented in Appendix B. By solving this constrained optimisation problem, all structural alternatives are evaluated, and the programme identifies the best solution. The installed capacities by mechanisation level and its production shares by crop and by country obtained by the optimisation model are displayed in Table 10.

Table 10. Existing capacity per mechanisation Level for China.

Crop	Installed Capacity PJ y^{-1} (GW)				Annual Avg. Yield [6] (kg ha^{-1})	Share of Production (%)			
	Traditional	Transitional	Modern	Modern Ren.		Traditional	Transitional	Modern	Modern Ren.
Maize	27.1 (0.9)	2395.5 (76.0)	153.3 (4.9)	135.6 (4.3)	5460.0	1.0%	88.3%	5.7%	5.0%
Soybean	42.3 (1.3)	236.5 (7.5)	2.8 (0.1)	0.0 (0.0)	1771.0	15.0%	84.0%	1.0%	0.0%
Wheat	0.0 (0.0)	681.5 (21.6)	952.1 (30.2)	0.0 (0.0)	4748.0	0.0%	41.7%	58.3%	0.0%
Rice	147.3 (4.7)	272.2 (8.6)	2261.4 (71.7)	265.2 (8.4)	6548.0	5.0%	9.2%	76.8%	9.0%

2.3. Scenarios

A set of three main scenarios have been developed to test the model:

- **Reference:** To drive service demands, the SSP2 narrative, which describes a middle-of-the-road development for mitigation and adaptation [71], has been considered. The scenario considers a carbon price that is endogenously calculated by MUSE.
- **Low development:** In this scenario, the SSP3 narrative, which describes a fragmented world, failing to achieve sustainable development goals, with little efforts in reducing fossil fuel utilisation and negative environmental effects has been considered. In this scenario, higher population growth (resulting in higher food and feed demand), as well as increases in technological costs and fuel prices, and a reduction in yield growth rates have been varied considering a 20% deviation from the Reference scenario. Additionally, no carbon price is considered.
- **High development:** In this scenario, the SSP1 narrative, which describes a sustainable pathway, with constant efforts in reaching development goals and reducing dependency on fossil fuels mainly driven by rapid technological development, has been considered. In this scenario, lower population growth (resulting in lower food and feed demand), as well as reduction in technological costs and fuel prices, and increase in yields growth rates have been varied considering a 20% deviation from the Reference scenario. Apart from a carbon price for CO_2 emissions, a specific tax has been imposed also on N_2O and CH_4. However, an imposed tax on these emissions needs to be weighted. Simulations have shown that a tax similar to the existing carbon level is too low, not only since N_2O and CH_4 are considerably more potent than CO_2 but also due to the lower level of emissions; therefore, a low tax would be insignificant. On the other hand, applying a tax that reflects the global warming potential (e.g., N_2O 298 times as strong as CO_2) pushes technology levels back down to traditional agricultural practices and thus has an important impact on predicted land use and food security. At high GHG price rates, it is of interest for a farmer to revert to cheap, traditional technologies and use significantly more land to make up for the emissions costs. Therefore, compared to CO_2 prices, a tax range between 0–300 times larger for N_2O and 0–50 times larger for CH_4 has been studied. After a sensitivity analysis, outputs suggest an optimal emission price of 10 times larger for CH_4 and 25 times larger for N_2O.

A linear decommissioning profile of existing technologies (existing mechanisation levels) has been applied. The imposed decommissioning profile implies that the capacity is totally renewed by 2040, 10 years ahead of the modelling horizon. Capital and operational costs of modern technologies are varied to understand the importance of price/subsidies in modern efficient technologies. Finally, yields annual growth rates are varied for both scenarios. A summary of each scenario is presented in Table 11.

Table 11. Summary of analysed scenarios for Chinese crops.

Scenario	Population/ Food Demand Growth by 2050	Fossil Fuel Price	Carbon Price	Modern Mechanisation Costs	Yields (Annual Growth)
Reference	SSP2 metrics [71]	IEA [72]/EIA [73] reference scenario	EMF27 [74] reference scenario	No changes	+0.8%
Low Development (high population growth/food demand)	SSP3 metrics +20%	−20%	No carbon tax	+20%	+0.5%
High Development (low population growth/food demand)	SSP1 metrics −20%	+20%	High price EMF 27 [74] 10 times larger tax—CH_4 25 times larger tax—N_2O	−20%	+1.3%

3. Results

3.1. Reference Scenario

3.1.1. Mechanisation Adoption and Investment

The reference scenario estimates that by 2050, China's overall demand (PJ y^{-1}) for the analysed crops is projected to increase by 71%. Based on the modelling outputs, the technological shift in Chinese agriculture by 2050 is illustrated in Figure 7. The shift towards technologies with higher efficiencies is exhibited for maize, wheat and rice. The overall crop supply increases constantly until 2035 when population starts declining and crop production stabilises.

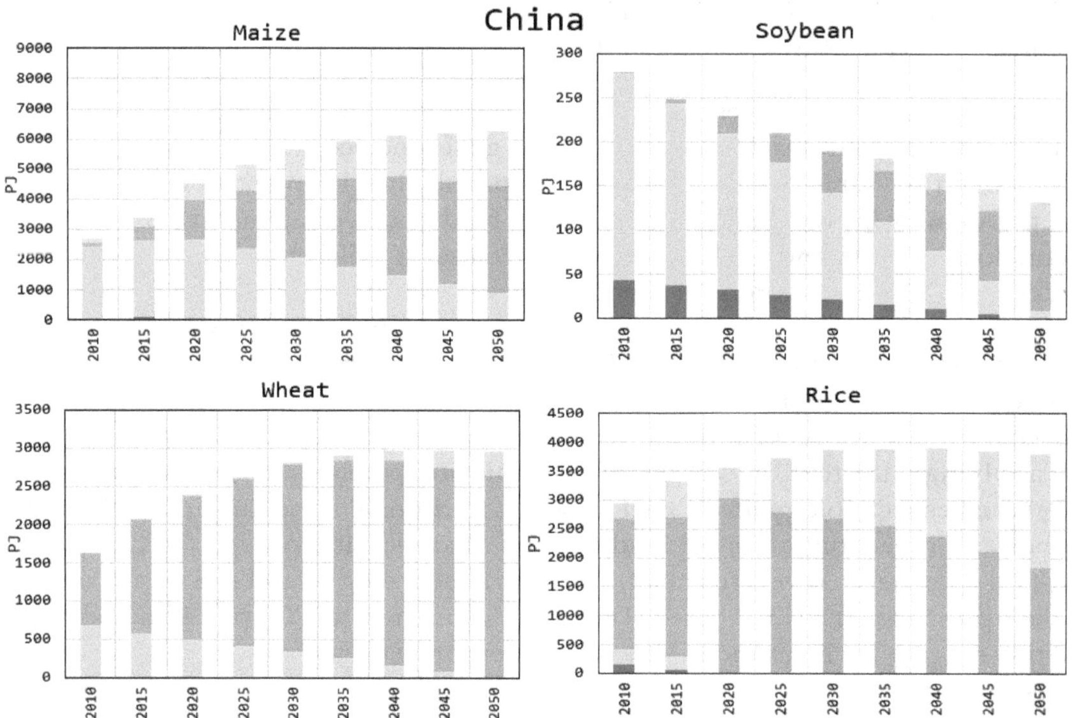

Figure 7. Technological diffusion for China in the reference scenario.

Overall, and considering all analysed crops, in the base year, "transitional" mechanisation is still responsible for 47.4% of the production, while "modern" and "modern renewable" for 44.5% and 5.3%, respectively. By 2050, "transitional" share decreases to 6.9%, overtaken by "modern" mechanisation with 61.9% of the production, followed by "modern renewable" with 31.1%. The forecast for soybean is rather different, but its low production compared to the rest of the crops, has barely any effect on the sector's performance. Nevertheless, soybean results show a steep decrease reducing crop production as GDP per capita improves. This is in line with data from the previous decade, where soybean reached a maximum production level of 17 Mt y^{-1} in 2004, followed by an average annual decrease rate of 2.7% y^{-1}. The main reason behind this was due to an increase in imports from Brazil, USA and Argentina in the last two decades [6]. It is possible that the country has been replacing more of its soybean production with other higher value crops. Additionally, in the short term it is not expected for soybean to play an important role in biodiesel production as currently the feedstock for Chinese biodiesel is mostly based on cotton, rapeseed and Jatropha [75].

In terms of investment and operational costs, the model suggests a necessary expenditure of around 319.4 billion 2017US$ by 2050, each crop requiring the following investment share: rice, 38.2%; maize, 35.3%; wheat, 25.2%; and soy, 1.3%. Modern technologies across the analysed crops would be responsible for around 95.8% of the total cost.

3.1.2. Energy/Fertiliser Demand and Related Emissions

Table 12 shows the regional fuel consumption aggregated for the four crops. Outputs show two main drivers that impact the future projections: (i) the high prevalence of coal in the sector and (ii) the current low mechanisation levels and high labour/draught inputs. In 2010, the analysed crops demand 525 PJ y^{-1} of conventional energy (fossil and renewable), representing 39.1% of the total national agriculture energy use of 1342 PJ y^{-1} [6]. Demand is similar to countries such as the USA; however, for China, labour and animal input is higher, as it is estimated that 12.4×10^9 h y^{-1} and 182.7×10^9 h y^{-1} (or 88.9 million workers) comes from draught and labour work, a value estimated to be 40 times higher than in the USA.

Table 12. Farm-related energy use and emissions in the reference scenario.

	Fuel Use (PJ y^{-1})								
Energy Commodity	**2010**	**2015**	**2020**	**2025**	**2030**	**2035**	**2040**	**2045**	**2050**
Electricity	141	171	195	209	221	225	229	228	227
Diesel	206	243	288	296	309	310	308	297	285
Natural Gas	1	1	2	2	3	3	4	4	5
Gasoline	29	43	49	52	55	56	57	57	56
Coal	145	211	227	227	223	211	197	179	161
Biogas	1	3	2	4	5	6	7	8	9
Biodiesel	2	5	4	8	9	11	13	15	18
Draught [10^9 hrs]	17.22	11.17	6.33	3.42	0.57	0.43	0.36	0.28	0.50
Labour [10^9 hrs]	182.65	164.80	110.20	108.16	97.45	85.71	78.06	66.33	56.63
	Total Emissions (Mt y^{-1}) ¥								
GHG (CO_2eq)	**2010**	**2015**	**2020**	**2025**	**2030**	**2035**	**2040**	**2045**	**2050**
CO_2	31	41	46	47	48	47	46	43	41
CH_4	109	109	102	105	107	106	105	102	100
N_2O	49	57	66	70	73	73	74	73	71

¥ CO_2 emission from electricity production are not considered towards the agricultural sector emissions.

Regarding to conventional energy sources, diesel, coal and electricity are the major fuels, representing 93.7% of the total fossil fuel inputs in the base year. By 2050, electricity and diesel retain their dominance of the overall fuel share, with coal reaching its maximum demand by 2025 and later being displaced by other energy sources used in modern technologies. Despite the growth in biofuels, their overall share remains small by 2050 (<5%), as "modern renewable" mechanisation is responsible of 31.1% compared to 61.9% of "modern" mechanisation that exclusively has fossil fuels as inputs. In terms of labour, it is expected a reduction of 69% reaching 56.6×10^9 h y^{-1} or 27.9 million workers. For emissions, CH_4 related emissions due to rice production represent the highest share, followed by N_2O from fertiliser use and CO_2 from fuel combustion. Overall emissions (CO_2eq) grow by 11.9%, with N-related emissions with the highest increase (+46.1%), followed by CO_2 (+30.2%). On the other hand, CH_4 emissions due to rice cultivation are reduced by 8.6%, mainly due to the installation of more efficient technologies. In terms of emission intensity, this is reduced from 0.025 to 0.016 $GtCO_2$eq $PJcrop^{-1}$ (−35.6%).

N-based fertiliser demand in Mt is shown in Figure 8. For comparison, a projected demand for the USA has been included. In China, the consumption has been calculated at 16.2 Mt y^{-1}, which represents 29.1% of the national demand. As seen, China's fertiliser demand is higher than in the USA due to differences in production. This difference becomes more pronounced as the Chinese agricultural sector intensifies. By mid-century, China reaches 23.0 Mt y^{-1} of fertiliser demand (after peaking in 2040 to 24.2 Mt y^{-1}).

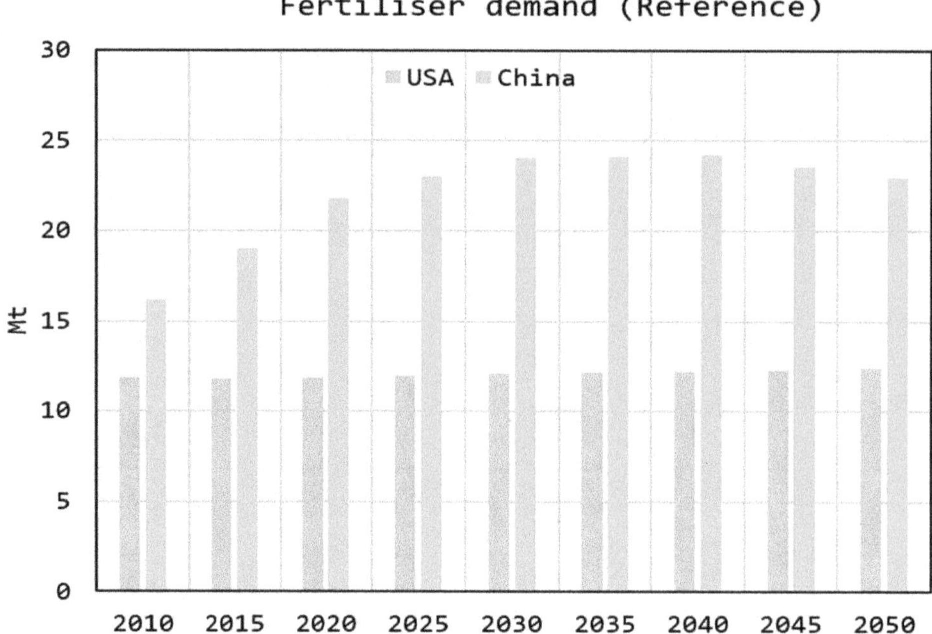

Figure 8. N-based fertiliser demand for both countries in the reference scenario.

3.1.3. Land Use

Figure 9 shows the land use modelling outputs for the four crops. In the base year, the analysed crops demand 106 Mha, with rice (33%) and maize (33%) as the as the largest croplands. Modelling results suggest that land demand will peak in 2030, increasing by 23%, to then shrink to 119 Mha by 2050, thus liberating land that could be used for other crops or for reforestation purposes. Apart from a reducing population at the end of the simulation period, the adoption of more efficient agricultural mechanisation processes with higher yields produces a decrease in land demand. The crop with the highest dynamism is maize, going from 35 Mha in 2010 to 51 Mha in 2050, while soybean gets reduced from 9 to 3 Mha.

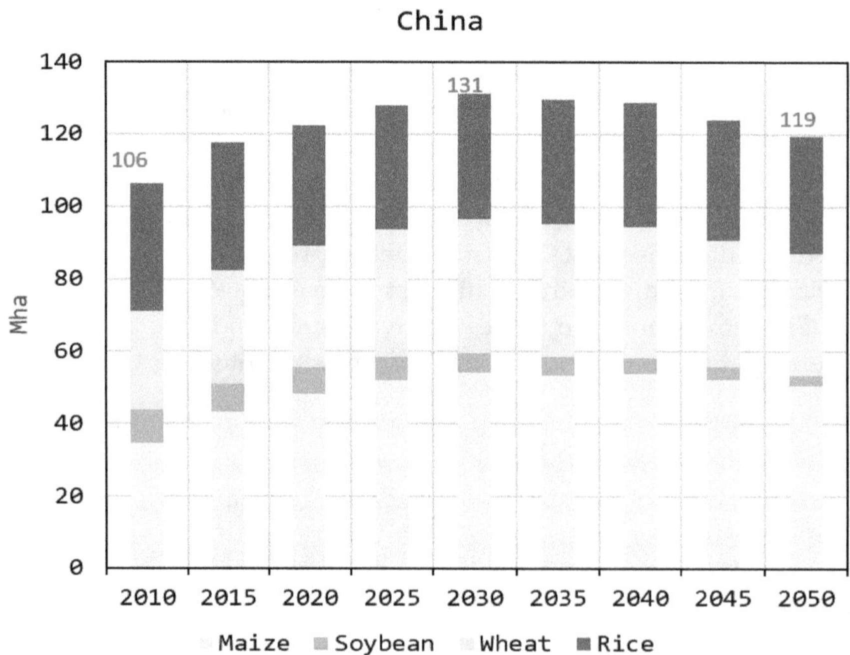

Figure 9. Land use projections by crop in the reference scenario.

3.2. "Low" and "High" Development Scenarios

3.2.1. Mechanisation Adoption and Investment

In the "Low" development scenario, food demand increases due to higher population growth and per capita food intake. Additionally, it is expected that fossil fuels are not hit by carbon prices and advanced more efficient technologies become more expensive. Productivity diminishes due to the decreased investment in modern efficient technologies, as both the NPV differential between modern and transitional technologies is reduced. In the "High" development scenario, food demand reduces due to lower population growth and food intake per capita. Additionally, fossil fuels become more expensive as carbon tax is implemented (including CH_4 and N_2O), and advanced technologies become more competitive, as capital cost reduces and fuel cost becomes cheaper than the traditional options.

As illustrated in Figure 10, China would take up to 2040 to decommission transitional mechanisation, especially for maize and soybean. The outputs showcase the rate of adoption of modern technologies, where "modern" mechanisation almost overtakes by 2050 in all analysed scenarios. This result is expected as current trends in mechanisation diffusion in the country is following modernisation, thus moving to a slight increase in efficiency and renewable fuel share. However, modern efficient technologies with renewable energy share are not as predominant as fossil-based modern, being the latter the preferred choice for investors. In 2030, renewable energy starts to enter the sector, especially for maize and rice crops.

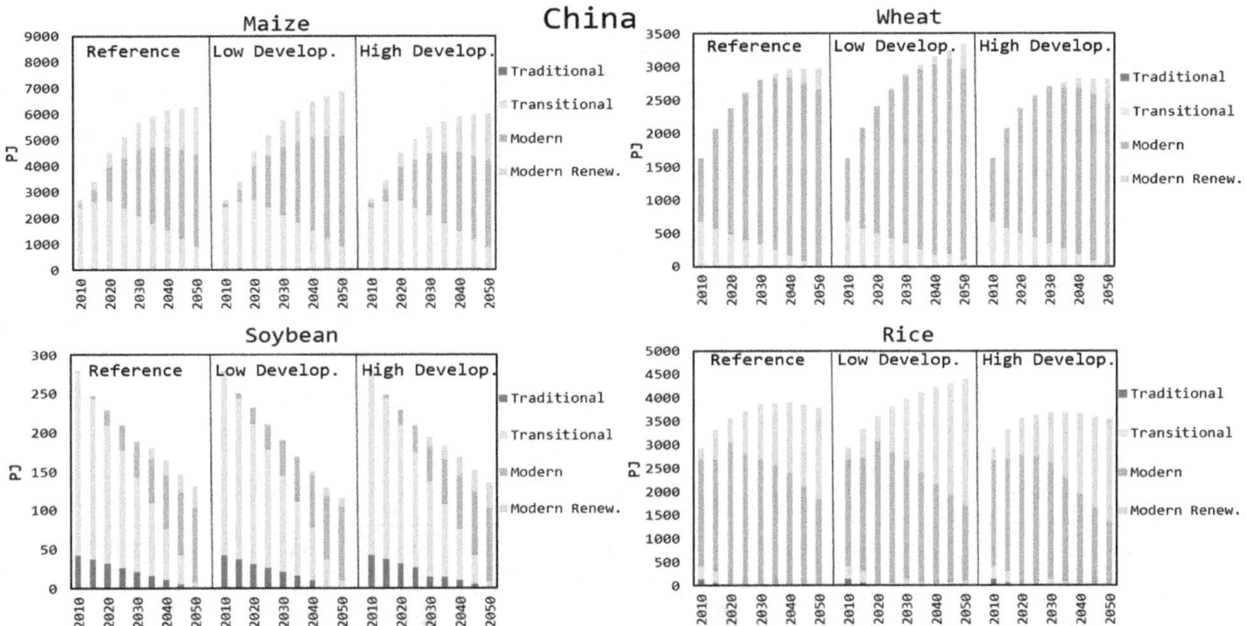

Figure 10. Technological diffusion for reference, low and high development scenarios in China.

Sector mechanisation and intensification is expected for all analysed crops. For the reference case in China, in the base year, "transitional" mechanisation is still responsible of 47.4% of the production, while "modern" is responsible of 44.5%. By 2050, a shift is expected as "transitional" will only be responsible of 6.9%, while "modern" and "modern renewable" will be responsible for 61.9% and 31.1% of the total production. In the "High" development scenario, the "modern renewable" share increases to 35.2%.

The 2010–2050 investment required for the "Low" development scenario has been calculated at around 361.6 billion 2017US$, while for the "High" development at 295.4 billion 2017US$. Apart from a reduced installed capacity in the "High" development scenario which directly explains the lower costs, other important cost difference is found in the higher investment in "transitional" mechanisation processes in the "Low" development scenario (22.4 against 11.7 billion 2017US$), mainly due to lower fossil fuel prices and high cost of modern technologies. Figure 11 presents a cost comparison among the analysed scenarios.

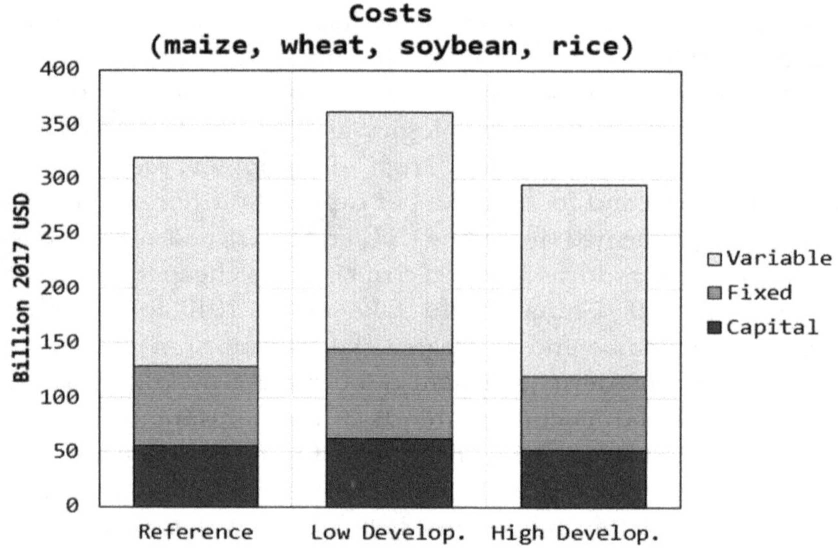

Figure 11. Necessary investment for each analysed scenario.

3.2.2. Energy/Fertiliser Demand and Related Emissions

While in the "Low" development scenario, fuel use increases by 33.9% by 2050 (Table 13), in the "High" development scenario, a reduction of 34.1% compared to the reference is observed. For the reference and "Low" development scenarios, coal still represents the major energy source; however, in the "Low" development scenario that coal has a predominant share. Electricity and diesel are expected to still be major energy sources, but it is gasoline that has the highest growth rates regardless the scenario. Moreover, outputs show a low share of biofuels in future Chinese agriculture. However, if policies are made that promote the use of biofuels in the country's energy system, this might slowly replace the demand of fossil fuels. For example, biodiesel or biomethane-fuelled agricultural technologies (e.g., farm machinery, farm equipment) could easily replace the demand of diesel in agriculture, which currently accounts for 8% of the national diesel consumption.

Table 13. Fuel use and emissions for reference, low and high development scenarios.

	Reference			Low Development			High Development		
Fuel (PJ)	**2010**	**2030**	**2050**	**2010**	**2030**	**2050**	**2010**	**2030**	**2050**
Electricity	141	221	227	141	225	262	141	207	215
Diesel	206	309	285	206	310	308	206	294	255
Natural Gas	1	3	5	1	2	3	1	3	6
Gasoline	29	55	56	29	56	65	29	52	53
Coal	145	223	161	145	277	321	145	165	118
Biogas	1	5	9	1	5	12	1	4	10
Biodiesel	2	9	18	2	11	24	2	9	20
Draught [10^9 h]	17.22	0.57	0.50	17.22	5.48	3.20	17.22	5.05	0.28
Labour [10^9 h]	182.65	97.45	56.63	182.65	120.41	70.92	182.65	114.80	47.96
Emissions (Mt CO_2eq)	**2010**	**2030**	**2050**	**2010**	**2030**	**2050**	**2010**	**2020**	**2050**
CO_2	31	48	41	31	53	58	31	41	34
CH_4	109	107	100	109	116	117	109	109	90
N_2O	49	73	71	49	74	79	49	70	67

In terms of total emissions, the model shows a range between 191.0 Mt y^{-1} and 211.2 Mt y^{-1}. Compared to the reference case base-year emission intensity of 0.025 $GtCO_2eq\ PJ_{crop}{}^{-1}$, even a "Low" development would decrease the index to 0.019 $GtCO_2eq\ PJ_{crop}{}^{-1}$, while a "High" development has the potential to reduce even further to 0.014 $GtCO_2eq\ PJ_{crop}{}^{-1}$.

Finally, the most striking output is the reduction of labour demand. In the "Low" development scenario, labour is reduced by 61.1% (reaching 70.92×10^9 h y^{-1} or 34.9 million workers), while in a highly mechanised sector, the reduction is in the order of 73.6% (employing 23.6 million workers). These values are important, as currently 20% of the Chinese workforce is employed in the agricultural sector [27]. To put this into perspective, the USA only employs 1.6% of its workforce for agricultural purposes.

Lastly, Table 14 shows the fertiliser demand for every scenario.

Table 14. Fertiliser demand for reference, low and high development scenarios.

	Fertiliser Demand (Mt)								
	Reference			Low Development			High Development		
	2010	**2030**	**2050**	**2010**	**2030**	**2050**	**2010**	**2030**	**2050**
China	16	24	23	16	25	26	16	23	21

3.2.3. Land Use Change

Figure 12 illustrates the total land use demand projections for both scenarios compared to the outputs from the reference case. As previously discussed, for the "Low" development case, yields annual growth declines from 0.8% to 0.5%, thus putting more pressure on land. On the other hand, a "High" development scenario explores a yield growth increase to 1.3%.

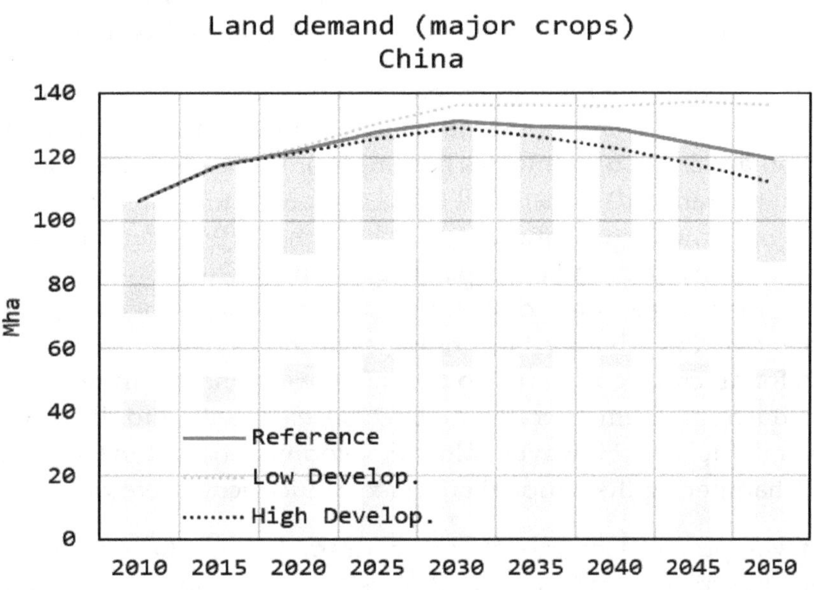

Figure 12. Land Use projections by crop for both countries for analysed scenarios.

Regardless of the scenario, it is expected an increase in land demand compared to the base year. However, the peak demand of land is expected between 2030 and 2035 for the reference case, in 2045 for the "Low" development, and in 2030 for the "High" development due to higher investment and more accessibility to modern technologies. At the end of the "High" scenario, land requirements for the four crops is expected to reach 112 Mha, compared to worst case of 136 Mha in the "Low" development. The "High" development scenario in turn could reduce emissions due to lower levels of deforestation and soil degradation.

4. Discussion

It is expected that developing economies will follow mechanisation diffusions similar to the past century trends experienced in developed economies. While some develop economies already have reached peak yields and technological diffusion in the majority of their crops, developing economies still have large potential for improvement, but also expect higher growth rates in food and feed demands due to the increase in population and per capita income. However, the greatest barrier for sustainable agricultural development is access to modern technologies, as most of the modern machinery is being developed in advanced economies thanks to larger and constant investment in R&D in the sector. As Safdar and Gevelt [76] discuss, developed economies are constantly strengthen its competitive advantage by combining investment and political lobbying. This has caused substantial barriers to local Chinese firms, that regardless of the state-support that the sector possesses, the technological gap of state-of-the-art machinery and processes between advanced economies and those from Chinese firms keeps widening.

Nevertheless, China keeps emphasising the development of the agricultural sector. The future development of the sector is not without related challenges. Among others, food demand security would require policies supporting smaller businesses and promoting advancements in production practices [77,78]. In addition to the economic implications, the introduction of advanced practices in agriculture becomes even more crucial when linked to emissions reduction from a perspective of low carbon transition.

Modelling outputs suggest that for all four crops analysed for the case of China, a shift towards modernisation is inevitable. The existence of some "transitional" agriculture is because in some rural regions, farmers are cultivating on marginal lands, where the access to more efficient technology might not be economically attractive. In this situation, it is the economic constraint that prevents the movement to the "modern" mechanisation levels, and not the lack of technological availability.

Analysis of variations over the reference scenario illustrate the dependence of the results on the made assumptions. In the "Low" development scenario, food demand is increased, fossil fuels are not hit by carbon taxes, and advanced technologies become more expensive. As expected, by mid-century, yields could decline in the range 10–14% for all crops compared to the reference scenario. This is due to the lower investment in modern technologies, as both the NPV differential between modern and transitional technologies is reduced, and the capacity gap is narrowed. On the other hand, in the "High" development scenario, food demand is reduced, fuels become more expensive, a tax on CH_4 and N_2O emissions is implemented, and advanced technologies become cheaper. In this case, yields improve in the range 16–24% for all crops compared to the reference scenario. In terms of direct energy use, despite the scenario, outcomes demonstrate that by 2050, electricity and diesel will still make up the largest portion of overall fuel use. However, if low development occurs in China, coal could still have a high share by 2050, hampering the national efforts to reduce economy-wide emissions.

The "High" development scenario shows a dramatic uptake of modern technologies in 2040—precisely when the current traditional and transitional technologies are fully decommissioned. It would be highly beneficial if policies promoting an early retirement of the existing technologies, when capital costs are outweighed by running costs, were put in place, accelerating decommissioning, namely if the resultant increase in capital costs is outweighed by the decrease in running costs.

Carbon taxes can have a significant impact on resource use and emissions. However, countries and relevant organisations are recommended to properly balance these taxes and subsidies. In the 2050 horizon, an optimal methane tax for controlling rice field emissions was observed at around USD 500/ton CH_4, specifically for Chinese agriculture, contributing to around 20.1% emission reduction compared to the base case. Due to the extremely high emissions linked with permanently flooded rice fields, the pathways to reduce these emissions focus mainly on reducing the flooded period as much as possible, close control of water levels on fields, new seeding techniques, and/or sequentially managing wetting and drying of patties to prevent methane build-up [79]. Erda et al. [80] investigated potential carbon reduction options in the Chinese agriculture sector. The author found a potential reduction of 4–40% of CH_4 emissions from better practice in rice, ruminants and animal waste, as well as reductions of 20% of N_2O from micro-dosing fertilisation.

New land use techniques such as precision agriculture [81], i.e., any form of agriculture that changes conventions by utilizing new technologies, are expected in the future. As Stafford [82] suggests, precision agriculture has the potential to reduce environmental impacts by providing better crop and land management while increasing economic returns, thus providing a stronger market. In the methodology, it was highlighted that fertiliser use and its subsequent emissions incur high investment and pollution costs. The emergence of precision agriculture and the optimised outputs it promises are, therefore, of high interest for policy makers. Considering the already high yields achieved by farms in some countries, it is reasonable to assume that when this technology is made globally available, a high increase in application efficiency could be observed. Although not analysed in this study, diet changes towards less meat-based products (which in turn would reduce feed crop demand) combined with waste minimisation could be the most cost-effective measure aiming at reducing sectoral emissions to lower levels [83].

5. Conclusions

In this paper, a specific crop model for the agricultural sector integrated in an energy systems model has been proposed. This study proposed a simulation framework capable of modelling the future energy use, production, fertiliser demand, land use and related emissions as a function of mechanisation diffusion for specific crops. Four main crops have been studied to demonstrate the model capabilities. The techno-economic inputs refined by crop type provided a deeper under-standing of the value of technological advances within the sector and their implication in the wider energy system and environment.

Agricultural and land use models are set to play a significant role in helping decision makers to make informed decisions to ensure a future for the energy system, farming and global food production in line with a sustainable development. Mechanisation planning allows for both a quantitative assessment of advancement and serves as a tool for analysis in countries where specific data are lacking.

The current study presents scenarios of mechanisation in agriculture using China as a case study, which represents about 20% of global crop production. The results obtained demonstrate the complex variations by crop in technology investment, yields, fuel use, land demand and GHG emissions. Additionally, it is expected that the energy demand in the sector will increase driven by projected higher food demand, especially if more efficient technologies are not deployed. This can be seen in the "Low" development scenario, where higher investments in "transitional" mechanisation processes instead of "modern" mechanisation have been found, mainly driven by lower fossil fuel prices and high cost of modern technologies. Although coal would remain as an important source for Chinese agriculture, there would be room for electrification in the sector. On the other hand, biofuels would merely increase their share from 0.5% to up to 4.4% by mid-century. Despite the increase in energy demand expected across scenarios, outputs suggest a reduction in sectoral emissions compared to the base year, mainly due to a larger share of renewable energy and more efficient use of fertilisers and land.

The model also revealed important socio-economic implications in the reduction in dedicated human labour. If the modelled trends of increased mechanisation are extrapolated to the entire sector [27], a reduction of up to 114.5 million workers by 2050 could be expected. A workforce that eventually would need to switch either the industry and/or the service economic sectors.

For future work, the characterisation of the majority of agricultural crops is envisioned, as well as the different types of animal-based food production, forestry products and energy crops. Additionally, we expect to integrate a spatially explicit model that will characterise land suitability based on several biophysical characteristics such as temperature, soil, or water availability, aiming at understanding its effects in crop productivity and reduce uncertainties in the outputs.

Author Contributions: Conceptualization, I.G.K. and S.G.; methodology, I.G.K.; software, I.G.K. and S.G.; validation, I.G.K., E.S. and M.T.; formal analysis, I.G.K., E.S. and M.T.; investigation, I.G.K., E.S. and M.T.; resources, A.H.; data curation, I.G.K., E.S. and M.T.; writing—original draft preparation, I.G.K.; writing—review and editing, I.G.K., S.G. and A.H.; visualization, I.G.K.; supervision, I.G.K. and A.H.; project administration, I.G.K and S.G.; funding acquisition, A.H. All authors have read and agreed to the published version of the manuscript.

Appendix A. Framework to Characterise Agricultural Processes Based on Qualitative and Quantitative Approaches

A pre-processing six step approach combining qualitative and quantitative methods has been developed.

Step 1. Data on total production by country has been collected for the main agricultural commodities (crops, meat, and forestry products) and converted into energy units. Following, land use demands have been obtained for each agricultural commodity per region.

Step 2. Finally, after aggregating total production and land use demand per type of commodity, it is possible to obtain yields in energy units per area (PJ Mha^{-1}). Distributions for each agricultural product on a global scale has been obtained.

Step 3. The outputs have been used to get a first qualitative definition of mechanisation levels depending on empirically observed yields (Step 3). In this case, as detailed in Section 2.1.3, three different levels of mechanisation have been defined using the quartiles calculated from the distributions. This technological classification is similar to mechanisation levels defined by the [6] (traditional, transitional, and modern) and by [84] (low level, intermediate level, and high level).

Step 4. To provide fuel share and land demand for each mechanisation process per agricultural product, a cluster analysis (k-means) has been implemented by locating countries with similar energy input/output ratios and yields characteristics per agricultural product.

Step 5. After grouping countries accordingly for each agricultural product, data on energy, fertiliser and land demand per unit product has been aggregated and probability distributions have been assigned to the different inputs to characterise those specific groups. Therefore, for each mechanisation level (traditional, transitional, and modern) and agricultural product (crops, meat, and forest) per region, fuel share, agrochemicals demand, yields and investment parameters have been defined with lower and upper bounds.

Step 6. To finalise the approach, it is necessary to calculate the total installed capacities by mechanisation level and by agricultural product for each region. For this, an optimisation problem (OP), based on integer linear programming (ILP) implemented in GAMS has been proposed, considering that only data on total sectoral energy demand and emissions by region are available.

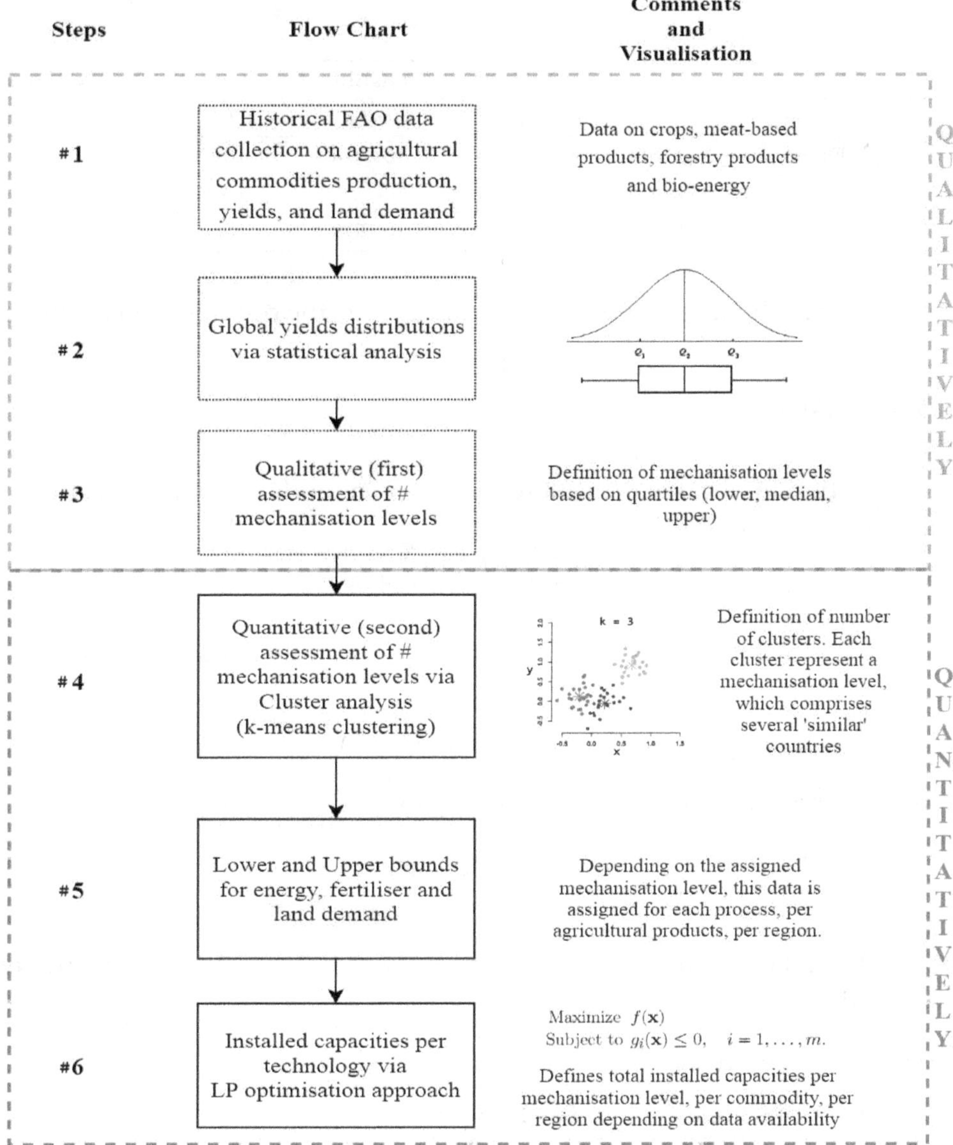

Figure A1. Framework to characterise agricultural processes based on qualitative and quantitative approaches.

Appendix B. Mechanisation Installed Capacity Optimisation Model

To quantitatively characterise mechanisation levels into modelling processes, an optimisation problem (OP), based on linear programming (LP), has been implemented in GAMS [71]. The OP model aims to minimise the difference between real emissions and estimated emissions values for 2010, based on a series of constraints. As main modelling assumptions, different shares of mechanisation level are allocated to every region based on regional yields and the level of economic development. This means every country and region has some level of mechanisation level to greater or lesser extent. Depending on the region's GDP agricultural contribution, different mechanisation share by level are characterize and used as bounds. Therefore, individual regions and crops would have to some extent a mechanisation level share. These bounds are shown in Table A1.

Table A1. Mechanisation share lower and upper bounds assumptions for different type of economies [59].

Type of Economy	Traditional	Transitional	Modern	Modern Renewable
	θ_{trad}	θ_{tran}	θ_{mod}	θ_{modern}
Least Developed (%GDPagr share > 0.16)	50–70%	10–20%	10–20%	1–2%
Emerging (0.02 < %GDPagr share < 0.16)	10–20%	50–70%	10–20%	3–5%
Developed (%GDPagr share < 0.02)	10–20%	10–20%	50–70%	5–10%

Along upper and lower bounds in other input parameters, the optimisation problem is formulated as follows:

Appendix B.1. Objective Function

The objective function minimises the difference between modelled emissions from real emissions in the sector:

$$minZ = \sum\nolimits_r |\Delta E_r| + \sum\nolimits_{r,f} |sl_{r,f}| \tag{A1}$$

where E_r is the difference in emissions, while the sum of $sl_{r,f}$ represents the aggregation of all calculated slack variables.

Appendix B.1.1. Emissions Difference

The difference in emissions (ΔE_r) can be obtained by:

$$\Delta E_r = R_E - M_E \tag{A2}$$

where R_E represent the real emissions (obtained from FAO or IEA), and M_E is the modelled emissions obtained by the model.

Appendix B.1.2. Emissions of the Model

The modelled emissions are calculated as follows:

$$M_E = \sum\nolimits_{f,t} CN_{r,f,t} x FE_f \tag{A3}$$

where $CN_{r,f,t}$ is the fuel consumption for region r and technology t, and FE_f is the emission factor for fuel type f.

Appendix B.2. Constraints

The OP is subject to the following quality and inequality constrains:

Appendix B.2.1. Mass Balance

First, the mass balance must be satisfied for every region, technology and agricultural service:

$$\sum\nolimits_{t \in TS_s} CP_{r,t} > DM_{r,s} \ \forall \in TS_s \tag{A4}$$

where CP refers to installed capacity, DM to service demand, r refers to region, t to technology mechanisation level, s to service type and TS_s are the technology mechanisation levels available for service s.

Appendix B.2.2. Service Demand

The demand per service must be met by the sum of the share of production per mechanisation levels:

$$DM_s = \sum_t \theta_{s,t} \quad \text{(A5)}$$

where θ refers to the demand share covered by technology mechanisation level t.

Appendix B.2.3. Mechanisation Level Share

The share of production per mechanisation level ($\theta_{s,t}$) is obtained by dividing the demand met by specific mechanisation level over the total demand for service s:

$$\theta_{s,t} = \frac{DM_{s,t}}{\sum_{t \in TS_s} DM_s} \ \forall \in TS_s \quad \text{(A6)}$$

The share sum of mechanisation levels is constrained by the following equality:

$$\sum \theta_t = \theta_{s,trad} + \theta_{s,tran} + \theta_{s,mod} + \theta_{s,mod\_ren} = 1 \quad \text{(A7)}$$

Appendix B.2.4. Fuel Balance

The capacity per region and technology ($CP_{r,t}$) is calculated from the total fuel consumption by technology in PJ multiplied by the technology yield or efficiency (PJ/PJ):

$$CP_{r,t} = \sum_f CN_{r,f,t} x\, Y_{t,s} \quad \text{(A8)}$$

where f refers to fuel type, $CN_{r,f,t}$ to fuel consumption per region and technology, and Y to efficiency of technology t for service s.

Appendix B.2.5. Fuel Constraint

The sum of fuel consumption by region and technology ($CN_{r,f,t}$) is constrained by the total fuel demand in the region:

$$\sum_t CN_{r,f,t} < (FD_{r,f} + sl_{r,f}) \quad \text{(A9)}$$

where $FD_{r,f}$ is fuel demand f in region r, and $sl_{r,f}$ is a slack variable added to make the problem feasible and fulfil the fuel constraints. The aim of adding a slack variable to an inequality constraint is to transform it into an equality constraint.

By solving the optimisation problem, all structural alternatives are evaluated, and the best solution is identified. The obtained capacities for each technology are used as an estimate of the base year stock.

References

1. Houghton, R.A.; Nassikas, A.A. Global and regional fluxes of carbon from land use and land cover change 1850–2015. *Glob. Biogeochem. Cycles* **2017**, *31*, 456–472. [CrossRef]
2. De Cara, S.; Houze, M.; Jayet, P.A. Methane and nitrous oxide emissions from agriculture in the EU: A spatial assessment of sources and abatement costs. *Environ. Resour. Econ.* **2005**, *32*, 551–583. [CrossRef]
3. Reay, D.S.; Davidson, E.A.; Smith, K.A.; Smith, P.; Melillo, J.M.; Dentener, F.; Crutzen, P.J. Global agriculture and nitrous oxide emissions. *Nat. Clim. Chang.* **2012**, *2*, 410–416. [CrossRef]
4. IPCC. Climate change 2014: Synthesis report. In *Contribution of Working Groups I, II and III to the 5th Assessment Report of the Intergovernmental Panel on Climate Change*; IPCC: Geneva, Switzerland, 2014.
5. Ray, D.K.; Mueller, N.D.; West, P.C.; Foley, J.A. Yield trends are insufficient to double global crop production by 2050. *PLoS ONE* **2013**, *8*, e66428. [CrossRef]
6. FAO. Faostat. 2017. Available online: http://www.fao.org/faostat/en/data (accessed on 15 January 2019).

7. Tilman, D.; Balzer, C.; Hill, J.; Befort, B.L. Global food demand and the sustainable intensification of agriculture. *Proc. Natl. Acad. Sci. USA* **2011**, *108*, 20260–20264. [CrossRef]
8. Valin, H.; Sands, R.D.; van der Mensbrugghe, D.; Nelson, G.C.; Ahammad, H.; Blanc, E.; Bodirsky, B.; Fujimori, S.; Hasegawa, T.; Havlik, P.; et al. The future of food demand: Understanding differences in global economic models. *Agric. Econ.* **2014**, *45*, 51–67. [CrossRef]
9. Bodirsky, B.L.; Rolinski, S.; Biewald, A.; Weindl, I.; Popp, A.; Lotze-Campen, H. Global food demand scenarios for the 21st century. *PLoS ONE* **2015**, *10*, e0139201. [CrossRef]
10. Thornton, P.K. Livestock production: Recent trends, future prospects. *Philos. Trans. R. Soc. Lond. B Biol. Sci.* **2010**, *365*, 2853–2867. [CrossRef]
11. He, G.; Zhao, Y.; Wang, L.; Jiang, S.; Zhu, Y. China's food security challenge: Effects of food habit changes on requirements for arable land and water. *J. Clean. Prod.* **2019**, *229*, 739–750. [CrossRef]
12. IEA. International Energy Agency Statistics. 2017. Available online: https://www.iea.org/statistics/ (accessed on 30 January 2019).
13. FAO. *World Agriculture: Towards 2015/2030 an FAO Perspective*; Earthscan Publications Ltd.: London, UK, 2003.
14. Heichel, G.H. Agricultural production and energy resources: Current farming practices depend on large expenditures of fossil fuels. How efficiently is this energy used, and will we be able to improve the return on investment in the future? *Am. Sci.* **1976**, *64*, 64–72.
15. Walker, L.P. A method for modelling and evaluating integrated energy systems in agriculture. *Energy Agric.* **1984**, *3*, 1–27. [CrossRef]
16. Uri, N.D. Motor gasoline and diesel fuel demands by agriculture in the United States. *Appl. Energy* **1989**, *32*, 133–154. [CrossRef]
17. Uri, N.D.; Gill, M. Agricultural demands for natural gas and liquefied petroleum gas in the USA. *Appl. Energy* **1992**, *41*, 223–241. [CrossRef]
18. Conforti, P.; Giampietro, M. Fossil energy use in agriculture: An international comparison. *Agric. Ecosyst. Environ.* **1997**, *65*, 231–243. [CrossRef]
19. Xu, B.; Lin, B. Factors affecting CO_2 emissions in china's agriculture sector: Evidence from geographically weighted regression model. *Energy Policy* **2017**, *104*, 404–414. [CrossRef]
20. Ozturk, I. The dynamic relationship between agricultural sustainability and food-energy-water poverty in a panel of selected sub-Saharan African countries. *Energy Policy* **2017**, *107*, 289–299. [CrossRef]
21. Robaina-Alves, M.; Moutinho, V. Decomposition of energy-related GHG emissions in agriculture over 1995-2008 for European countries. *Appl. Energy* **2014**, *114*, 949–957. [CrossRef]
22. Li, T.; Balezentis, T.; Makuteniene, D.; Streimikiene, D.; Krisciukaitiene, I. Energy-related CO_2 emission in European Union agriculture: Driving forces and possibilities for reduction. *Appl. Energy* **2016**, *180*, 682–694. [CrossRef]
23. Arnell, N.W.; Lowe, J.A.; Challinor, A.J. Global and regional impacts of climate change at different levels of global temperature increase. *Clim. Chang.* **2019**, *155*, 377–391. [CrossRef]
24. Woods, J.; Williams, A.; Hughes, J.K.; Black, M.; Murphy, R. Energy and the food system. *Philos. Trans. R. Soc. Lond. Ser. B* **2010**, *365*, 2991–3006. [CrossRef]
25. Camargo, G.G.T.; Ryan, M.R.; Richard, T.L. Energy use and greenhouse gas emissions from crop production using the farm energy analysis tool. *BioScience* **2013**, *63*, 263–273. [CrossRef]
26. Pimentel, D. Energy inputs in food crop production in developing and developed nations. *Energies* **2009**, *2*, 1–24. [CrossRef]
27. World-Bank. World Bank Open Data. 2017. Available online: https://data.worldbank.org/ (accessed on 30 January 2019).
28. Riahi, K.; van Vuuren, D.P.; Kriegler, E.; Edmonds, J.; O'Neill, B.C.; Fujimori, S.; Bauer, N.; Calvin, K.; Dellink, R.; Fricko, O.; et al. The shared socioeconomic pathways and their energy, land use, and greenhouse gas emissions implications: An overview. *Glob. Environ. Chang.* **2017**, *42*, 153–168. [CrossRef]
29. Djanibekov, U.; Gaur, V. Nexus of energy use, agricultural production, employment and incomes among rural households in Uttar Pradesh, India. *Energy Policy* **2018**, *113*, 439–453. [CrossRef]
30. Schmitz, A.; Moss, C. Mechanized agriculture: Machine adoption, farm size, and labor displacement. *AgBioForum* **2015**, *18*, 278–296.
31. Olesen, J.E. Socio-economic Impacts—Agricultural Systems. In *North Sea Region Climate Change Assessment. Regional Climate Studies*; Springer: Berlin/Heidelberg, Germany, 2016. [CrossRef]

32. Ren, C.; Liu, S.; van Grinsven, H.; Reis, S.; Jin, S.; Liu, H.; Gu, B. The impact of farm size on agricultural sustainability. *J. Clean. Prod.* **2019**, *220*, 357–367. [CrossRef]
33. Appel, F.; Ostermeyer-Wiethaup, A.; Balmann, A. Effects of the German renewable energy act on structural change in agriculture—The case of biogas. *Util. Policy* **2016**, *41*, 172–182. [CrossRef]
34. Baruah, D.C.; Bora, G.C. Energy demand forecast for mechanized agriculture in rural India. *Energy Policy* **2008**, *36*, 2628–2636. [CrossRef]
35. European Commission. Eurostat: Agri-Environmental Indicators. Available online: https://ec.europa.eu/eurostat/web/agriculture/agri-environmental-indicators (accessed on 2 November 2020).
36. Ozkan, B.; Akcaoz, H.; Fert, C. Energy input-output analysis in Turkish agriculture. *Renew. Energy* **2004**, *29*, 39–51. [CrossRef]
37. Singh, G. Estimation of a mechanisation index and its impact on production and economic factors—A case study in India. *Biosyst. Eng.* **2006**, *93*, 99–106. [CrossRef]
38. Mileusnic, Z.I.; Petrovic, D.V.; Devic, M.S. Comparison of tillage systems according to fuel consumption. *Energy* **2010**, *35*, 221–228. [CrossRef]
39. Dalgaard, T.; Halberg, N.; Porter, J.R. A model for fossil energy use in Danish agriculture used to compare organic and conventional farming. *Agric. Ecosyst. Environ.* **2001**, *87*, 51–65. [CrossRef]
40. Nkakini, S.O.; Ayotamuno, M.J.; Ogaji, S.O.T.; Probert, S.D. Farm mechanization leading to more effective energy-utilizations for cassava and yam cultivations in rivers state, Nigeria. *Appl. Energy* **2006**, *83*, 1317–1325. [CrossRef]
41. Alluvione, F.; Moretti, B.; Sacco, D.; Grignani, C. EUE (energy use efficiency) of cropping systems for a sustainable agriculture. *Energy* **2011**, *36*, 4468–4481. [CrossRef]
42. Veiga, J.P.S.; Romanelli, T.L.; Gimenez, L.M.; Busato, P.; Milan, M. Energy embodiment in Brazilian agriculture: An overview of 23 crops. *Sci. Agric.* **2015**, *72*, 471–477. [CrossRef]
43. Garcia Kerdan, I.; Giarola, S.; Hawkes, A. Implications of future natural gas demand on sugarcane production, land use change and related emissions in Brazil. *J. Sustain. Dev. Energy Water Environ. Syst.* **2020**, *8*, 304–327. [CrossRef]
44. Garcia Kerdan, I.; Giarola, S.; Jalil-Vega, F.; Hawkes, A. Carbon sequestration potential from large-scale reforestation and sugarcane expansion on abandoned agricultural lands in Brazil. *Polytechnica* **2019**, *2*, 9–25. [CrossRef]
45. De Oliveira, L.L.; Garcia Kerdan, I.; de Oliveira Ribeiro, C.; do Nascimento Oller, C.A.; Rego, E.E.; Giarola, S.; Hawkes, A. Modelling the technical potential of bioelectricity production under land use constraints: A multi-region Brazil case study. *Renew. Sustain. Energy Rev.* **2020**, *123*, 109765. [CrossRef]
46. Yang, J.; Huang, Z.; Zhang, X.; Reardon, T. The rapid rise of cross-regional agricultural mechanization services in China. *J. Agric. Econ.* **2013**, *95*, 1245–1251. [CrossRef]
47. Wise, M.; Calvin, K.; Kyle, P.; Luckow, P.; Edmonds, J. Economic and physical modeling of land use in GCAM 3.0 and an application to agricultural productivity, land, and terrestrial carbon. *Clim. Chang. Econ.* **2014**, *5*, 1450003. [CrossRef]
48. FAO. *"Energy-Smart" Food for People and Climate*; Food and Agriculture Organization of the United Nations: Rome, Italy, 2011.
49. Uri, N.D. Energy substitution in agriculture in the United States. *Appl. Energy* **1988**, *31*, 221–237. [CrossRef]
50. Elobeid, A.; Tokgoz, S.; Dodder, R.; Johnson, T.; Kaplan, O.; Kurkalova, L.; Secchi, S. Integration of agricultural and energy system models for biofuel assessment. *Environ. Model. Softw.* **2013**, *48*, 1–16. [CrossRef]
51. Miljkovic, D.; Ripplinger, D.; Shaik, S. Impact of biofuel policies on the use of land and energy in U.S. agriculture. *J. Policy Model.* **2016**, *38*, 1089–1098. [CrossRef]
52. Rochedo, P. Development of a Global Integrated Energy Model to Evaluate the Brazilian Role in Climate Change Mitigation Scenarios. Ph.D. Thesis, Federal University of Rio de Janeiro, Rio de Janeiro, Brazil, 2016.
53. Al-Mansour, F.; Jejcic, V. A model calculation of the carbon footprint of agricultural products: The case of Slovenia. *Energy* **2017**, *136*, 7–15. [CrossRef]
54. Daioglou, V.; Doelman, J.; Wicke, B.; Faaij, A.; van Vuuren, D.P. Integrated assessment of biomass supply and demand in climate change mitigation scenarios. *Glob. Environ. Chang.* **2019**, *54*, 88–101. [CrossRef]
55. Wu, W.; Hasegawa, T.; Ohashi, H.; Hanasaki, N.; Liu, J.; Matsui, T.; Fujimori, S.; Masui, T.; Takahashi, K. Global advanced bioenergy potential under environmental protection policies and societal transformation measures. *GCB Bioenergy* **2019**, *11*, 1041–1055. [CrossRef]

56. Li, M.; Fu, Q.; Singh, V.P.; Liu, D.; Li, T. Stochastic multi-objective modeling for optimization of water-food-energy nexus of irrigated agriculture. *Adv. Water Resour.* **2019**, *127*, 209–224. [CrossRef]
57. Elkadeem, M.R.; Wang, S.; Sharshir, S.W.; Atia, E.G. Feasibility analysis and techno-economic design of grid-isolated hybrid renewable energy system for electrification of agriculture and irrigation area: A case study in Dongola, Sudan. *Energy Convers. Manag.* **2019**, *196*, 1453–1478. [CrossRef]
58. Jones, J.W.; Antle, J.M.; Basso, B.; Boote, K.J.; Conant, R.T.; Foster, I.; Godfray, H.C.J.; Herrero, M.; Howitt, R.E.; Janssen, S.; et al. Brief history of agricultural systems modelling. *Agric. Syst.* **2017**, *155*, 240–254. [CrossRef]
59. Garcia Kerdan, I.; Giarola, S.; Hawkes, A. A novel energy systems model to explore the role of land use and reforestation in achieving carbon mitigation targets: A Brazil case study. *J. Clean. Prod.* **2019**, *232*, 796–821. [CrossRef]
60. Paris Reinforce. The ModUlar Energy System Simulation Environment (MUSE). Available online: http://paris-reinforce.epu.ntua.gr/detailed_model_doc/muse (accessed on 9 December 2020).
61. USDA. USDA Food Composition Databases. 2017. Available online: https://www.ers.usda.gov/data-products/commodity-costs-and-returns/ (accessed on 15 February 2019).
62. Pradhan, P.; Lüdeke, M.; Reusser, D.; Kropp, J. Embodied crop calories in animal products. *Environ. Res. Lett.* **2013**, *8*, 044044. [CrossRef]
63. Giarola, S.; Budinis, S.; Sachs, J.A.; Hawkes, A. Long-Term Decarbonisation Scenarios in the Industrial Sector, International Energy Workshop. 2017. Available online: http://events.pnnl.gov/images/IEW%202017/IEW2017_abstracts/Longterm%20decarbonisation%20scenarios%20in%20the%20industrial.pdf (accessed on 15 February 2019).
64. Shahzad, S.J.H.; Hernandez, J.A.; Al-Yahyaee, K.H.; Jammazi, R. Asymmetric risk spillovers between oil and agricultural commodities. *Energy Policy* **2018**, *118*, 182–198. [CrossRef]
65. Van Ruijven, B.J.; van Vuuren, D.P.; Boskaljon, W.; Neelis, M.L.; Saygin, D.; Patel, M.K. Long-term model-based projections of energy use and CO_2 emissions from the global steel and cement industries. *Resour. Conserv. Recycl.* **2016**, *112*, 15–36. [CrossRef]
66. Hartigan, J.A.; Wong, M.A. Algorithm AS 136: A K-Means Clustering Algorithm. *J. R. Stat. Soc. C Appl.* **1979**, *28*, 100–108. [CrossRef]
67. Rousseeuw, P.J. Silhouettes: A graphical aid to the interpretation and validation of cluster analysis. *J. Comput. Appl. Math.* **1987**, *20*, 53–65. [CrossRef]
68. Calabi-Floody, M.; Medina, J.; Rumpel, C.; Condron, L.M.; Hernandez, M.; Dumont, M.; de la Luz Mora, M. Smart Fertilizers as a Strategy for Sustainable Agriculture. *Adv. Agron.* **2018**, *147*, 119–157. [CrossRef]
69. IPCC. *2006 IPCC Guidelines for National Greenhouse Gas Inventories*; UNEP: Nairobi, Kenya, 2006.
70. GAMS. General Algebraic Modeling System (Gams) Release 24.2.1. 2013. Available online: https://www.gams.com/ (accessed on 15 February 2019).
71. Fricko, O.; Havlik, P.; Rogelj, J.; Klimont, Z.; Gusti, M.; Johnson, N.; Kolp, P.; Strubegger, M.; Valin, H.; Amann, M.; et al. The marker quantification of the shared socioeconomic pathway 2: A middle-of-the-road scenario for the 21st century. *Glob. Environ. Chang.* **2017**, *42*, 251–267. [CrossRef]
72. IEA. *Energy Technology Perspectives 2017: Catalysing Energy Technology Transformations*; OECD/IEA: Paris, France, 2017.
73. EIA. US Energy Information and Administration—Annual Energy Outlook 2017. 2017. Available online: https://www.eia.gov/outlooks/aeo/ (accessed on 15 February 2019).
74. EMF. Energy Modelling Forum—EMF 27: Global Model Comparison Exercise. 2017. Available online: https://emf.stanford.edu/projects/emf-27-global-model-comparison-exercise (accessed on 3 March 2019).
75. Xu, Y.J.; Li, G.X.; Sun, Z.Y. Development of biodiesel industry in China: Upon the terms of production and consumption. *Renew. Sustain. Energy Rev.* **2016**, *54*, 318–330. [CrossRef]
76. Safdar, M.T.; van Gevelt, T. Catching Up with the 'Core': The Nature of the Agricultural Machinery Sector and Challenges for Chinese Manufacturers. *J. Dev. Stud.* **2020**, *56*, 1349–1366. [CrossRef]
77. Huang, J.; Otsuka, K.; Rozelle, S. *The Role of Agriculture in China's Development: Past Failures, Present Successes, and Future Challenges*; Stanford University: Stanford, CA, USA, 2007.
78. OECD. Review of Agricultural Policies—China. Available online: https://www.oecd.org/china/oecdreviewofagriculturalpolicies-china.htm (accessed on 9 December 2020).

79. Sibayan, E.B.; Samoy-Pascual, K.; Grospe, F.S.; Casil, M.E.D.; Tokida, T.; Padre, A.T.; Minamikawa, K. Effects of alternate wetting and drying technique on greenhouse gas emissions from irrigated rice paddy in central Luzon, Philippines. *J. Soil Sci. Plant Nutr.* **2017**, *64*, 39–46. [CrossRef]
80. Erda, L.; Yue, L.; Hongmin, D. Potential GHG mitigation options for agriculture in China. *Appl. Energy* **1997**, *56*, 423–432. [CrossRef]
81. Cisternas, I.; Velásquez, I.; Caro, A.; Rodríguez, A. Systematic literature review of implementations of precision agriculture. *Comput. Electron. Agric.* **2020**, *176*, 105626. [CrossRef]
82. Stafford, J. Implementing Precision Agriculture in the 21st Century. *J. Agric. Eng. Res.* **2000**, *76*, 267–275. [CrossRef]
83. FAO. Annex 3: Agricultural Policy and Food Security in China. Available online: http://www.fao.org/3/ab981e/ab981e0c.htm#bm12.3.7 (accessed on 9 December 2020).
84. IRENA. *Agriculture and Environment in EU-15—The IRENA Indicator Report*; European Environment Agency: Copenhagen, Denmark, 2005.

3

Yeast Microbiota during Sauerkraut Fermentation and its Characteristics

Paweł Satora [1,*], Magdalena Skotniczny [1], Szymon Strnad [1] and Katarína Ženišová [2]

[1] Department of Fermentation Technology and Microbiology, Faculty of Food Technology, University of Agriculture in Krakow, Balicka 122, 30-149 Krakow, Poland; magdalena.skotniczny@urk.edu.pl (M.S.); szymon.strnad@urk.edu.pl (S.S.)

[2] Department of Microbiology, Molecular Biology and Biotechnology, National Agricultural and Food Centre, Food Research Institute, Priemyselna 4, P.O. Box 25, 824 75 Bratislava, Slovakia; katarina.zenisova@nppc.sk

* Correspondence: pawel.satora@urk.edu.pl

Abstract: Sauerkraut is the most important fermented vegetable obtained in Europe. It is produced traditionally by spontaneous fermentation of cabbage. The aim of this study was to determine biodiversity of yeasts present during fermentation of eight varieties of cabbages (Ambrosia, Avak, Cabton, Galaxy, Jaguar, Kamienna Głowa, Manama and Ramco), as well as characterize obtained yeast isolates. WL Nutrient Agar with Chloramphenicol was used to enumerate yeast. Isolates were differentiated using RAPD-PCR and identified by sequencing of the 5.8S-ITS rRNA gene region. The volatiles production was analyzed using SPME-GC-TOFMS. Our research confirmed that during sauerkraut fermentation there is an active growth of the yeasts, which begins in the first phases. The maximal number of yeast cells from 1.82 to 4.46 log CFU g^{-1} occurred after 24 h of fermentation, then decrease in yeast counts was found in all samples. Among the isolates dominated the cultures *Debaryomyces hansenii*, *Clavispora lusitaniae* and *Rhodotorula mucilaginosa*. All isolates could grow at NaCl concentrations higher than 5%, were relatively resistant to low pH and the presence of lactic acid, and most of them were characterized by killer toxins activity. The highest concentration of volatiles (mainly esters and alcohols) were produced by *Pichia fermentans* and *D. hansenii* strains.

Keywords: sauerkraut; yeast microbiota; 5.8S-ITS rRNA gene; *Debaryomyces*; resistance for stress conditions; volatile profile

1. Introduction

Poland is the fourth largest producer of sauerkraut in Europe, and its annual consumption in our country is about 3 kg per capita [1]. This fermented vegetable is produced by spontaneous fermentation of cabbage leaves, mainly by lactic acid bacteria [2]. As a result of the fermentation process, the fermenting sugars found in the vegetable are transformed into lactic and acetic acid, ethanol, CO_2, mannitol and other compounds [3].

The quality of the product depends mainly on the microbiota contained on the crude vegetable [4]. The development of microorganisms in the appropriate sequence is necessary to obtain the taste and aroma characteristic of sauerkraut [5].

In the initial phase of fermentation of vegetable, among the large variety of microorganism, yeasts represent a large group. Sometimes it even happens that they begin to dominate and control the fermentation process. In the fermentation of vegetable silages, the negative activity of yeast leads to an increase in pH, which can be caused by one of two processes. The first is related to the fact that the yeasts present in the silage carry out ethanol fermentation using sugars, which should be fermented by the bacteria to lactic acid. The second process is related to the development of *Geotrichum candidum* and

Candida mycoderma yeasts on the silage surface under aerobic conditions. The above microorganisms use organic acids stabilizing the product as a carbon source. This reduces the acidity, increases the pH and creates conditions for the development of spoilage bacteria. Yeast is also responsible for the production of a large amount of gases [6]. Too soft silage structure may be associated with insufficient salt concentration. It should be added to the freshly shredded cabbage in such quantity as to stimulate the growth of the desired lactic bacteria in the right order. Lactic acid and salt determine the correct hardness of the product by inhibiting pectinolytic enzymes [7]. Very large economic losses in the industrial production of sauerkraut causes the appearance of a defective pink color associated with the development of yeast *Rhodotorula*. They produce carotenoids, which give the silage pink color. This disadvantage is usually associated with too high concentration of salt added, which exceeds 3% [8].

Disadvantages in the form of strange aftertaste and aroma, which are unspecific for pickled products and are associated more with the yeast industry, are most often caused by a too fast fermentation process carried out at high temperature. These conditions favor the growth of aerobic microorganisms, mainly yeasts and molds, which produce metabolites responsible for undesirable taste and smell. Sauerkraut in which yeast and mold develops is characterized by a high content of esters responsible for the aroma of raw cabbage, while the amount of lactic acid is much lower in it than in products in which the contaminating microbiota did not occur [7].

Most of the so far published articles related to the microbiology of the sauerkraut concern LAB (Lactic Acid Bacteria). There is very little information regarding the presence of yeast during fermentation of sauerkraut and their characteristics. The aim of this study was to determine biodiversity of yeasts present during fermentation of different varieties of cabbages as well as characterize obtained yeast isolates. For the experiments, eight varieties of cabbages, most often used by Polish farmers for sauerkraut production were used.

2. Results

2.1. Yeast Population Kinetics

The isolation was done in 0., 1., 2., 3., 7., 10. and 14. day of fermentation (Figure 1). Quantitative analysis showed differences in total yeast contents during fermentation of cabbages of different varieties. The fresh cabbages contained from 0.60 (Cabton) to 3.74 (Manama) log CFU of yeasts per gram. For the majority of trials in the following day, there was an increase in the number of yeast cells (Manama, Ramco, Ambrosia, Jaguar, Cabton), reaching their maximum. In the sauerkraut of the Kamienna Głowa, Galaxy and Avak varieties, from the very beginning of fermentation, a gradual decrease in the number of yeast cells has been reported.

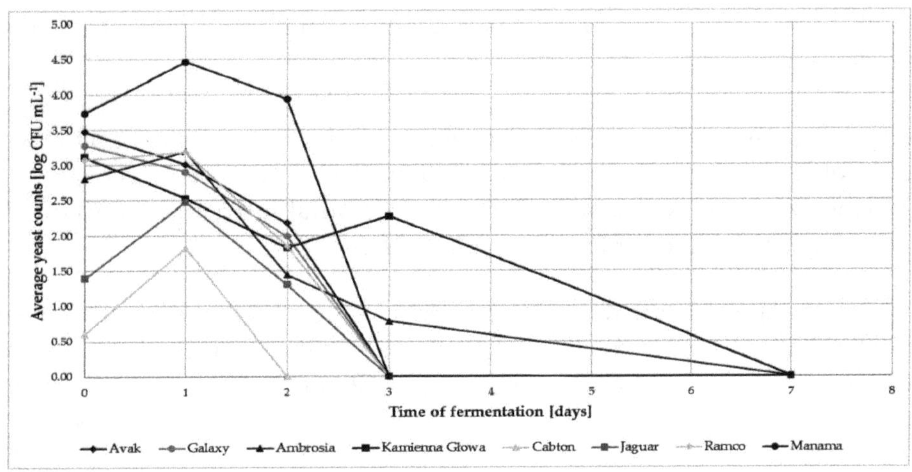

Figure 1. The amount of yeast cells during sauerkraut fermentation of different varieties (standard deviations do not exceed 5%).

After 24 h of fermentation, the decrease in yeast counts was found in all samples. After two days of the process in Cabton cabbage (with the smallest initial number of yeast cells), the presence of these microorganisms was not observed; a similar tendency was observed in the majority of the other samples after 72 h of the process. The exceptions were the Kamienna Głowa and Ambrosia varieties, in which on the third day of fermentation, 2.28 and 0.78 log CFU g^{-1} were still detected. During sampling in 7th day of fermentation no yeast was found.

2.2. Biodiversity of Yeasts during Fermentation

A total of 246 pure yeast cultures were isolated from various stages of fermented sauerkraut from eight varieties of cabbages.

The RAPD-PCR method was used to differentiate isolates and reduce their number before further analysis. Isolated yeasts were classified into 25 groups based on distinct electrophoretic patterns (Figure 2). Representatives of each groups of RAPD patterns were analyzed by 5.8S-ITS (internal transcribed spacer) PCR-RFLP (polymerase chain reaction-restriction fragment length polymorphism) (Table 1) and were identified by 5.8S-ITS rRNA gene region sequencing.

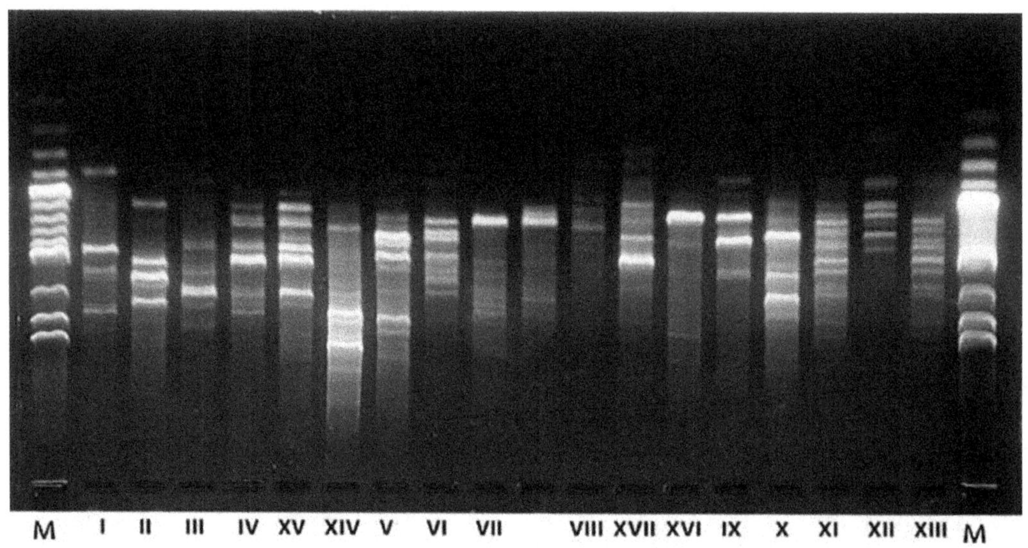

Figure 2. Results of electrophoresis of RAPD-PCR products with M13 starter of yeast isolates from fermenting sauerkraut.

The RFLP analysis showed that the isolates belong to nine yeast species. This was confirmed by sequencing, with 25 electrophoretic patterns obtained by RAPD (Figure 2), classified into 17 different strains by comparing obtained sequences with those available in the GenBank NCBI database, considering an identity threshold of at least 98% (Table 1).

Among the isolates dominated the cultures *Debaryomyces hansenii* and *Clavispora lusitaniae*—four strains each, and *Rhodotorula mucilaginosa*—three strains. The presence of single strains of *Cryptococcus macerans, Nakazawaea holstii, Meyerozyma guilliermondii, Candida sake, Pichia fermentans* and *Tausonia pullulans* were also found.

Figure 3 present participation of individual yeast strains identified during spontaneous fermentation of different cabbage varieties at various fermentation stages.

Table 1. Identified species of yeast based on the results of the RFLP analysis and 5.8S ITS rRNA gene sequencing.

Strain	5.8S-ITS rRNA Gene [bp]	Restriction Fragments [bp]			Species Identification by 5.8S-ITS rRNA Gene Sequencing	Restriction Pattern	Accession No.
		*Cfo*I	*Hinf*I	*Hae*III			
I	650	260 + 200 + 95 + 95	270 + 200 + 175	440 + 80 + 70 + 60	*Cryptococcus macerans*	1	MK312605
II	650	590	310 + 310	570 + 80	*Nakazawaea holstii*	2	MK312606
III	640	320 + 240 + 80	340 + 225 + 75	425 + 215	*Rhodotorula mucilaginosa*	3	MK312607
IV	650	300 + 300 + 50	325 + 325	420 + 150 + 90	*Debaryomyces hansenii*	4	MK312608
V	370	210 + 180	180 + 160	370	*Clavispora lusitaniae*	7	MK312609
VI	625	300 + 265 + 60	320 + 300	400 + 115 + 90	*Meyerozyma guilliermondii*	8	MK312610
VII	450	250 + 200	230 + 220	450	*Candida sake*	9	MK312611
VIII	640	320 + 240 + 80	340 + 225 + 75	425 + 215	*Rhodotorula mucilaginosa*	11	MK312612
IX	640	320 + 240 + 80	340 + 225 + 75	425 + 215	*Rhodotorula mucilaginosa*	14	MK312613
X	370	210 + 180	180 + 160	370	*Clavispora lusitaniae*	15	MK312614
XI	370	210 + 180	180 + 160	370	*Clavispora lusitaniae*	16	MK312615
XII	650	300 + 300 + 50	325 + 325	420 + 150 + 90	*Debaryomyces hansenii*	17	MK312616
XIII	370	210 + 180	180 + 160	370	*Clavispora lusitaniae*	18	MK312617
XIV	450	170 + 100 + 100 + 80	250 + 200	340 + 80 + 30	*Pichia fermentans*	21	MK312618
XV	650	300 + 300 + 50	325 + 325	420 + 150 + 90	*Debaryomyces hansenii*	22	MK312619
XVI	650	300 + 300 + 50	325 + 325	420 + 150 + 90	*Debaryomyces hansenii*	23	MK312620
XVII	500	300 + 100	300 + 150	500	*Tausonia pullulans*	25	MK312621

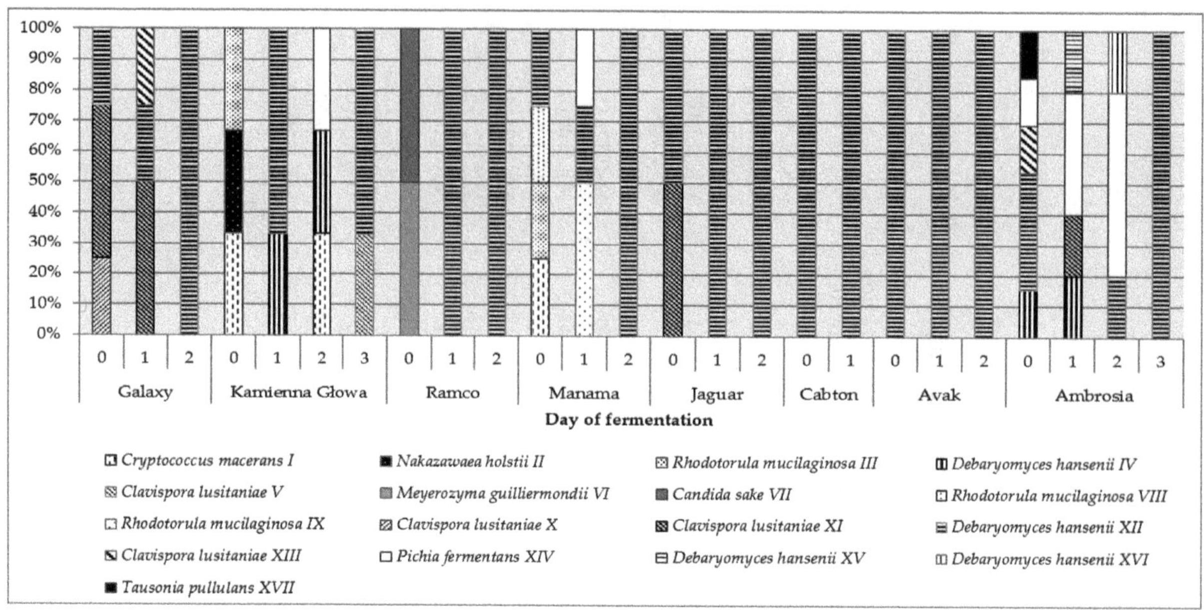

Figure 3. Distribution of yeast strains (%) isolated from the initial stage of different cultivars of cabbage fermentation.

In the case of 4 analyzed varieties (Galaxy, Kamienna Głowa, Manama, Ambrosia) a large diversity of the yeast population was found, in the others much smaller. The yeast population in terms of species and strains was the most diverse at the beginning of fermentation, in the course of time, selected cultures remained.

In the prepared for fermentation cabbage of the Galaxy variety, *C. lusitaniae* X, XI and *D. hansenii* XII strains prevailed. A similar composition of yeast microbiota was found on the cabbage of the Jaguar variety. Kamienna Głowa and Manama samples were similar and contained cultures of *Cry. macerans* I and *R. mucilaginosa* III. In addition, in the first case *N. holstii* II was detected, and the second—*D. hansenii* XII and *R. mucilaginosa* VIII. The Ambrosia cabbage was characterized by the most diverse yeast population—*D. hansenii* (strains IV, XII), *C. lusitaniae* (XIII), *P. fermentans* (XIV) as well as *T. pullulans* (XVII) were identified. Similarly, in the structure of yeast microbiota, Ramco was distinguished, on the surface of which *M. guilliermondii* VI and *C. sake* VII were found. Cabbage of Avak cultivars were the least diverse in this respect, in which only yeast *D. hansenii* XII was isolated.

After 24 h of fermentation the growth of *Basidiomycota* yeast was significantly reduced and cultures with fermentative metabolism, such as *D. hansenii* (strains IV, XII, XV, XVI), *C. lusitaniae* (XI, XIII) and *P. fermentans* (XIV) started to prevail.

In subsequent days of the process, at the time of a significant decrease in the amount of yeast (Figure 1), the strain *D. hansenii* XII began to dominate, regardless of the variety of the cabbage subjected to fermentation.

2.3. Production of Volatile Components by Isolates

A total of 31 different aroma components were detected in the samples (Figure 4). Alcohols and esters constituted the most numerous groups. Based on the cluster analysis and heat map, the strains analyzed were divided into three groups. The first was cultures forming small amounts of volatile compounds, such as *R. mucilaginosa* (strains III, VIII, IX), *Cry. macerans* (I), *N. holstii* (II), *C. lusitaniae* (V, XIII), *C. sake* (VII) and *T. pullulans* (XVII), the second strains producing average amounts of volatile compounds—*C. lusitaniae* (X, XI), *M. guillermondii* (VI) and *D. hansenii* (IV), and to the third largest producer of volatile compounds—*D. hansenii* (XII, XV, XVI) and *P. fermentans* (XIV). The most volatile components were formed by *P. fermentans* XIV—producing almost all analyzed components, some of them such as ethyl acetate were produced in more than 10 times larger quantities than in the case of other isolates. The presence of high concentrations of other esters, such as isobutyl acetate, isoamyl acetate, 2-phenylethyl acetate, ethyl dodecanoate and others were also characteristic of these samples. Several other volatile compounds were also unique to selected cultures, e.g., ethyl 2-hydroxypropanoate was only present in the samples of *M. guillermondii* VI, cyclohexanol—*C. lusitaniae* XI, or cyclohexanone—*C. lusitaniae* V and XI. Isolates producing smaller amounts of volatile components mainly produced fewer volatile substances, such as fatty acid esters, alcohols with more than eight carbon atoms per molecule.

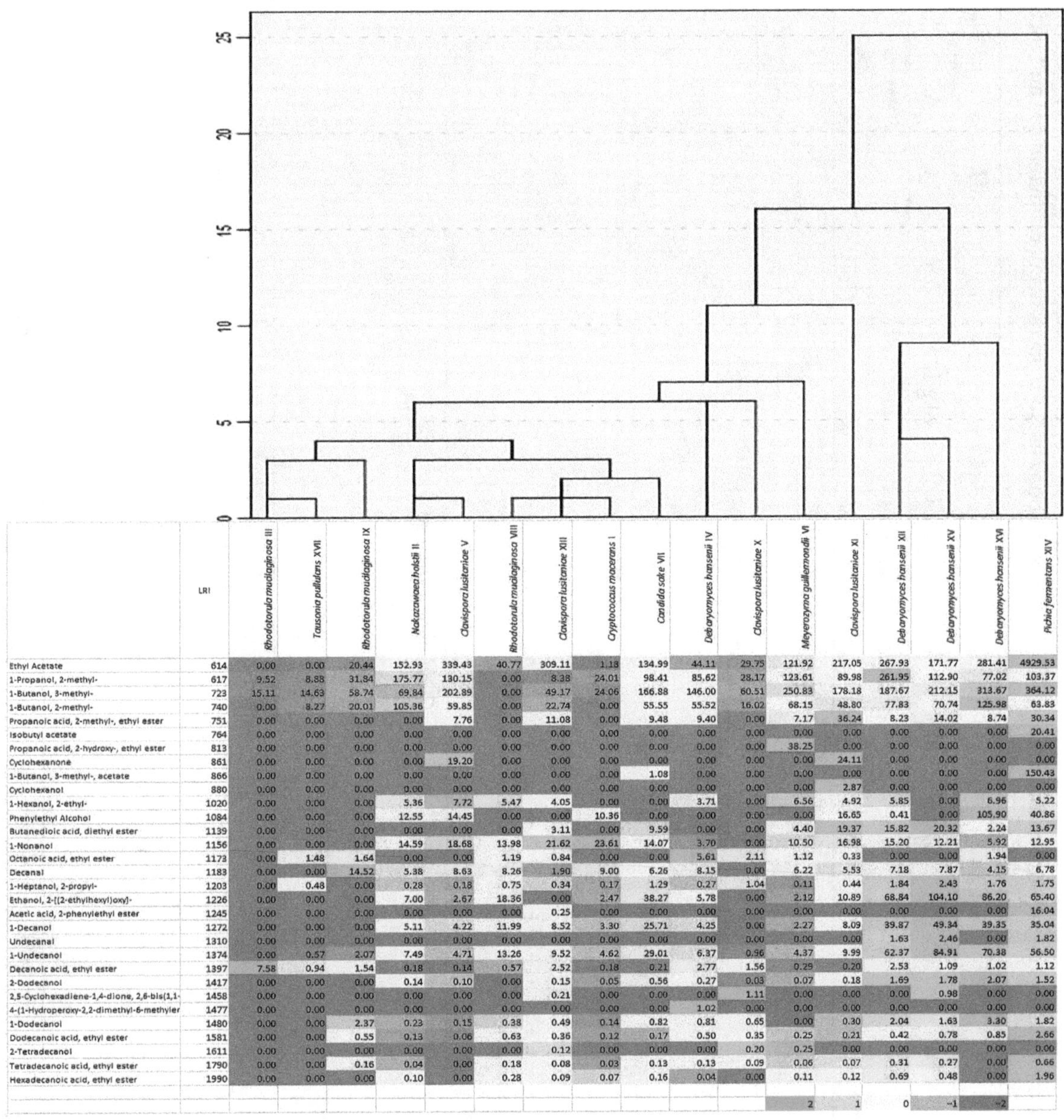

	LRI	*Rhodotorula mucilaginosa* III	*Tausonia pullulans* XVII	*Rhodotorula mucilaginosa* IX	*Nakazawaea holstii* II	*Clavispora lusitaniae* V	*Rhodotorula mucilaginosa* VIII	*Clavispora lusitaniae* XIII	*Cryptococcus macerans* I	*Candida sake* VII	*Debaryomyces hansenii* IV	*Clavispora lusitaniae* X	*Meyerozyma guilliermondii* VI	*Clavispora lusitaniae* XI	*Debaryomyces hansenii* XII	*Debaryomyces hansenii* XV	*Debaryomyces hansenii* XVI	*Pichia fermentans* XIV
Ethyl Acetate	614	0.00	0.00	20.44	152.93	339.43	40.77	309.11	1.18	134.99	44.11	29.75	121.92	217.05	267.93	171.77	281.41	4929.53
1-Propanol, 2-methyl-	617	9.52	8.88	31.84	175.77	130.15	0.00	8.38	24.01	98.41	85.62	28.17	123.61	89.98	261.95	112.90	77.02	103.37
1-Butanol, 3-methyl-	723	15.11	14.63	58.74	69.84	202.89	0.00	49.17	24.06	166.88	146.00	60.51	250.83	178.18	187.67	212.15	313.67	364.12
1-Butanol, 2-methyl-	740	0.00	8.27	20.01	105.36	59.85	0.00	22.74	0.00	55.55	55.52	16.02	68.15	48.80	77.83	70.74	125.98	63.83
Propanoic acid, 2-methyl-, ethyl ester	751	0.00	0.00	0.00	0.00	7.76	0.00	11.08	0.00	9.48	9.40	0.00	7.17	36.24	8.23	14.02	8.74	30.34
Isobutyl acetate	764	0.00	0.00	0.00	0.00	0.00	0.00	0.00	0.00	0.00	0.00	0.00	0.00	0.00	0.00	0.00	0.00	20.41
Propanoic acid, 2-hydroxy-, ethyl ester	813	0.00	0.00	0.00	0.00	0.00	0.00	0.00	0.00	0.00	0.00	0.00	38.25	0.00	0.00	0.00	0.00	0.00
Cyclohexanone	861	0.00	0.00	0.00	0.00	19.20	0.00	0.00	0.00	0.00	0.00	0.00	0.00	24.11	0.00	0.00	0.00	0.00
1-Butanol, 3-methyl-, acetate	866	0.00	0.00	0.00	0.00	0.00	0.00	0.00	0.00	1.08	0.00	0.00	0.00	0.00	0.00	0.00	0.00	150.43
Cyclohexanol	880	0.00	0.00	0.00	0.00	0.00	0.00	0.00	0.00	0.00	0.00	0.00	0.00	2.87	0.00	0.00	0.00	0.00
1-Hexanol, 2-ethyl-	1020	0.00	0.00	0.00	5.36	7.72	5.47	4.05	0.00	0.00	3.71	0.00	6.56	4.92	5.85	0.00	6.96	5.22
Phenylethyl Alcohol	1084	0.00	0.00	0.00	12.55	14.45	0.00	0.00	10.36	0.00	0.00	0.00	0.00	16.65	0.41	0.00	105.90	40.86
Butanedioic acid, diethyl ester	1139	0.00	0.00	0.00	0.00	0.00	0.00	3.11	0.00	9.59	0.00	0.00	4.40	19.37	15.82	20.32	2.24	13.67
1-Nonanol	1156	0.00	0.00	0.00	14.59	18.68	13.98	21.62	23.61	14.07	3.70	0.00	10.50	16.98	15.20	12.21	5.92	12.95
Octanoic acid, ethyl ester	1173	0.00	1.48	1.64	0.00	0.00	1.19	0.84	0.00	0.00	5.61	2.11	1.12	0.33	0.00	0.00	1.94	0.00
Decanal	1183	0.00	0.00	14.52	5.38	8.63	8.26	1.90	9.00	6.26	8.15	0.00	6.22	5.53	7.18	7.87	4.15	6.78
1-Heptanol, 2-propyl-	1203	0.00	0.48	0.00	0.28	0.18	0.75	0.34	0.17	1.29	0.27	1.04	0.11	0.44	1.84	2.43	1.76	1.75
Ethanol, 2-[(2-ethylhexyl)oxy]-	1226	0.00	0.00	0.00	7.00	2.67	18.36	0.00	2.47	38.27	5.78	0.00	2.12	10.89	68.84	104.10	86.20	65.40
Acetic acid, 2-phenylethyl ester	1245	0.00	0.00	0.00	0.00	0.00	0.00	0.25	0.00	0.00	0.00	0.00	0.00	0.00	0.00	0.00	0.00	16.04
1-Decanol	1272	0.00	0.00	0.00	5.11	4.22	11.99	8.52	3.30	25.71	4.25	0.00	2.27	8.09	39.87	49.34	39.35	35.04
Undecanal	1310	0.00	0.00	0.00	0.00	0.00	0.00	0.00	0.00	0.00	0.00	0.00	0.00	0.00	1.63	2.46	0.00	1.82
1-Undecanol	1374	0.00	0.57	2.07	7.49	4.71	13.26	9.52	4.62	29.01	6.37	0.96	4.37	9.99	62.37	84.91	70.38	56.50
Decanoic acid, ethyl ester	1397	7.58	0.94	1.54	0.18	0.14	0.57	2.52	0.18	0.21	2.77	1.56	0.29	0.20	2.53	1.09	1.02	1.12
2-Dodecanol	1417	0.00	0.00	0.00	0.14	0.10	0.00	0.15	0.05	0.56	0.27	0.03	0.07	0.18	1.69	1.78	2.07	1.52
2,5-Cyclohexadiene-1,4-dione, 2,6-bis(1,1-	1458	0.00	0.00	0.00	0.00	0.00	0.00	0.21	0.00	0.00	0.00	1.11	0.00	0.00	0.00	0.98	0.00	0.00
4-(1-Hydroperoxy-2,2-dimethyl-6-methyler	1477	0.00	0.00	0.00	0.00	0.00	0.00	0.00	0.00	0.00	1.02	0.00	0.00	0.00	0.00	0.00	0.00	0.00
1-Dodecanol	1480	0.00	0.00	2.37	0.23	0.15	0.38	0.49	0.14	0.82	0.81	0.65	0.00	0.30	2.04	1.63	3.30	1.82
Dodecanoic acid, ethyl ester	1581	0.00	0.00	0.55	0.13	0.06	0.63	0.36	0.12	0.17	0.50	0.35	0.25	0.21	0.42	0.78	0.85	2.66
2-Tetradecanol	1611	0.00	0.00	0.00	0.00	0.00	0.00	0.12	0.00	0.00	0.00	0.20	0.25	0.00	0.00	0.00	0.00	0.00
Tetradecanoic acid, ethyl ester	1790	0.00	0.00	0.16	0.04	0.00	0.18	0.08	0.03	0.13	0.13	0.09	0.06	0.07	0.31	0.27	0.00	0.66
Hexadecanoic acid, ethyl ester	1990	0.00	0.00	0.00	0.10	0.00	0.28	0.09	0.07	0.16	0.04	0.00	0.11	0.12	0.69	0.48	0.00	1.96
													2	1	0	-1	-2	

Figure 4. A heat map and cluster analysis results of 31 volatiles [μg L^{-1}] produced by yeast isolates from sauerkraut fermentation. The highest content is in the darkest green and the lowest content is in the darkest red.

2.4. *Resistance of Isolates to Selected Stress Factors Present during Sauerkraut Fermentation and Killer Activity*

Most of the tested isolates can be classified as halophiles, because they grew very well even in an environment containing 10% sodium chloride. *N. holstii* II, *C. lusitaniae* X, *P. fermentans* XIV and *D. hansenii* XVI strains were an exception and their growth were visibly inhibited on media with 6% NaCl and more (Table 2).

Table 2. Growth of sauerkraut isolates on substrates containing different concentrations of NaCl, lactic acid or different pH and their killer activity.

Isolate	Control (pH 5.6)	NaCl				Lactic Acid			pH			Killer Activity [mm]
		5%	6%	8%	10%	6 g L^{-1}	8 g L^{-1}	10 g L^{-1}	3.6	3.4	3.2	
						Cryptococcus macerans						
I	+ + + +	+ + + +	+ + +	+ + +	+	+ +	+	-	+ +	-	-	16
						Nakazaweae holstii						
II	+ + + +	+ +	-	-	-	-	-	-	+ +	-	-	-
						Rhodotorula mucilaginosa						
III	+ + + +	+ + + +	+ + +	+ + +	-	+ +	+ +	-	+	-	-	-
VIII	+ + + +	+ + +	+ + +	+ +	+	+ +	+	-	+ + +	+	-	-
IX	+ + + +	+ + + +	+ + +	+ + +	+ +	+	+	-	+	-	-	-
						Debaryomyces hansenii						
IV	+ + + +	+ + +	+ + +	+ + +	+ +	+ + +	+ + +	+ +	+ +	-	-	16
XII	+ + + +	+ + + +	+ + + +	+ + +	+ +	+ +	-	-	+	-	-	24
XV	+ + + +	+ + + +	+ + + +	+ + + +	+ + +	+	+	-	+	+	-	14
XVI	+ + + +	+	-	-	-	+	-	-	+	-	-	11
						Clavispora lusitaniae						
V	+ + + +	+ + + +	+ + +	+ + +	+ +	+ + +	+ +	-	+ + +	+	-	22
X	+ + + +	+	-	-	-	+ +	+	-	+	-	-	16
XI	+ + + +	+ + + +	+ + +	+ +	+	+ +	+	-	+ +	+	-	6
XIII	+ + + +	+ + + +	+ + +	+ +	+ +	+ +	+	-	+ +	-	-	-
						Meyerozyma guilliermondii						
VI	+ + + +	+ + + +	+ + +	+ + +	+ + +	+ +	+ +	-	+ + +	+ +	-	-
						Candida sake						
VII	+ + + +	+ + + +	+ + +	+ +	+	+ +	+ +	-	+ + +	+	-	22
						Pichia fermentans						
XIV	+ + + +	+	-	-	-	+	+	-	+	+	-	18

(+ + + +) growth diameter over 6 mm, (+ + +) growth diameter 4–6 mm, (+ +) growth diameter 2–4 mm, (+) growth diameter over less than 2 mm, (-) no growth or inhibition detected.

Differences in salt tolerance were also found within a species, e.g., in the case of *D. hansenii* XVI and *C. lusitaniae* X strains, proliferation was significantly inhibited with the addition of 6% NaCl, while the other strains analyzed showed good growth even at 10% salt concentration (Table 2).

Based on the results obtained (Table 2), it was shown that 10 g L^{-1} lactic acid inhibits the growth of almost all tested yeasts. In the case of the *N. holstii* II strain, no growth was observed with the addition of 6 g L^{-1} acid, whereas the strains resistant to high acid concentration were strains *R. mucilaginosa* III, *C. lusitaniae* V, *M. guilliermondii* VI, *C. sake* VII and *D. hansenii* IV growing at a concentration of up to 10 g L^{-1} lactic acid.

Decreasing the pH of the environment to 3.4 significantly inhibited the growth of most yeasts, which explains why, at the final stage of sauerkraut fermentation, these microorganisms are practically absent. Only in case of *M. guilliermondii* VI, *C. sake* VII, *D. hansenii* XV and XVI strains, weak colony growth was noted at pH 3.4 (Table 2).

Eleven of the seventeen analyzed strains showed killer activity (Table 2). The highest killer activity was observed in the yeasts *D. hansenii* XII, *C. lusitaniae* V and *C. sake* VII. All analyzed *D. hansenii* strains were able to inhibit the growth of a sensitive culture. The highest differentiation in this trait was found in the case of the yeast *C. lusitaniae*, among which there were both cultures with strong killer activity, such as strains V and X, and with low or no activity—XI and XIII. The strain with the highest killer activity—*D. hansenii* XII, was predominant during sauerkraut fermentation and was found in the fermenting sauerkraut samples of each variety (Figure 3).

3. Discussion

Previous studies of sauerkraut fermentation process included mainly the determination of the quantitative and qualitative composition of bacterial microbiota [5,9–12], content and formation of selected bioactive components (such as biogenic amines, glucosinolates) [13–16], the influence of selected physicochemical factors on the above [10,17], as well as management of post-fermentation brine as a waste material for the production of components of biotechnological importance [18,19]. Very little attention was paid to the presence of yeast during this process, and available publications come from many years ago, or contain only mention of yeasts as spoilage organisms [6,8].

In this work we have proved that with properly conducted fermentation process, with anaerobic conditions, yeast growth is typical for sauerkraut fermentation and occurs at the initial period of fermentation (Figure 1). Their high number may affect the sensory characteristics of the product being produced without deterioration in its quality. It is dependent on the salt concentration in brine, as well as the use of the starter culture [10]. In our research, we used a 2.5% NaCl solution, used by Polish producers of sauerkraut, which creates better conditions for the process and more strongly limits the growth of undesirable microbiota than the concentrations used in other regions of the world [10].

Cabbage cultivar strongly influenced the quantitative and qualitative composition of yeast microbiota at the beginning and during its fermentation (Figures 1 and 4). We have not shown a relationship between the content of sugars and dry matter in the cabbage [20] and the amount of yeast. Although the Manama cabbage contained the most sugars (21.7 g kg^{-1} of fructose and 25.8 g kg^{-1} glucose) and the highest number of yeast cells was found on it, a similar trend was not confirmed in the other analyzed samples. Therefore, it should be assumed that other factors, i.e., agrotechnical treatments used during cultivation, as well as climatic conditions present during its development (late or mid-late varieties with a similar harvest time, intended for processing, were used for experiments) could determine the population of yeasts. Similar factors affect the composition of microbiota during spontaneous fermentation of other vegetables/fruits [21,22].

The sauerkrauts were dominated by *D. hansenii* strains (Table 1, Figure 3). Using the RAPD-PCR method and the sequencing of 5.8S-ITS rRNA gene region, four different strains of this species were found, with strain XII occurring in all samples, IV in the Kamienna Głowa and Ambrosia cabbages, and the other two were specific to the Ambrosia variety (Figure 3). Isolated strains significantly differed in terms of sensitivity to stress factors occurring during sauerkraut fermentation, killer activity but also in terms of volatile compounds formed (Table 2, Figure 4). Culture more resistant such as strain IV produced less volatiles. Strain XII was present in most sauerkraut samples (regardless of the variety of cabbage used), formed relatively large amounts of alcohols and esters; it can be assumed that its positive participation in sensory features of sauerkraut could be significant. *D. hansenii* is a halo-, osmo-, and xerotolerant species of yeast. It occurs in many habitats with low water activity, it was also isolated from wine, beer, fruit, cheese, meat and soil as well as from high-sugar products [23]. The generation of volatile and aroma compounds by *D. hansenii* is considered an important contribution to the ripening process of cheeses and fermented sausages [24–26]. *D. hansenii* can prevent the formation of lipid oxidation products in fermented sausages, and contributes to improve sensory components, principally flavors, such as ethyl esters [26]. Unfortunately, this yeast can also spoil brine-preserved foods, such as gherkins [23]. Deak and Beuchat [27] report that yeast *D. hansenii* develops only at the beginning of the fermentation process of cabbage and their population gradually dies off. Not without significance is the induction of fermentation properties of *D. hansenii* in the presence of salt [28], which can increase the survival of cultures under anaerobic conditions and stimulate the cells to form larger amounts of volatile compounds positively affecting the silage aroma. However, confirmation of the positive impact of *D. hansenii* cultures on the process of sauerkraut fermentation still requires further research.

The second most common yeast species detected in our analyses was *C. lusitaniae*. This is the first report regarding the presence of these yeast in fermented vegetable. It is a saprobial, fermentative yeast, that has been associated with clinical specimens from immunocompromised patients and is now recognized as an opportunistically infectious organism. It has also been isolated from food and food production processes, including fermentation of agave must and sap, Lager brewing process in Thailand [29], traditional Egyptian dairy products [30], dates, Kopanisti cheese ripening [31] and others. This microorganism is most often found with *D. hansenii* and lactic acid bacteria, but there is no references on the properties of cultures, as well as resistance to stress factors occurring during sauerkraut fermentation, such as the presence of NaCl, lactic acid, low pH, etc. Our research is also the first to determine the ability of these yeasts to show killer activity and produce volatile compounds. The analyzed isolates significantly differed in this respect; they were classified into groups of microorganisms producing small or medium amounts of volatile compounds, forming specific components not obtained from other cultures.

Other yeasts identified during sauerkraut fermentation are also detected in various food products and during their production. *P. fermentans* is responsible for the formation of the cocoa beans' aroma [32]. Isolated in our research from sauerkraut fermentation strain of *P. fermentans* were the strongest producer of volatiles, especially esters. *Meyerozyma caribbica* (closely related to *M. guilliermondii*) and *N. holstii* are often found together, for example they are the predominant yeasts present in the crushed olives, olive pomace, and fresh table olive [33]. *Cry. macerans* was found in sauerkraut obtained on farms in the south of Poland [34].

Some of the isolates obtained in this work may have interesting properties which can positively shape the quality of food. *D. hansenii* can produce volatile compounds, such as alcohols, esters, carbonyl and sulphur compounds [35]. Furthermore, recent reports on the potential probiotic properties of *Candida lusitaniae* (anamorph *Clavispora lusitaniae*) and *Meyerozyma caribbica* cultures seem very interesting [36]. The finding of such abilities in yeast isolates from sauerkraut could confirm that vegetable silages (also containing numerous LAB cultures) can be a rich source of positively acting microbiota on our organism.

Our preliminary research of commercial and farm-made sauerkraut showed the largest amount of yeast was found in the sauerkrauts produced in the farms located in the Muszyna commune (3.3–4.2 log CFU g^{-1}) and in one commercial product (3.4 log CFU g^{-1}). In other commercial sauerkraut products analyzed, no yeast was found. Representatives of two species: *Cry. macerans* and *D. hansenii* (3 different strains) predominated among the isolates [34]. These cultures correspond to the strains *Cry. macerans* I and *D. hansenii* IV, XII and XV isolated in this work (Table 1). Their presence in the finished product indicates that single yeast cells can survive the fermentation process, being resistant to the stress factors prevailing in it, such as the presence of salt, anaerobic conditions, increasing lactic acid concentration and decreasing pH. This fact has been confirmed in our research (Table 2). Initially, osmotic pressure associated with the presence of salt and rapid depletion of oxygen in the environment affect microorganisms derived from the raw material. Among the isolates, we only detected cultures that all can grow at NaCl concentrations higher than 5%. Aerobic cultures, such as *Cry. macerans*, *R. mucilaginosa* i *T. pullulans* [32], occurred only in the initial period of fermentation, then they were replaced by yeast with fermentative metabolism (Figure 3). Strains resistant to low pH and the presence of lactic acid produced by lactic acid bacteria were the longest present. Lactic acid has been found to have a weaker inhibitory effect on yeast cells than inorganic acids (8 g L^{-1} lactic acid corresponds to pH 3.4). Earlier studies Hassan et al. [37], showed a similar phenomenon, the effect of eight organic acids (among others acetic and lactic acids) as antifungal agents on the growth of four fungal species (*Aspergillus flavus*, *Penicillium purpurogenum*, *Rhizopus nigricans* and *Fusarium oxysporum*) was different. It has been found that, there is no relationship between the efficiency of organic acid and its final pH.

Another factor influencing the presence of yeasts during fermentation of sauerkraut was their ability to produce killer toxins. Most of the analyzed isolates showed killer phenomena. These antimicrobial substances can inhibit growth of different genera of yeast and play an important role in colonization of the environment [38]. *D. hansenii* has been reported to produce strong and active toxic proteins or glycoproteins, as killer toxins. The research showed that the optimum inhibitory effect of killer toxin was in the presence of NaCl in the environment [39]. The above dependence may explain why during the sauerkraut fermentation, regardless of the cabbage variety used, as the fermentation progressed, the strain of *D. hansenii* XII characterized by the highest killer activity, dominated. A similar phenomenon could be associated with other killer cultures, confirming that in addition to stress factors such as NaCl concentration, lack of oxygen, low pH, and killer activity affects the growth of yeast during fermentation of sauerkraut.

Summarizing, our research confirmed that during properly conducted fermentation process of sauerkraut, with anaerobic conditions, yeast growth occurs always and are present in the initial period of fermentation. We found a large similarity in the composition of yeast microbiota between samples regardless of the cabbage variety used. Among the isolates, we only detected cultures that all can grow at NaCl concentrations higher than 5%, and relatively resistant to low pH and the presence of lactic acid. Most of the cultures were characterized by killer toxins activity. Yeast *D. hansenii* and *C. lusitaniae* predominated in the analyzed samples, because of their beneficial properties may directly affect the sensory characteristics of the finished product, but also increase its pro-health value. However, further research is required to determine the impact of individual cultures on the quality of sauerkraut.

4. Materials and Methods

4.1. Cabbages and Fermentation of Sauerkraut

Eight cultivars of cabbage (*Brassica oleracea* var. *capitata* f. *alba*)—Ambrosia, Avak, Cabton, Galaxy, Jaguar, Kamienna Głowa, Manama and Ramco—from planters of southern Poland (harvested from August to October 2016) were used in this study. The obtained cabbage was kept for 24 h before fermentation in room temperature, in dark. After preliminary cleansing from dirt, the core as well as outer leaves were removed. Then, the cabbage was shredded with a sterile slicer MA-GO 612 P to obtain shavings of thickness of 3.5 mm and 4 kg of shredded cabbage was placed in layers a 5-L jar and 2.5% NaCl (*w*/*w*) was added. The experiment was done in triplicate.

4.2. Yeast Enumeration and Isolation

Fermentation was carried out in the temperature of 20 °C for two weeks. During the process (0., 1., 2., 3., 7., 10. and 14. day of fermentation), 5 g of sauerkraut and 5 mL of brine were collected from the middle of the jar. Samples were placed in sterile stomacher bags (BagPage 400 mL; Interscience, Woburn, MA, USA) and homogenized for 5 min. Ringer's solution (sodium chloride 2.25 g L^{-1}, anhydrous calcium chloride 0.12 g L^{-1}, sodium bicarbonate 0.05 g L^{-1}; POCH S.A., Gliwice, Poland) was used for serial decimal dilutions. The appropriate dilutions were plated in triplicate on Petri dishes and poured with WL Agar (Wallerstein Laboratory) (BIOCORP, Warsaw, Poland) supplemented with 100 mg L^{-1} of chloramphenicol.

After the incubation at 20 °C for 5 days (WL Agar) the colonies were enumerated, representative 10 cultures were isolated from each cabbage variety from subsequent fermentation days, based on their colony morphologies (size, shape, color), pure cultures were obtained by streaking on Sabouraud Dextrose with Chloramphenicol LAB-AGAR (BIOCORP, Warsaw, Poland) and identified.

4.3. *DNA Extraction and RAPD-PCR Analysis*

The total yeast genomic DNA was extracted from isolates using a commercial kit, Yeast Genomic Mini AX Spin (A&A Biotechnology, Gdynia, Poland), following the manufacturer's instructions.

The RAPD-PCR analysis was done according to the method described by Cioch-Skoneczny et al. [22].

4.4. *PCR-RFLP Analysis and 5.8 S-ITS rRNA Gene Region Sequencing*

ITS1 (5′ TCCGTAGGTGAACCTGCGG-3′) and ITS4 (5′-TCCTCCGCTTATTGATATGC-3′) primers were used for the 5.8S-ITS rRNA gene region amplification. The PCR-RFLP analysis and 5.8 S-ITS rRNA gene region sequencing were conducted according to the method described by Cioch-Skoneczny et al. [22]. Sequences were deposited in the GenBank NCBI database with the accession numbers: MK312605-312621.

4.5. *Production of Volatile Components by Isolates (SPME-GC-TOFMS)*

Identified yeast isolates were growing overnight in Sabouraud Dextrose Broth (BIOCORP, Warsaw, Poland). The cells were then centrifuged (735× *g*), resuspended in Ringer's solution and 10^6 cells mL^{-1} were inoculated into YNB solution (Yeast Nitrogen Base; Sigma-Aldrich, St. Louis, MO, USA) with 0.5% of glucose and 0.5% of fructose as a carbon source. After 10 days of incubation (25 °C), the cells were removed by centrifugation (735× *g*), and the supernatants were analyzed by SPME-GC-TOFMS, according to the method described by Zdaniewicz et al. [40]. Detected volatiles were identified using the NIST database (http://webbook.nist.gov/chemistry/) and determined semi-quantitatively (μg L^{-1}) by measuring the relative peak area of each identified component, in relation to that of the internal standards (4-methyl-2-pentanol and ethyl nonanoate). All tests were carried out three times.

4.6. *Resistance of Isolates to Selected Stress Factors Present during Sauerkraut Fermentation*

Sterile Petri dishes with Sabouraud Dextrose LAB-Agar medium, supplemented with NaCl (2.5, 5, 6, 8 or 10%), lactic acid (6, 8 or 10 g L^{-1}) or hydrochloric acid (pH 3.2, 3.4 or 3.6) were prepared. The control was a pH 5.6 medium (without adjusted pH value), with 2.5% NaCl. The plates were divided into 4 parts, and then in each quarter, in five repetitions, 30 μL of yeast suspension in 0.85% NaCl (8 log cells mL^{-1}, growing over-night) was placed. The cultures were kept at a temperature of about 20 °C for 48 h. The diameter of the grown colonies was then measured.

4.7. *Killer Activity*

Killer activity was assayed using seeded-agar-plate technique. Killer sensitive strain of *Saccharomyces cerevisiae* (DBPVG 6500) was suspended in Ringer's solution (~5 log cells mL^{-1}) and inoculated into YEPD-MB agar [38]. Then, wells (5 mm) were sterilely cut in the YEPD-MB agar and the potential killer strains were seeded in the wells at 100 μL of yeast inoculum per well. The plates were incubated at 20 °C for up to 7 days. If the tested strain was surrounded by a zone of inhibition fringed with blue color, it was recorded as killer. Killer activity was measured by subtracting diameter of the well from diameter of the inhibition zone.

4.8. Statistical Analysis

A heat map and cluster analysis were prepared using the statistical package SPSS 18.0 (SPSS Inc., Chicago, IL, USA).

Author Contributions: Conceptualization, P.S.; methodology, P.S., M.S.; software, P.S.; validation, P.S., M.S. and K.Ž.; formal analysis, P.S., M.S., S.S. and K.Ž.; investigation, M.S., S.S. and K.Ž.; resources, M.S., S.S.; data curation, P.S.; writing—original draft preparation, P.S.; writing—review and editing, P.S., M.S.; visualization, P.S., M.S.; supervision, P.S.; project administration, P.S.; funding acquisition, P.S. All authors have read and agreed to the published version of the manuscript.

References

1. Filipiak, T.; Maciejczak, M. Vegetable production in Poland and selected countries of the European Union. *Econ. Sci. Rural Dev.* **2011**, *2*, 30–39.
2. Beganović, J.; Kos, B.; Pavunc, A.L.; Uroić, K.; Jokić, M.; Šušković, J. Traditionally produced sauerkraut as source of autochthonous functional starter cultures. *Microbiol. Res.* **2014**, *169*, 623–632. [CrossRef] [PubMed]
3. Kusznierewicz, B.; Śmiechowska, A.; Bartoszek, A.; Namieśnik, J. The effect of heating and fermenting on antioxidant properties of white cabbage. *Food Chem.* **2008**, *108*, 853–861. [CrossRef] [PubMed]
4. Peñas, E.; Frias, J.; Gomez, R.; Vidal-Valverde, C. High hydrostatic pressure can improve the microbial quality of sauerkraut during storage. *Food Control* **2010**, *21*, 524–528. [CrossRef]
5. Plengvidhya, V.; Breidt, F., Jr.; Lu, Z.; Fleming, H.P. DNA fingerprinting of lactic acid bacteria in sauerkraut fermentations. *Appl. Environ. Microbiol.* **2007**, *73*, 7697–7702. [CrossRef] [PubMed]
6. Li, K.-Y. Fermentation. In *Handbook of Food and Beverage Fermentation Technology*; Hui, Y.H., Meunier-Goddik, L., Hansen, Å.S., Josephsen, J., Nip, W.K., Stanfield, P.S., Toldrá, F., Eds.; Marcel Dekker Inc.: New York, NY, USA, 2004; pp. 595–610.
7. Hang, Y.D. Sauerkraut. In *Handbook of Food and Beverage Fermentation Technology*; Hui, Y.H., Meunier-Goddik, L., Hansen, Å.S., Josephsen, J., Nip, W.K., Stanfield, P.S., Toldrá, F., Eds.; Marcel Dekker Inc.: New York, NY, USA, 2004; pp. 669–676.
8. Shih, C.T.; Hang, Y.D. Production of carotenoids by *Rhodotorula rubra* from sauerkraut brine. *LTW-Food Sci. Technol.* **1996**, *29*, 570–572.
9. Du, R.; Ge, J.; Zhao, D.; Sun, J.; Ping, W.; Song, G. Bacterial diversity and community structure during fermentation of Chinese sauerkraut with *Lactobacillus casei* 11MZ-5-1 by Illumina Miseq sequencing. *Lett. Appl. Microbiol.* **2018**, *66*, 55–62. [CrossRef] [PubMed]
10. Müller, A.; Rösch, N.; Cho, G.S.; Meinhardt, A.K.; Kabisch, J.; Habermann, D.; Böhnlein, C.; Brinks, E.; Greiner, R.; Franz, C.M.A.P. Influence of iodized table salt on fermentation characteristics and bacterial diversity during sauerkraut fermentation. *Food Microbiol.* **2018**, *76*, 473–480. [CrossRef]
11. Touret, T.; Oliveira, M.; Semedo-Lemsaddek, T. Putative probiotic lactic acid bacteria isolated from sauerkraut fermentations. *PLoS ONE* **2018**, *13*, e0203501. [CrossRef]
12. Zhou, Q.; Zang, S.; Zhao, Z.; Li, X. Dynamic changes of bacterial communities and nitrite character during northeastern Chinese sauerkraut fermentation. *Food Sci. Biotechnol.* **2017**, *27*, 79–85. [CrossRef]
13. Beljaars, P.R.; Van Dijk, R.; Jonker, K.M.; Schout, L.J. Liquid chromatographic determination of histamine in fish, sauerkraut, and wine: Interlaboratory study. *J. AOAC Int.* **1998**, *81*, 991–998. [CrossRef] [PubMed]
14. Hallmann, E.; Kazimierczak, R.; Marszałek, K.; Drela, N.; Kiernozek, E.; Toomik, P.; Matt, D.; Luik, A.; Rembiałkowska, E. The Nutritive Value of Organic and Conventional White Cabbage (*Brassica Oleracea* L. Var. *Capitata*) and Anti-Apoptotic Activity in Gastric Adenocarcinoma Cells of Sauerkraut Juice Produced Therof. *J. Agric. Food Chem.* **2017**, *65*, 8171–8183. [CrossRef] [PubMed]
15. Martinez-Villaluenga, C.; Peñas, E.; Frias, J.; Ciska, E.; Honke, J.; Piskula, M.K.; Kozlowska, H.; Vidal-Valverde, C. Influence of fermentation conditions on glucosinolates, ascorbigen, and ascorbic acid content in white cabbage (*Brassica oleracea* var. *capitata cv. Taler) cultivated in different seasons. J. Food Sci.* **2009**, *74*, C62–C67. [PubMed]
16. Palani, K.; Harbaum-Piayda, B.; Meske, D.; Keppler, J.K.; Bockelmann, W.; Heller, K.J.; Schwarz, K. Influence of fermentation on glucosinolates and glucobrassicin degradation products in sauerkraut. *Food Chem.* **2016**, *190*, 755–762. [CrossRef]

17. Harbaum-Piayda, B.; Palani, K.; Schwarz, K. Influence of postharvest UV-B treatment and fermentation on secondary plant compounds in white cabbage leaves. *Food Chem.* **2016**, *197 Pt A*, 47–56. [CrossRef]
18. Hang, Y.D.; Woodams, E.E. Lipase production by *Geotrichum candidum* from sauerkraut brine. *World J. Microbiol. Biotechnol.* **1990**, *6*, 418–421. [CrossRef]
19. Ku, M.A.; Hang, Y.D. Effect of inulin on yeast inulinase production in sauerkraut brine. *World J. Microbiol. Biotechnol.* **1994**, *10*, 354–355. [CrossRef]
20. Satora, P.; Skotniczny, M.; Strnad, S.; Piechowicz, W. Chemical composition of sauerkraut produced of different cabbage varieties. *LWT-Food Sci. Technol.* **2021**, *136*, 110325. [CrossRef]
21. Kiai, H.; Hafidi, A. Chemical composition changes in four green olive cultivars during spontaneous fermentation. *LWT-Food Sci. Technol.* **2014**, *57*, 663–670. [CrossRef]
22. Cioch-Skoneczny, M.; Satora, P.; Skotniczny, M.; Skoneczny, S. Quantitative and qualitative composition of yeast microbiota in spontaneously fermented grape musts obtained from cool climate grape varieties 'Rondo' and 'Regent'. *FEMS Yeast Res.* **2018**, *18*, 1–11. [CrossRef]
23. Breuer, U.; Harms, H. *Debaryomyces hansenii*—An extremophilic yeast with biotechnological potential. *Yeast* **2006**, *23*, 415–437. [CrossRef] [PubMed]
24. Masoud, W.; Jakobsen, M. Surface ripened cheeses: The effects of *Debaryomyces hansenii*, NaCl and pH on the intensity of pigmentation produced by *Brevibacterium linens* and *Corynebacterium flavescens*. *Int. Dairy J.* **2003**, *13*, 231–237. [CrossRef]
25. Pitt, J.I.; Hocking, A.D. Yeasts. In *Fungi and Food Spoilage*; Pitt, J.I., Hocking, A.D., Eds.; Springer: Boston, MA, USA, 2009; pp. 357–382.
26. Ramos, J.; Melero, Y.; Ramos-Moreno, L.; MichÁn, C.; Cabezas, L. *Debaryomyces hansenii* strains from valle de los pedroches iberian dry meat products: Isolation, identification, characterization, and selection for starter cultures. *J. Microbiol. Biotechnol.* **2017**, *27*, 1576–1585. [CrossRef] [PubMed]
27. Deak, T.; Beuchat, L.R. *Handbook of Food Spoilage Yeasts;* CRC Press: New York, NY, USA, 1996.
28. Calahorra, M.; Sánchez, N.S.; Peña, A. Activation of fermentation by salts in *Debaryomyces hansenii*. *FEMS Yeast Res.* **2009**, *9*, 1293–1301. [CrossRef]
29. Pérez-Brito, D.; Magaña-Alvarez, A.; Lappe-Oliveras, P.; Cortes-Velazquez, A.; Torres-Calzada, C.; Herrera-Suarez, T.; Larqué-Saavedra, A.; Tapia-Tussell, R. Genetic diversity of *Clavispora lusitaniae* isolated from Agave fourcroydes Lem, as revealed by DNA fingerprinting. *J. Microbiol.* **2015**, *53*, 14–20. [CrossRef]
30. El-Sharoud, W.M.; Belloch, C.; Peris, D.; Querol, A. Molecular identification of yeasts associated with traditional Egyptian dairy products. *J. Food Sci.* **2009**, *74*, M341–M346. [CrossRef]
31. Kaminarides, S.E.; Anifantakis, E.M. Evolution of the microflora of Kopanisti cheese during ripening. Study of the yeast flora. *Le Lait INRA Ed.* **1989**, *69*, 537–546. [CrossRef]
32. Boekhout, T. *Yeasts in Food: Beneficial and Detrimental Aspects*; Behr's: Hamburg, Germany, 2003.
33. Romo-Sánchez, S.; Alves-Baffi, M.; Arévalo-Villena, M.; Ubeda-Iranzo, J.; Briones-Pérez, A. Yeast biodiversity from oleic ecosystems: Study of their biotechnological properties. *Food Microbiol.* **2010**, *27*, 487–492. [CrossRef]
34. Satora, P.; Celej, D.; Skotniczny, M.; Trojan, N. Identyfikacja drożdży obecnych w kiszonej kapuście komercyjnej i otrzymywanej w gospodarstwach rolnych. *ŻYWNOŚĆ Nauka Technol. Jakość* **2017**, *4*, 27–36.
35. Cano-García, L.; Flores, M.; Belloch, C. Molecular characterization and aromatic potential of *Debaryomyces hansenii* strains isolated from naturally fermented sausages. *Food Res. Int.* **2013**, *52*, 42–49. [CrossRef]
36. Amorim, J.C.; Piccoli, R.H.; Duarte, W.F. Probiotic potential of yeasts isolated from pineapple and their use in the elaboration of potentially functional fermented beverages. *Food Res. Int.* **2018**, *107*, 518–527. [CrossRef]
37. Hassan, R.; Sand, M.I.; El-Kadi, S.M. Effect of some organic acids on fungal growth and their toxins production. *J. Agric. Chem. Biotechn. Mansoura Univ.* **2012**, *3*, 391–397. [CrossRef]
38. Satora, P.; Tuszynski, T. Biodiversity of yeasts during plum Wegierka Zwykla spontaneous fermentation. *Food Technol. Biotechnol.* **2005**, *43*, 277–282.
39. Al-Qaysi, S.A.S.; Al-Haideri, H.; Thabit, Z.A.; Al-Kubaisy, W.H.A.A.-R.; Ibrahim, J.A.A.-R. Production, Characterization, and Antimicrobial Activity of Mycocin Produced by *Debaryomyces hansenii* DSMZ70238. *Int. J. Microbiol.* **2017**, 1–9. [CrossRef] [PubMed]
40. Zdaniewicz, M.; Satora, P.; Pater, A.; Bogacz, S. Low Lactic Acid-Producing Strain of *Lachancea thermotolerans* as a New Starter for Beer Production. *Biomolecules* **2020**, *10*, 256. [CrossRef]

4

Waste Management in Dairy Cattle Farms in Aydın Region: Potential of Energy Application

Gürel Soyer [1] and Ersel Yilmaz [2,*]

1 11th Regional Directorate, General Directorate of State Hydraulic Works, Ministry of Agriculture and Forestry, 22100 Edirne, Turkey; gurel.soyer@dsi.gov.tr
2 Department of Biosystems Engineering, Aydın Adnan Menderes University, 09020 Aydın, Turkey
* Correspondence: eyilmaz@adu.edu.tr

Abstract: In this paper, the dairy cattle waste management systems on farms in Aydın region in Turkey were investigated. Number of farms and livestock herd size, type of barn, type of machinery and farm labour force were studied. The collection, management and storage systems of manure produced in dairy cattle farms were taken into consideration. Additionally, biogas amount, which is produced from animal waste, was calculated for all districts of Aydın by using the number of livestock animals and various criteria such as the rate of dry matter. Results show that the typical and representative farm in the Aydın region is facility with a total head over 100 heads. 89.6% of the farms have heads in the range of 100 to 200. The amount of biogas that can be produced from all manure collected in Aydın region in the biogas plants is approximately 160,438 m^3/day (based on 0.5 m^3/day biogas per cattle), which would produce around 100 GWh/year that can be used for own needs of farms owners.

Keywords: animal waste; biogas; dairy cattle farms; energy potential; waste management

---

1. Introduction

Nowadays, expansion and intensification of large-scale animal feeding operations has resulted in an increase in the size of farms and in the amount of waste produced from farms causing serious problems such as a negative impact on environment and public health in rural areas.

By the end of 2014, according to FAOSTAT [1], 24.99 billion animals were produced on farms all over the World. The livestock sector is one of the fastest growing parts of the agricultural economy. In recent years there has been an increasing demand for cattle production. The large cattle producers are Brazil about 218 million, India 186 million and China 83 million heads [1].

In Turkey, the greatest livestock production belongs to cattle farms, with about 17 million heads of cattle being bred in 2018, resulting in an increase of 33% compared to 2010. Dairy cattle produced about 22 million tonnes of milk and 1 million tonnes of meat in 2018 [2]. Table 1 presents the total amount of animal production from species across years in Turkey.

The breeding and agricultural activities, especially livestock production on an industrial scale, are seen as one of the main sources of natural environment pollution [3,4]. Depending on the farming system, animal farms generate solid (dung) and liquid (liquid manure) animal excrement. In this day and age, no-mulch systems are becoming more and more popular, particularly for livestock production on a large scale. The excrement in this system is so-called liquid manure, i.e., liquid, or a semiliquid mixture of faeces, urine, water and feed leftovers.

Table 1. Total amount of animal production in Turkey by species [1].

Animal	Unit	2008	2010	2012	2014
Buffaloes	Head	84,705	87,207	107,435	121,826
Camels	Head	1057	1041	1315	1442
Cattle	Head	11,036,753	10,723,958	13,914,912	14,223,109
Sheep	Head	25,462,292	21,794,508	27,425,233	31,140,244
Chickens	Head ×1000	269,368	229,969	253,712	293,728
Goats	Head ×1000	6,286,358	5,128,285	8,357,286	10,344,936
Turkeys	Head ×1000	2675	2755	2761	2990

It is estimated that the cattle residues produced in Turkey reached the value of 1.3×10^6 tonnes/year in 2012 [5]. The amount of wet manure from animals could be a major problem for farms. If the wet manure cannot be utilized properly, it can create pollution risk with a potentially disastrous impact on the environment.

Manure management depends on many factors such as the size of the herd and type of manure, as well as available labour, soil type, climate and region [6,7]. Additionally, intensive animal production can be significantly problematic with respect to manure storage and removal [3].

Effluents of unproperly stored manure can flow directly or indirectly into surface waters in open lagoons. As a result, gaseous emissions and odours can also be released upon decomposition of manure, with negative consequences for farmers' fields and livestock farms [8,9]. Fangueiro [10] reported that greenhouse gas (NH_3, N_2O, CH_4) emissions during storing depend on type of manure, i.e., emission from separated solids, are typically higher than from liquid or unseparated manure. Animal manure contains a wide range of micro-organisms which could be a source of hazards to humans and animals. These micro-organisms can cause food contamination and epidemics and are dangerous to public health [3,11]. Therefore, sustainable manure management systems on farms must minimize risks for the environment associated with storage, handling and utilization of manure.

Animal manure contains essential nutrients such as nitrogen, phosphorus, potassium and can be applied to land as a natural fertilizer [7,8], which is the most common method of manure application. Organic matter improves the physical and biological properties of soil, as well as aeration and soil water infiltration [12].

However, in recent years, we have observed a large problem of environmental pollution caused by nitrates connected with irrational use of natural fertilizers in agriculture [13,14]. The manure contains large amount of N in organic form and converted to inorganic form through mineralization process which is ultimately a serious risk to the environment. Manure is applied to the soil at one time (usually by spreading out on the field), so more leaching occurs as compared to chemical fertilizer and the N content may reach the ground and surface waters [15].

Animal manure can also be used as substrate for biogas production in the process anaerobic digestion [16–18]. Biogas is a product of methane fermentation of organic fraction of many types of biomass.

The methane fermentation process consists of four phases (hydrolysis, acidogenesis, acetogenesis and methanogenesis) [19]. The main stages of anaerobic digestion are presented in Figure 1.

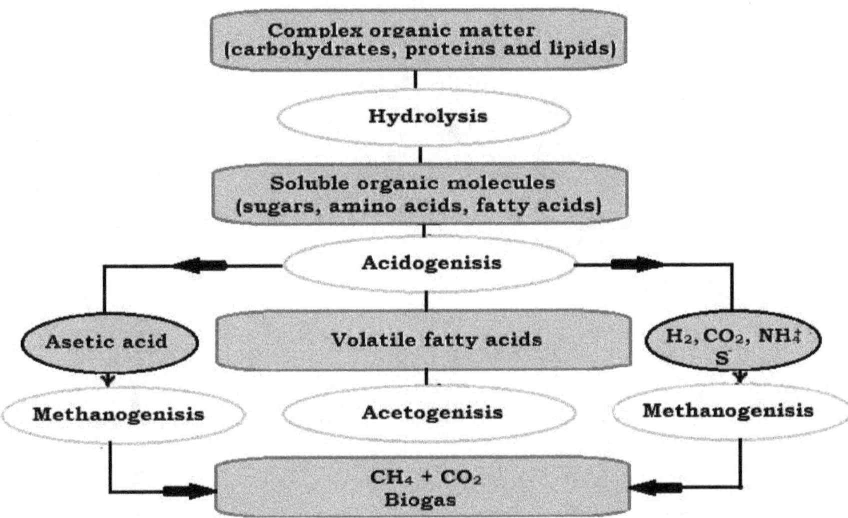

Figure 1. Stages of anaerobic digestion (methane fermentation process) [20].

The composition of biogas is different and depends on the applied substrates; however, typically it consists mostly of CH_4 (40%–70%) and CO_2 (15%–60%), as well as other compounds in small amounts: H_2O (2%–7%), N_2 (2%–5%), O_2 (0%–2%) < 1% H_2, NH_3 (0%–1%) and H_2S (0.005%–2%) [8,21].

In the process of biochemical transformations in the absence of oxygen instead of biogas is also produces the nutrient rich organic fertilizer which is easy assimilated by the plants, with a reduction in the odours and the disease-causing agents [19].

The biogas energy potential of Turkey was found to be 2.18 billion m^3 based on the animal numbers in the last agricultural census. The total biogas potential originates from 68% cattle, 5% small ruminants, and 27% poultry. The biogas energy equivalence of Turkey is approximately 49 PJ [5]. After comparing the biogas potential for animal manure of Turkey with that for different countries (Germany 20.6 billion m^3, Poland 6.4 billion m^3, Italy 1.9 billion m^3 and Sweden 7.04 billion m^3) [22], Turkey has a high biogas potential, which is associated with the increasing production in the livestock sector.

As of now, only 7% of this potential is used. There are 19 biogas power plants that produce electricity from animal manure in Turkey. The total installed power capacity of the biogas power plants is 43.41 MW_e. The range of the installed power capacity is from 0.33 to 6.40 MW_e [23].

The collection, storage and utilization of animal manure are the major problems for local livestock farmers. Problems and strategies with respect to manure management should be taken care of on a local scale and adapted to the existing conditions in a given area. There are several studies focused on cattle in Turkey [24–27]. However, region-based studies are few and limited [28,29].

The aim of this study was to investigate the collection and management of manure in the cattle farms in Aydın region. Number of farms, livestock herd size, type of barns, type of machinery for collecting manure, farm labour force and manure management were also studied to evaluate the possibility of using manure as a feedstock for biogas production for energy generation.

2. Materials and Methods

2.1. Study Area

Aydın province is located in Aegean Region of western Turkey (Figure 2). The Aegean Region has a typical Mediterranean climate with hot-dry summer and warm-rainy winter. The average annual temperature is 17.6 °C, 26.77 °C in summer and 9.33 °C in winter.

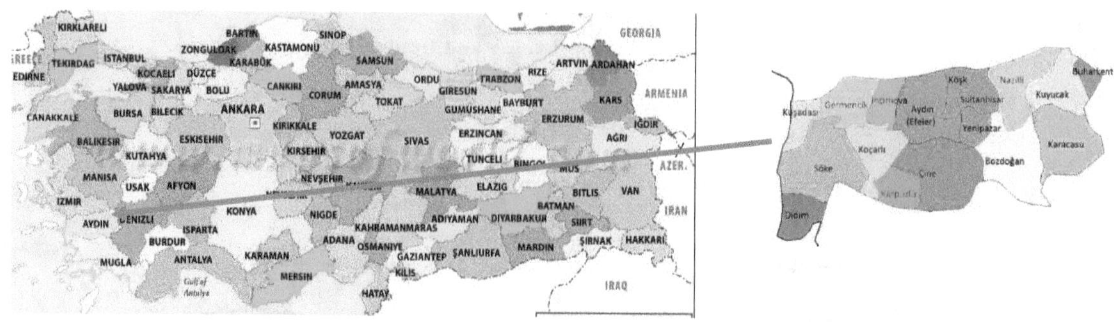

Figure 2. Turkey and Aydın province [30].

The relative humidity of the air is between 48% and 55% and the average rainfall is on the level of 647 mm [31].

The area of Aydın province with 17 districts is 8007 km^2. The population was 1,097,746 with density of 140 people/km^2 in 2018. The cultivated area is about 395,494 ha corresponding to 49.3% of soil sources of Aydın and 75,000 ha of cultivated area is used fo cereal production. The main agricultural products in Aydın province are fig, olive, chestnut, cotton and fruits [32].

2.2. Methods

The aim of this study was to investigate manure collection and management in cattle farms in the Aydın region of Turkey and determine the energy potential of the waste generated on farms.

In this study, a survey was conducted by interviewing the owners of 87 farms located in 17 districts of Aydın province and each farm was photographically documented.

The survey included general topics presented below:

- education level of farmers and the possibility of using new technologies,
- livestock herd size,
- type of barns,
- type of machinery for collecting manure,
- manure storage systems,
- methods of manure application.

In addition, the energy parameters of manure waste as a potential substrate for the biogas production were also examined. The tests were carried out in accordance with the following standards:

- moisture—EN ISO 18134:2015,
- ash—EN ISO 18122:2015,
- organic matter—EN 12176:2004,
- high heating value (HHV)—EN 14918:2009, ISO 18125:2015 using IKA C 5000 Calorimeter,
- elementary analysis (C, H, N, S, O)—EN ISO 16948:2015 using Elementary Vario Macro Cube analyser.

Based on the data obtained from 87 farms livestock size in the study region, potential of biogas and electricity production were calculated using equation 1 and 2 below:

Biogas production (BP) [33]:

$$BP = N_c \times C, \tag{1}$$

where: N_c—the number of cattle, C—production of manure per day/cattle (on the basis of an assumption of 0.4 m^3/day/cattle [34].

Electricity production (EP) [21]:

$$EP = BP \times LHV x \%CH_4, \tag{2}$$

where: $\%CH_4$—methane content in the biogas (on the basis of an assumption it is 62%), LHV_{CH4}—low heating value of methane (21 MJ/m^3)—1.7 kWh in cogeneration process: 1.7 kWh electricity and 2 kWh heat) [34].

3. Results and Discussion

In Aydın province, many of farms are located in the districts: Efeler (18 farms), Cine and Kuyucak (12 farms) and Söke (10 farms). Table 2 presents number of cattle in districtes of Aydın.

Table 2. Total amount of cattle in Aydın's districts [35].

Aydın's Districts	Number of Cattle, Head
Efeler	34,300
Bozdoğan	26,244
Buharkent	2025
Çine	62,376
Didim	3047
Germencik	19,144
İncirliova	9048
Karacasu	10,219
Karpuzlu	27,027
Koçarlı	23,953
Köşk	8757
Kuşadası	1283
Kuyucak	21,713
Nazilli	26,000
Söke	24,145
Sultanhisar	4595
Yenipazar	17,000
Total	**320,876**

Farms in Aydın province usually have more than 100 cattle and the number of animals in the 89.6% of the farms range between 100 and 200 heads.

The study shows that 69.55% of the farmers are under age 50. The ages of the youngest and oldest farmers are 28 and 74, respectively. The percentage of owners who have a university degree was 14.9%, whereas most of the owners have an elementary school degree (43.7%). Only farmers with higher education showed an interest in application of new technologies.

Manure storage type is of importance in terms of impacts on gaseous emissions and the flexibility it offers for land application and hence the potential for nutrient losses to ground and surface waters.

Generally, owners of farms have noticed the problem of disposal of manure, the facility must minimize the impact on water quality, especially on groundwater and surface water. It is indicated that the manure storage facility should be located at least 100 m away from the water resources [36].

In the Aydın region, the distance between open-air manure storage and water resources, as well as source of drinking water supplies, is 96 m on average. Çayır and Atılgan [37] examined about 74 farms in Burdur province and determined the distances to be 1–10 m in 39 farms, 11–20 m in 20 farms and 21–30 m in 10 farms, and 31 m or more in 5 of 74 farms. According to Mutlu [38], Jacopson et al. [39] and Nizam et al. [40], this distance should be much longer.

In the study area, manure storage facilities are located in the open area. The most common type of manure storage is midden (60%), and 30% and 10.3% of farms store the manure on flat ground and on leak-proof pits, respectively. Figure 3 presents the types of manure storage used in Aydın region. Manure stored on flat ground is shown in Figure 4.

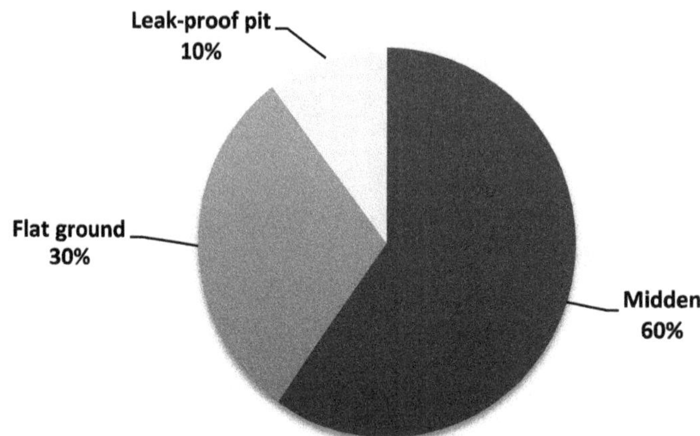

Figure 3. Types of storage used in Aydın region.

Figure 4. Manure storage on flat ground.

Study shows that 66.6% of farms do not have manure storage pits on protected ground (concrete floor). Therefore, there is a danger of contamination of ground water with nitrogen compounds.

For example, according to the survey by Smith et al. [6], manure is stored in concrete floor compounds (40%) and temporary field heaps (60%) in England. Loyon [41] stated that 23% of facilities for storing manure are covered in France.

Manure is usually stored for many months, and during its decomposition, manure emits unpleasant gases such as ammonia and hydrogen sulphide and impacts the health and comfort of surrounding people. Another problem is to minimize odour from manure storage locations as well as from open barns, which depends on the size of the intensive livestock operation, the type of livestock or manure management system and storage time.

The results of this study show that 48.2% of farms have closed-wall barns, 41.4% semi-open barns and rest of them have open sheltered barns. Semi-open barns are shown in Figure 5.

The conventional method of handling manure has been to use sufficient bedding to keep the manure relatively dry and then to move it out of the confinement area and deposit it into a manure pile [42].

In large production units, manure is handled both mechanically and hydraulically. Mechanical removal of the wastes is normally done with tractors, manure spreaders or scrapers with permanently installed equipment, such as shuttle conveyors, floor augers or pumps.

The information collected from the dairy farms assessed in this study showed that 67.8% of the farms used tractor shovels for the collection of manure produced in barns. The percentage of manual collection was 14.9%, and there were only 9 farms (10.4% of the farms evaluated) in which the manure was collected with scrapers equipped with chain. 89.7% of the farms do not have any impermeable manure pits.

Figure 5. Semi-open barns.

The most common waste management strategy on farms is to apply the manure to the land. Atılgan et al. [43] divided it according to content of solids, i.e., above 25% solid fertilizer; 10–20% semi-solid and 0–10% content of soils is called liquid manure. All produced manure in the studied farms is used in agriculture as fertilizer, mostly for their own purposes, and only 12.6% of farms sell it. The studied farms utilize only the solid manure, which provides minimum benefit because of the loss of organic nitrogen content during long storage, while it can also cause serious environmental pollution.

It is stated that the main source of nitrate contamination in groundwater is agricultural fertilizers. In the Aydın region the high level of nitrite (0–124 mg/L) [44] in groundwater used for the irrigation was noted. The average level of nitrite in the surface water is in the range of 0.01–0.7 mg/L, nitrate 1.20–3.70 mg/L, ammonia 004–5.20 mg/L [45]. According to the World Health Organization (WHO) [46], the standard nitrate level in drinking water is 50 mg/L. Groundwater is used as irrigation (from 8–32 m deep) and drinking (>32 m deep). Elevated nitrate concentrations in groundwater can cause public health problems. Now Turkey has updated regulations aiming to combat agricultural nitrate pollution in rivers and soil. The revised rules include procedures and principles for determining, reducing and eliminating nitrate pollution [47].

Due to the increase in the Turkish population, and therefore the increase in demand from the animal sector, contamination from pollutants may also cumulatively increase in the next years. Regulations are required in order to control manure management, especially the localization of manure storage and type of floor construction for its temporary storage, as well as the limits for the use of manure as fertilizer.

Baytekin [33] claims that, under normal conditions, a healthy cow produces 40–45 kg of manure per day. According to this value, the total manure amount obtained from the research area is as presented in Table 3.

Table 3. Total amount of produced manure in farms.

	Daily Tonnes	Weekly Tonnes	Monthly Tonnes	Annual Tonnes
Production of Manure *	**13,637**	**95,460**	**409,116**	**4,977,588**

* calculated from obtained data.

The production of biogas from manure, in particular, is one of the alternative utilization methods of organic wastes that can be implemented in this region. This study also attempts to identify the biogas potential of the Aydın region basing on obtained data of animal manure production.

As a first step for this application, the energy parameters of manure, as a potential feedstock for the biogas production were tested. Obtained energy parameters of manure is presented in Table 4.

Table 4. Energy properties of manure.

Parameter	Moisture % wt	Dry mass % wt	Ash %wt	Organic Matter %wt	C % dm	H % dm	N % dm	S % dm	O * % dm	HHV MJ/kg
Value	79.9	20.1	11.50	74.62	48.09	7.13	2.14	0.27	42.66	16.48

wt—weight percentage; dm—dry mass; HHV—High Heating Value; *—calculated on the basis of the obtained difference.

One of the key parameters in terms of the efficiency of biogas production is associated with the high content of organic matter in wastes, and this determines the course of the fermentation process and the volume of the biogas [21]. The tested samples have a high content of organic matter (74.62%), which is comparable with data obtained by Zue et al. [48], it may range from 68% to 76%. For methane fermentation, especially for the growth of microorganisms, one of the important factors is the ratio of C:N (optimum 20–30:1) [49]. The ratio of C:N in the waste (24:1) is adequate in this respect.

For the Aydın province, based on the amount of produced waste, it is possible to obtain about 160,438 m^3 biogas/day, assuming 0.5 m^3/day biogas per cattle. Table 5 contains total amount of produced biogas in Aydın and LPG equivalent.

Table 5. Total amount of produced waste and biogas in Aydın and its LPG equivalent.

Production of Biogas *	Daily m^3	Weekly m^3	Monthly m^3	Annual m^3	Equivalent of LPG m^3/year
	160,438	**1,123,066**	**4,813,140**	**58,559,870**	**93,895,792**

* calculated from obtained data.

It gives production of electricity on level 99,552 MWh annually. In Aydın in 2012 electricity consumption was 1,860,667 MWh. In the case of the use of biogas, which can be substituted for conventional fuels, 5.4% of electricity can be covered by biogas.

There is only one biogas power plant—Efeler BGEPP—in the Aydin region, with a max installed power of 4.8 MW_e. Because of the distributed allocation of small-capacity livestock farms in the region, and also due to the low interest of farmers in installing their own small biogas installations, it is proposed to build centralized facilities.

Considering the use of only half of the manure generated in the region, it is possible to install 7 biogas power plants with a capacity 4.8 MW_e in different locations. Because of the topography and the distances between farm locations, the Aydın region can be divided into four districts, in which 3-4 biogas power plants can be installed with capacities similar to the Efeler BGEPP facility.

There are some funds that support investments in renewable energy sources in Turkey such as: *Renewable Energy Resources Support Mechanism (YEKDEM)*, coordinated by the Energy and Natural Resources Ministry, the regional scale *Agriculture and Rural Development Support Institution (TKDK)*, supported by the Agriculture and Forestry Ministry, and also local development agencies subordinate to teh Ministry of Development, for example, the South Aegean Development Agency (GEKA) serving, i.a., the Aydın region.

According to the Turkish National Energy and Mining Strategy, it is a top priority for Turkey to generate 30% of its electricity from local and renewable resources by 2023. The costs of achieving this target by 2023 are estimated to require investment in renewable energy generation of around 21 billion USD (1,5 billion USD/year) [50]. In the case of Turkey, which is a net energy importer, 73% of its energy needs come from foreign suppliers, and investments in a local and secure energy supply is main pillar of the energy sector.

Biogas, biomass, and geothermal energy resources are expected to comprise a considerable part of RES with the rapid growth in utilization of these resources in the market [51].

Biogas can be used for heating and electricity production, providing local autonomy for the region in the face of the increasing cost of fossil fuels.

Manure storage facilities on farms should be considered to be a temporary solution, and farmers should have knowledge about the negative influence on the environment caused by improper treatment of manure. Education and financial support in changing the approach to animal waste management can be a key factor.

The conversion of animal waste to biogas through anaerobic digestion processes can provide added value to manure as an energy resource and reduce the environmental problems associated with animal waste. It is worth mentioning that dairy cattle manure is endowed with considerable biogas production, offering numerous benefits with respect to environmental, agricultural and socio-economic standards.

Author Contributions: Conceptualization, E.Y.; methodology, E.Y. formal analysis, E.Y.; investigation, G.S.; resources, G.S.; writing—original draft preparation, G.S.; writing—review and editing, E.Y.; supervision, E.Y. All authors have read and agreed to the published version of the manuscript.

References

1. FAOSTAT Website for Statistics. Available online: http://faostat3.fao.org/home (accessed on 20 January 2020).
2. TUIK (Turkish Statistical Institute). *Livestock Statistics*; TUIK (Turkish Statistical Institute): Ankara, Turkey, 2019.
3. Malomo, G.A.; Madugu, A.S.; Bolu, S.A. *Sustainable Animal Manure Management Strategies and Practices*; Intech Open: London, UK, 2018.
4. Delgad, J.A.; Nearing, M.A.; Rice, C.W. Chapter Two—Conservation practices for climate change adaptation. *Adv. Agron.* **2013**, *121*, 47–115.
5. Acaroglu, M.; Aydogan, H. Biofuels energy sources and future of biofuels energy in Turkey. *Biomass Bioenergy* **2012**, *3*, 69–76. [CrossRef]
6. Smith, K.A.; Williams, A.G. Production and management of cattle manure in the UK and implications for land application practice. *Soil Use Manag.* **2016**, *32*, 73–82. [CrossRef]
7. Mac-Safley, L.M.; Boyd, W.H.; Schmidt, A.R. Agricultural waste management systems. In *Agricultural Waste Management Field Handbook*; Hickman, D., Owens, L., Pierce, W., Self, S., Eds.; USDA: Washington, WA, USA, 2011; pp. 9–31.
8. Font-Palma, C. Methods for the treatment of cattle manure—A review. *J. Carbon Res.* **2019**, *5*, 27. [CrossRef]
9. Chandra, M.; Manna, M.; Naidux, R.; Sahu, A.; Bhattacharjya, S.; Wanjari, R.H.; Patra, A.K.; Chaudharijj, S.K.; Majumdar, K.; Khanna, S.S. Bio-waste management in subtropical soils of India: Future challenges and opportunities in agriculture. *Adv. Agron.* **2018**, *152*. [CrossRef]
10. Fangueiro, D.; Dave, J.C.; Chadwick, R.; Trindade, H. Effect of cattle slurry separation on greenhouse gas and ammonia emissions during storage. *J. Environ. Qual.* **2008**, *37*, 2322–2331. [CrossRef]
11. Manyi-Loh, C.E.; Mamphweli, S.N.; Meyer, E.L.; Makaka, G.; Simon, M.I.; Okoh, A.I. An overview of the control of bacterial pathogens in cattle manure. *Int. J. Environ. Res. Public Health* **2016**, *13*, 843. [CrossRef]
12. Francis, J.; Larney, D.; Angers, A. The role of organic amendments in soil reclamation: A review. *Can. J. Soil Sci.* **2012**, *92*, 913–938.
13. European Council. European Council Directive of 12 December 1991 concerning the protection of waters against pollution caused by nitrates from agricultural sources (91/676/EEC). *Off. J. Eur. Commun. L.* **1991**, *375*, 1–8.
14. Hokeem, K.R.; Akhtor, J.; Sobir, M. *Soil Science: Agricultural and Environmental Prospective*; Springer: Berlin/Heidelberg, Germany, 2016.
15. Webb, J.; Pain, B.F.; Bittman, S.; Morgan, J. The impacts of manure application methods on emissions of ammonia, nitrousoxide and on crop response—A review. *Agric. Ecosyst. Environ.* **2010**, *137*, 39–46. [CrossRef]
16. Abdeshahian, P.; Lim, J.S.; Hoa, W.S.; Hashim, H. Potential of biogas production from farm animal waste in Malaysia. *Renew. Sustain. Energy Rev.* **2016**, *60*, 714–723. [CrossRef]
17. Sun, Q.; Li, H.; Yan, J.; Liu, L.; Yu, Z.; Yu, X. Selection of appropriate biogas upgradingtechnology-a review of biogas cleaning, upgrading and utilisation. *Renew. Sustain. Energy Rev.* **2015**, *51*, 521–532. [CrossRef]

18. Cucui, G.; Ionescu, C.A.; Goldbach, I.R.; Coman, M.D.; Marin, E.L.M. Quantifying the economic effects of biogas installations for organic waste from agro-industrial sector. *Sustainability* **2018**, *10*, 2582. [CrossRef]
19. Gould, M.C. Bioenergy and anaerobic digestion. In *Bioenergy*; Elsevier Inc.: Amsterdam, The Netherlands, 2015; Volume 18, pp. 297–317.
20. Rameshprabu, R.; Natthawud, D. Biological purification processes for biogas using algae cultures: A review. *Int. J. Sust. Green Energy* **2015**, *4*, 20–32.
21. Pawlita-Posmyk, M.; Wzorek, M. Assessment of application of selected waste for production of biogas. In Proceedings of the E3S Web of Conferences, International Conference Energy, Environment and Material Systems (EEMS 2017), Polanica-Zdrój, Poland, 13–15 September 2017; Volume 19, p. 02017. [CrossRef]
22. Karaca, C. Determination of biogas production potential from animal manure and GHG emission abatement in Turkey. *Int. J. Agric. Biol. Eng.* **2019**, *11*, 205–210. [CrossRef]
23. Available online: https://www.enerjiatlasi.com/elektrik-uretimi (accessed on 20 January 2020).
24. Avcioğlu, A.O.; Türker, U. Status and potential of biogas energy from animal wastes in Turkey. *Renew. Sustain. Energy Rev.* **2012**, *16*, 1557–1561. [CrossRef]
25. Acaroglu, M. The potential of biomass and animal waste of Turkey and the possibilities of these as fuel in thermal generating stations. *Energy Source* **1999**, *21*, 339–346.
26. Kaygusuz, K.; Türker, M.F. Biomass energy potential in Turkey. *Biomass Bioenergy* **2002**, *26*, 661–678. [CrossRef]
27. Balat, M. Use of biomass sources for energy in Turkey and a view to biomass potential. *Biomass Bioenergy* **2005**, *29*, 32–41. [CrossRef]
28. Kizilaslan, H.; Onurlubas, H.E. Potential of production of biogas from animal origin waste in Turkey (Tokat Provincial Example). *J. Anim. Vet. Adv.* **2004**, *9*, 1083–1087. [CrossRef]
29. Eryilmaz, T.; Yesilyurt, M.K.; Gokdogan, O.; Yumak, B. Determination of biogas potential from animal waste in Turkey: A case study for Yozgat Province. *Eur. J. Sci. Technol.* **2015**, *2*, 106–111.
30. World Map. Available online: https://www.mapsofworld.com (accessed on 20 January 2020).
31. Meteoroloji Genel Müdürlüğü. Available online: https://meteor.gov.tr (accessed on 26 January 2020).
32. TÜİK. *Selected Indicators in Aydın. No 4038:189*; TÜİK: Ankara, Turkey, 2016.
33. Baytekin, H. *Bitkisel Üretimde Çiftlik Gübresi ve Biyogaz Kompostu Kullanımının Yaygınlaştırılması. Türk—Alman Biyogaz Projesi*; Çevre ve Şehircilik Bakanlığı: Ankara, Turkey, 2012.
34. Demirer, G.N.; Chen, S. Anaerobic digestion of dairy manure in a hybrid reactor with biogas recirculation. *World J. Microb. Biot.* **2015**, *21*, 1509–1514. [CrossRef]
35. Soyer, G. *Aydin ili süt sığırcılığı işletmelerinde elde edilen gübrenin değerlendirilmesi üzerine bir çalışmà, Yüksek Lisans Tezi*; Fen Bilimleri Enstitüsü, Adnan Menderes Üniveritesi: Aydın, Turkey, 2014.
36. Camberato, J.; Lıppert, B.; Chastaın, J.; Plank, O. Land Application of Animal Manure. In *Agricultural uses of by-Products and wastes American Chemical Society*; Washington, DC, USA, 1996; Available online: http://hubcap.clemson.edu (accessed on 26 January 2020).
37. Çayır, A.; Atılgan, A. Büyükbaş Hayvan Barınaklarındaki Gübrelikler ve Su Kaynaklarına Olan Durumlarının İncelenmesi. *Süleyman Demirel Üniversitesi Ziraat Fakültesi Dergisi* **2012**, *7*, 1–9.
38. Mutlu, A. *Adana İli ve Çevresindeki Hayvancılık Tesislerinde ortaya Çıkan Atıkların Yarattığı Çevre KirliliğiÜzerinde Bir Çalışma. Yüksek Lisans Tezi, Ç.Ü*; Fen Bilimleri Enstitüsü: Adana, Turkey, 1999.
39. Jacobson, L.D.; Moon, R.; Bicudo, J.; Yanni, K.; Noll, S. *Generic Environmental Impact Statement on Animal Agriculture. A Summary of the Literature Related to Air Quality and Odor of Animal Science*; University of Minnesota: Twin cities, MN, USA, 1999.
40. Nizam, S.; Armağan, G. Aydın İlinde Pazara Yönelik Süt Sığırcılığı İşletmelerinin Verimliliklerinin Belirlenmesi. *ADÜ Ziraat Fakültesi Dergisi* **2006**, *3*, 53–60.
41. Loyon, L. Overview of manure treatment in France. *Waste Manag.* **2017**, *61*, 516–520. [CrossRef]
42. *Managing Manure Nutrients at Concentrated Animal Feeding Operations*; U.S. Environmental Protection Agency: Austin, TX, USA, 2004.
43. Atılgan, A.; Erkan, M.; Saltuk, B. Akdeniz Bölgesindeki Hayvancılık İşletmelerinde Gübrenin Yarattığı Çevre Kirliliği. *Ekoloji* **2006**, *15*, 1–7.
44. Öztürk, S. *Determination of Ground Water Pollution Degrees in Some Pilot Areas Where Intensive Irrigated Agriculture is Practiced in Söke Plain of Aydın*; Enstitute of Applied Science, Adnan Menderes Üniversitesi: Aydın, Turkey, 2009.

45. Yildiz, B. *A Research on Determination of Polluter Parameters according to Receiver Environmental Conditions;* Yüksek Lisans Tezi, Enstitute of Applied Science, Adnan Menderes Üniversitesi: Aydın, Turkey, 2020.
46. WHO: World Health Organization. *Guidelines for Drinking-Water Quality,* 4th ed.; WHO: Geneva, Switzerland, 2011.
47. Evci, B. *Implementation of the Nitrate Directive in Turkey;* Ministry of Agriculture and Rural Affairs: Ankara, Turkey, 2007. Available online: https://www.slideshare.net/iwlpcu/implementation-of-the-nitrate-directive-in-turkey-evci (accessed on 15 February 2020).
48. Zhu, L.; Yan, C.; Li, Z. Microalgal cultivation with biogas slurry for biofuel production. *Bioresour. Technol.* **2016**, *220*, 629–636. [CrossRef]
49. Matheri, A.N.; Ndiweni, M.; Belaid, N.S.; Muzenda, E.; Hubert, R. Optimising biogas production from anaerobic co-digestion of chicken manure and organic fraction of municipal solid waste. *Renew. Sustain. Energy Rev.* **2017**, *80*, 756–764. [CrossRef]
50. Uğurlu, A. An overview of Turkey's renewable energy trend. *J. Energy Syst.* **2017**, *1*, 148–157. [CrossRef]
51. Guide to Investing in Turkish Renewable Energy Sector, Presidency of the Republic of Turkey Investment Office. Available online: https://www.invest.gov.tr (accessed on 15 February 2020).

5

Lyophilized Protein Structures as an Alternative Biodegradable Material for Food Packaging

Katarzyna Kozłowicz [1], Sybilla Nazarewicz [1], Dariusz Góral [1,*], Anna Krawczuk [2] and Marek Domin [1]

[1] Department of Biological Bases of Food and Feed Technologies, University of Life Sciences in Lublin, Głęboka 28, 20-612 Lublin, Poland; katarzyna.kozlowicz@up.lublin.pl (K.K.); sybilla_klap.94@o2.pl (S.N.); marek.domin@up.lublin.pl (M.D.)

[2] Department of Machinery Exploitation and Management of Production Processes, University of Life Sciences in Lublin, Akademicka 13, 20-950 Lublin, Poland; anna.krawczuk@up.lublin.pl

* Correspondence: dariusz.goral@up.lublin.pl

Abstract: Considering the need for sustainable development in packaging production and environmental protection, a material based on lyophilized protein structures intended for frozen food packaging was produced and its selected thermophysical properties were characterized. Analyses of density, thermal conductivity and thermal diffusivity were performed and strength tests were carried out for lyophilized protein structures with the addition of xanthan gum and carboxymethyl cellulose. Packagings were made of new materials for their comparative assessment. Then, the surface temperature distribution during thawing of the deep-frozen product inside the packaging was tested. In terms of thermal insulation capacity, the best properties were obtained for sample B4 with a thermal conductivity of $\lambda = 0.06$ W·(mK)$^{-1}$), thermal capacity $C = 0.29$ (MJ·(m^{3}K)$^{-1}$) and thermal diffusivity $a = 0.21$ (mm^2·s^{-1}). The density and hardness of the obtained lyophilized protein structures were significantly lower compared to foamed polystyrene used as a reference material. Thermal imaging analysis of the packaging showed the occurrence of local freezing. Lyophilized protein structures obtained from natural ingredients meet the needs of consumers and are environmentally friendly. These were made in accordance with the principles of sustainable development and can be an alternative material used for the production of frozen food packaging.

Keywords: packaging; biodegradable material; lyophilized protein structure

---

1. Introduction

Packaging is an important factor in maintaining food quality and commercial attractiveness, facilitating its transport at the same time. Nowadays, packaging, in addition to protecting, informing or marketing functions, should show additional functional properties and meet many different requirements. Innovations play a special role in modifying packaging functionality and improving its barrier, strength and aging resistance properties while avoiding the use of environmentally harmful materials. This contributes to the creation of a new generation of packagings that allow to maintain and even improve the quality of the packaged product, which is a very desirable feature, especially in the case of food packaging [1,2]. Legal regulations, legislative pressure from governments and non-governmental organizations dealing with environmental protection [3,4], as well as consumers themselves make food producers more and more interested in providing new pro-ecological and sustainable solutions in production. At the same time, packaging made in accordance with the principles of sustainable development must be safe for health, life and the environment [5–7].

One of the directions of development of food packaging technology that meets the requirements of sustainable development is obtaining fully composted and biodegradable materials with the addition of natural fillers [8]. An alternative that can reduce carbon footprint, pollution risks and greenhouse gas emissions caused by the use of conventional polymers is the use of biopolymers from agroindustrial sources that are renewable and low cost [9]. Recent studies have shown that starch can be used to obtain foams [10–12] retaining its biodegradable character when converted to a thermoplastic material. Research on the use of natural fillers such as: plantain flour, wood fiber, sugarcane bagasse, asparagus peel fiber [13,14], sunflower protein and cellulose fiber [15], plant protein, palm oil [16] and grape stalks [1] help to improve the physical, chemical, mechanical and technological properties of the material. For example, protein derivatives are the most attractive biopolymers for edible film formulations because they provide high nutritional value, superior mechanical properties and exhibit the oxygen barrier [17]. Carboxymethylcellulose based films are easily water soluble as it contains a hydrophobic polysaccharide backbone and many hydrophilic carboxyl groups [18]. Carboxymethylcellulose improves protein film mechanical properties by increasing thermal stability and the elasticity modulus [19]. Gelatin films blended with xanthan gum characterize a transparent film with excellent ultraviolet light resistance, low total soluble matter and moisture content, low water vapour permeability, improved mechanical properties and thermal stability [20]. Xanthan gum is cross-linker to be blended with various materials; it may dissolves directly in many highly acidic, alkaline, alcoholic systems containing different components. It is also compatible with commercially available thickeners such as sodium alginate, carboxymethylcellulose and starch [21].

Biodegradable packaging, due to the possibility of full compostability, does not pose a threat to the environment. On the contrary, it can enrich the soil with nutrients. Despite significant technological progress, there are no packaging that meets all the requirements. For each group of products, especially chilled and frozen ones, the most rational packaging is selected taking into account technical, economic and legal conditions [22–24].

This paper discusses the need for sustainable development in packaging production to protect the environment, and a material based on lyophilized protein structures for frozen food has been produced and its selected thermophysical properties have been characterized. Hence, the present work used xanthan gum and carboxymethylcellulose as a crosslinking agents to form a potentially new natural and biodegradable materials. For this purpose, prototype packagings were made from the obtained lyophilized structures. The thawing kinetics of food stored in these packagings and in packaging made of foamed polystyrene were compared.

2. Materials and Methods

2.1. Research Material

The research material was lyophilized protein structures prepared on the basis of foams obtained from powdered albumin with high foaming activity containing 84.3% protein (*Basso*), modified by addition of carboxymethylcellulose (*Agnex, Białystok*) and xantan gum (*Agnex, Białystok*) in various percentage. Preparation of foams consisted of whipping albumin in distilled water for 5 minutes. Powdered carboxymethylcellulose and xanthan gum were added to the resulting foams and mixed for another 2–3 min. Four different foam variants with different percentages of individual components were prepared: 1-distilled water 88.0%, albumin 10.0% and carboxymethyl cellulose 2.0% (B1); 2-distilled water 88.0%, albumin 10.0% and xanthan gum 2.0% (B2); 3-distilled water 88.0%, albumin 6.0% and xanthan gum 6.0% (B3); 4-distilled water 88.0%, albumin 8.0% and xanthan gum 4.0% (B4). The obtained protein foams were transferred into 0.125 × 0.125 m plastic moulds. The samples prepared in this way were frozen in an air blast freezer at −30.0 °C. Then, after freezing, the sample was lyophilized for 72 h at a pressure of 20 Pa and an ice condensation temperature of −64 °C (ALPHA 2–4LD Plus freeze-dryer, Christ, Osterode am Harz, Germany) [25]. The obtained lyophilized protein structures with a thickness of 0.027 m are shown in Figure 1.

Figure 1. Cross-section of lyophilized protein structures: (**a**) B1, (**b**) B2, (**c**) B3, (**d**) B4.

2.2. Determination of Protein Structure Density

The density of the obtained samples (ρ) was calculated as the ratio of the mass of the samples to their volume. The volume of samples was determined using the formula for the cuboid volume (multiplying the dimensions: length, height and width).

2.3. Determination of the Strength of Protein Structures

For determination of the strength (hardness) of lyophilized protein structures, a fracture test was performer using a symmetrical knife with dimensions: blade thickness 0.003 m, blade angle 30° (texture analyzer LFRA 4500, Brookfield, Middleboro, MA, USA). Parameters of the device operation: knife speed 0.5 $mm{\cdot}s^{-1}$ and measurement accuracy 0.02 N. The dimensions of the tested samples were 0.10 × 0.04 m. The test for each sample was performed in 3 replications [26].

2.4. Measurement of Thermophysical Properties of Protein Structures

Lyophilized samples were analyzed for thermophysical properties such as thermal conductivity, heat capacity and thermal diffusivity using a KD2 Pro meter (Decagon Devices, Pullman, WA, USA) with the SH-1 probe. The measurement was carried out in 8 replications under the same conditions [27].

2.5. Making a Packaging Prototype

The packaging was made of 6 even cuboid walls with dimensions of 0.115 × 0.115 m and thickness of 0.18 m. The walls were joined with a specialist adhesive approved for food contact (*LOCTITE 454*). Only samples B3 and B4 were used for packaging. The remaining samples were rejected due to the soft and fragile structure that made them impossible to test. The reference packaging was made of foamed polystyrene (XPS) (Figure 2).

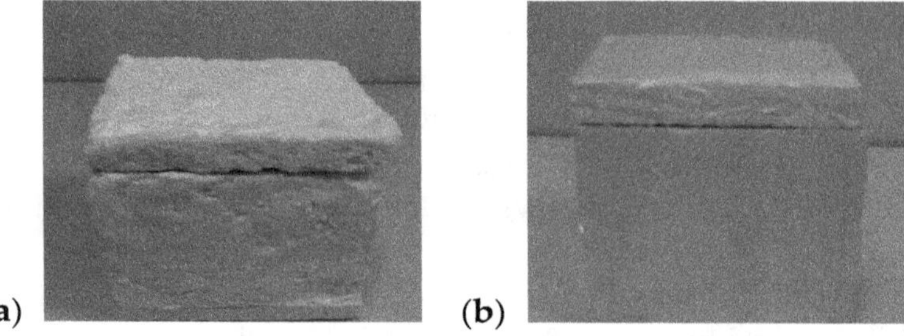

Figure 2. View of lyophilized protein structure of the sample B3 (**a**) and foamed polystyrene (**b**) packagings.

2.6. Thermovision Analysis of Lyophilized Protein Structures

For comparative assessment of the samples, the temperature fields recorded on their surface during thawing of the deep-frozen product (carrot with green peas) placed inside the packaging were tested. Thawing was carried out at an ambient temperature of 25 °C until the thawed product inside the packaging reached 23 °C. The temperature analyzer LB-515P cooperating with mini temperature and humidity data logger (type 23) was used to measure and record temperature. The mini data logger recorded temperature in a frozen product inside the packaging. The data obtained were used to compare thawing kinetics. For analysis of changes occurring during thawing of products on the surface of packaging walls, a Testo 882 thermal imaging camera (detector resolution 320 × 240 pixels, measuring accuracy ±2 °C, thermal sensitivity < 50 mK), cooperating with the Testo IRSoft program Version 3.4., was used. Thermovision analysis of lyophilized structures was based on the detection of thermal bridges, i.e., points on the surface of the packaging wall with the highest temperature and histograms of temperature distribution along the line passing through the point with the lowest temperature (CS).

2.7. Statistical Analysis of Results

The results were evaluated with statistical analysis methods (determination of mean values, standard deviation) using the *Statistica* 13.1 program and analysis of variance (ANOVA) at a at the 95% confidence level. To verify the significance of differences between the average values, the Student's *t*-test was used, where the distribution of variables, followed a normal distribution.

3. Results and Discussion

3.1. Physical Properties of Obtained Lyophilized Protein Structures

Materials used for food packaging should have appropriate density. This parameter is closely related to the thermal conductivity coefficient. This relationship determines the decrease in the thermal conductivity coefficient as the material density increases. Material density is also important when mechanical properties are formed [28]. Table 1 shows the density and hardness of lyophilized protein structures modified by the addition of carboxymethylcellulose and xanthan gum in varying percentages. It was found that the density of lyophilized protein structures defined as the ratio of mass to volume increases with the increase of xanthan gum share in the samples B2, B3 and B4. The highest density (28.22 kg·m^{-3}), was noted for the sample B1 containing 10.0% albumin and 2.0% carboxymethyl cellulose, while the sample B2, which contained 10.0% albumin and 2.0% xanthan gum, had the lowest density (20.41 kg·m^{-3}). The percentage and type of modification used had a statistically significant impact on the density of lyophilized protein structures. The values of density of the obtained materials were significantly lower compared to foamed polystyrene which served as the reference sample. The density of the material is determined by the volume and dimensions of air bubbles which in turn depend on the viscosity of the environment and the conditions for the formation and dispersion of the gaseous phase [29].

Table 1. Density and bending force of lyophilized samples and foamed polystyrene.

Properties	B1	B2	B3	B4	XPS
Density (kg·m^{-3})	28.22 ± 0.28 ^{b}	20.41 ± 0.07 ^{c}	25.01 ± 0.11 ^{d}	24.51 ± 0.66 ^{d}	36.51 ± 1.17 ^{a}
p-value	0.01	0.001	0.003	0.002	
Bedning force (N)	0.68 ± 0.20 ^{b}	1.00 ± 0.21 ^{b}	4.49 ± 0.53 ^{c}	1.55 ± 0.11 ^{d}	20.84 ± 0.52 ^{a}
p-value	0.0001	0.0004	0.001	0.0004	

^{a,b,c,d} Means in the same line indicated by different letters were significantly different (p value < 0.05). The results are expressed as mean ± SD (n = 3).

The conducted fracture test showed a much greater hardness of lyophilized protein structures with xanthan gum than those with carboxymethyl cellulose. The sample B3 containing 6.0% albumin and 6.0% xanthan gum had the highest hardness (4.49 N), whereas the lowest hardness (0.68 N) was noted for the sample B1, which contained 10.0% albumin and 2.0% carboxymethyl cellulose. The obtained lyophilized protein structures had a statistically significantly lower hardness compared to the reference material, which was foamed polystyrene (20.84 N). Hazirah et al. [30] showed that physical and mechanical properties of gelatin-carboxymethylcellulose films were best improved with 5% xanthan gum added. The use of xanthan gum a non-gelling nature, as an alternative crosslinking agent in gelatin/carboxymethylcellulose film blend have formed a blend of composite film and improved several physical and mechanical properties of gelatin/carboxymethylcellulose film blend alone. Results Lima et al. [31] demonstrated that films containing higher content of xanthan gum show the highest tensile strength and the lowest elongation. Xanthan gum addition did not affect the water vapor permeability, solubility, and moisture of films.

3.2. Characteristics of Thermophysical Properties of Obtained Protein Structures

Table 2 presents thermophysical properties (coefficient of thermal conductivity—λ, heat capacity—C, thermal diffusivity—a) of lyophilized protein structures modified with a different percentage of carboxymethyl cellulose and xanthan gum. The coefficient of thermal conductivity (λ) characterises the material's ability to conduct heat. It is defined as the heat flux per unit area of the material with a temperature gradient of 1 $K \cdot m^{-1}$ [28]. The study demonstrated that modification of material composition affected statistically significantly ($p < 0.05$) the value of thermal conductivity coefficient. The lowest value of this parameter (0.06 $W \cdot (mK)^{-1}$) was noted for protein structures obtained from 8.0% albumin and 4.0% xanthan gum (B4) and it was comparable to foamed polystyrene (0.04 $W \cdot (mK)^{-1}$), which is widely used as a material with insulating properties. Similar values of thermal conductivity were reported by Kozłowicz et al. [25]. The thermal conductivity of lyophilized gelatin structures ranged from 0.045 to 0.063 $W \cdot (mK)^{-1}$. The values of thermal conductivity determined for lyophilized gelatin structures modified with hydrated paper pulp, ground extruded starch and hydrogel balls were in the range of 0.047–0.081 $W \cdot (mK)^{-1}$ [32].

Table 2. Thermal properties of lyophilized samples and foamed polystyrene.

Properties	B1	B2	B3	B4	XPS
Thermal conductivity λ ($W \cdot (mK)^{-1}$)	0.12 ± 0.01^{b}	0.13 ± 0.01^{b}	0.12 ± 0.01^{b}	0.06 ± 0.00^{c}	0.04 ± 0.00^{a}
p-value	0.000	0.000	0.000	0.000	
Heat capacity C ($MJ \cdot (m^3K)^{-1}$)	0.40 ± 0.02^{b}	0.41 ± 0.03^{b}	0.41 ± 0.02^{b}	0.29 ± 0.02^{c}	0.26 ± 0.02^{a}
p-value	0.000	0.000	0.000	0.02	
Thermal diffusivity *a* ($mm^2 \cdot s^{-1}$)	0.30 ± 0.01^{b}	0.30 ± 0.01^{b}	0.28 ± 0.01^{b}	0.21 ± 0.01^{c}	0.18 ± 0.01^{a}
p-value	0.000	0.000	0.000	0.000	

[a,b,c] Means in the same line indicated by different letters were significantly different (p value < 0.05). The results are expressed as mean ± SD ($n = 3$).

Lyophilized protein structures modified with carboxymethylcellulose 2% (B1) and xanthan gum 2% (B2) and 6% (B3) had a significantly higher heat capacity C (0.40 $MJ \cdot m^{-3} \cdot K^{-1}$ and 0.41 $MJ \cdot m^{-3} \cdot K^{-1}$) than lyophilized protein structures with addition of 4% xanthan gum (B4) (0.29 $MJ \cdot m^{-3} \cdot K^{-1}$) and foamed polystyrene (0.26 $MJ \cdot m^{-3} \cdot K^{-1}$). Heat capacity is defined as the ratio of the amount of heat absorbed by the material to the resulting increase in temperature. Materials with high heat capacity require more heat than those with low heat capacity to achieve the same effect of a temperature rise. In packaging, it is preferable to use materials with high heat capacity. Packagings made of such materials will maintain a constant temperature of the product inside.

Another value characterizing a material in terms of its thermal properties is thermal diffusivity a, which determines the ability of a given material to transfer heat within it and reduce temperature gradients. Thermal diffusivity in lyophilized protein structures, modified with carboxymethylcellulose and xanthan gum, ranged from 0.21–0.30 $mm^2 \cdot s^{-1}$. The obtained thermal diffusivity values were significantly higher ($p < 0.05$) than those noted for foamed polystyrene (0.18 $mm^2 \cdot s^{-1}$). This value is determined by the structure, chemical composition and temperature describing the speed of heat conduction in the material [28].

3.3. *Analysis of Thawing Kinetics*

The thawing curve is a widely used data source for determining changes of the frozen product's temperature during thawing (Figure 3).

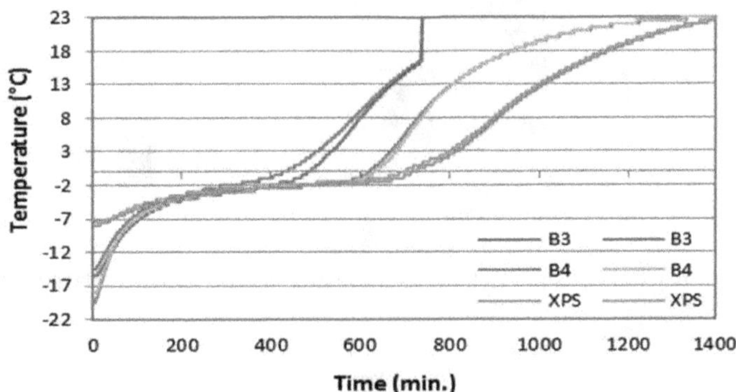

Figure 3. Thawing curves for carrots with green peas in packaging (B3—6.0% albumin, 6.0% xanthan gum, B4—8.0% albumin, 4.0% xanthan gum, XPS—foamed polystyrene).

Considering the temperature changes over time in the form of thawing curves summarized in diagrams (Figure 3), it was found that in a packaging made of lyophilized protein structures B3, the thawed product kept the temperature below 0 °C for the shortest time (temperature measured after 7.5 h). The temperature of 0 °C was reached by the thawed product in a packaging made of lyophilized protein structures B4 after 10 h. It was the time of thawing compared to that recorded for the product placed in the packaging made of foamed polystyrene (11.7 h).

3.4. *Thermovision Analysis of Designed Packaging*

The operation of the thermal imaging camera is based on the phenomenon of infrared radiation. The temperature distribution on the surface of the tested packaging is presented in the form of colored isotherms, where the individual color corresponds to points having the same temperature [33].

The packagings were subjected to thermovision analysis when testing changes in temperature during defrosting of the product. Figure 4 presents a thermovision image of packaging made of lyophilized protein structures of the B3 sample with characteristic parameters. The point CS1 specifying the lowest temperature occurring on the packaging wall (20.1 °C) and temperature profiles on the straight lines P1 and P2 are presented. A histogram was prepared for the thermovision image shown in Figure 5. The histogram is used to present the empirical distribution of features, so it is possible to use it to present the results obtained for certain quantitative variables. It was found that temperatures in the range of 21.4–22.5 °C had the largest share of the entire area, constituting about 65%. Analysis of the temperature distribution profile on the line P1 shows a minimum temperature of 20.1 °C (uneven material structure, uneven aeration) and a maximum temperature of 22.5 °C. The temperature distribution on the line P2 shows the minimum value of 21.8 °C and the maximum value of 24.6 °C (Figure 6).

Figure 7 presents a thermovision image of a packaging made of lyophilized protein structures (8.0% albumin, 4.0% xanthan gum) of the sample B4. At CS1, the lowest packaging surface temperature was 19.9 °C. The largest share in the whole area, constituting 82.0% of the total individual temperatures, was recorded for the range of temperature 23.3–25.0 °C (Figure 8). Analysis of the temperature distribution profile on the line P2 shows the temperature with minimum of 20.8 °C and maximum of 25.2 °C, while the temperature distribution on the line P3 shows the minimum value of 20.0 °C and the maximum value of 24.8 °C (Figure 9).

Figure 4. Infrared image of packaging surface with analysis of temperature profile and its extreme values (B3).

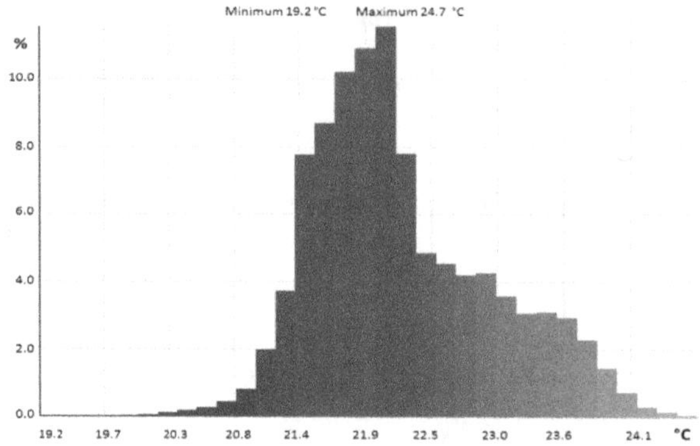

Figure 5. The histogram of temperature distribution on packaging surface (B3).

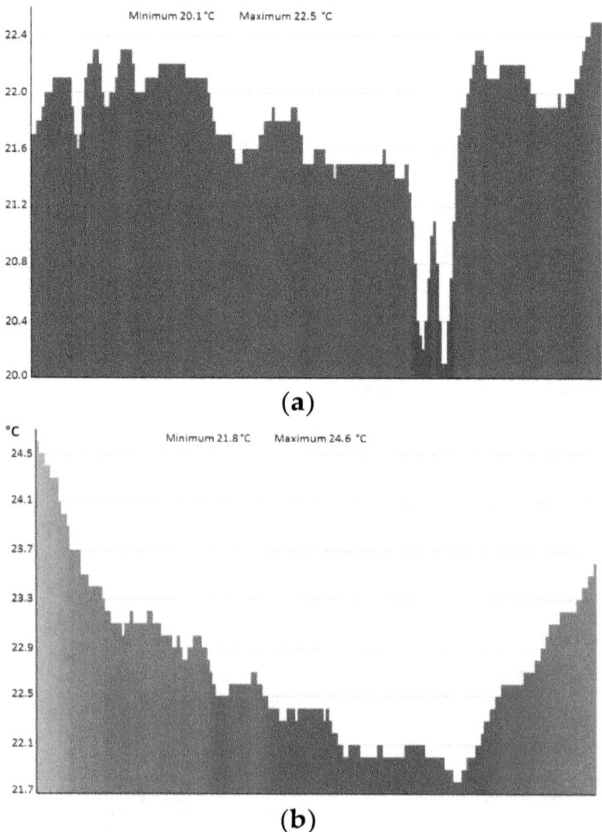

Figure 6. The temperature distribution profile of pack surface samples B3 in line (**a**) P1, (**b**) P2.

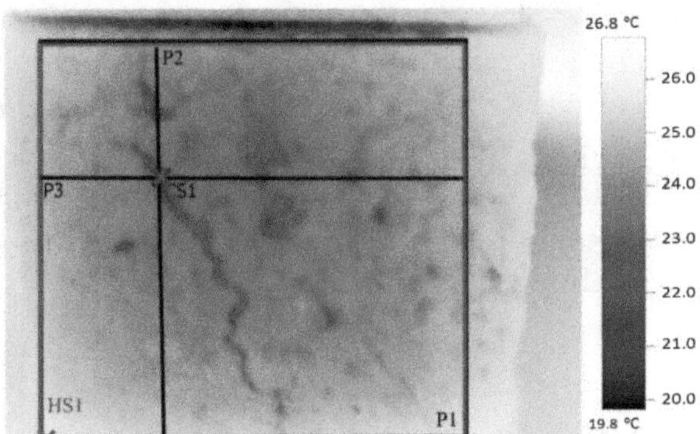

Figure 7. Infrared image of packaging surface with analysis of temperature profile and its extreme values (B4).

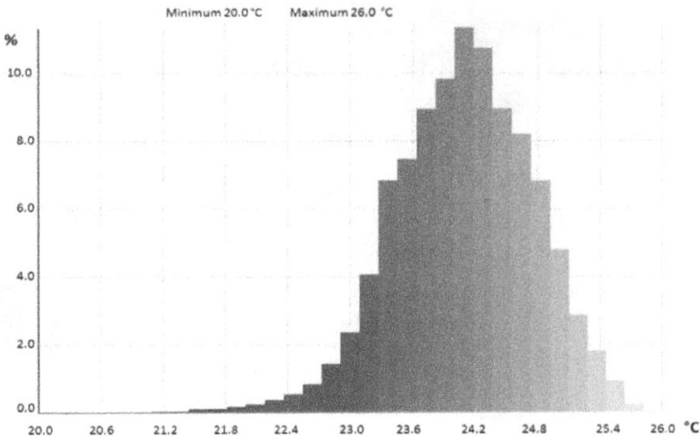

Figure 8. The histogram of temperature distribution on packaging surface (B4).

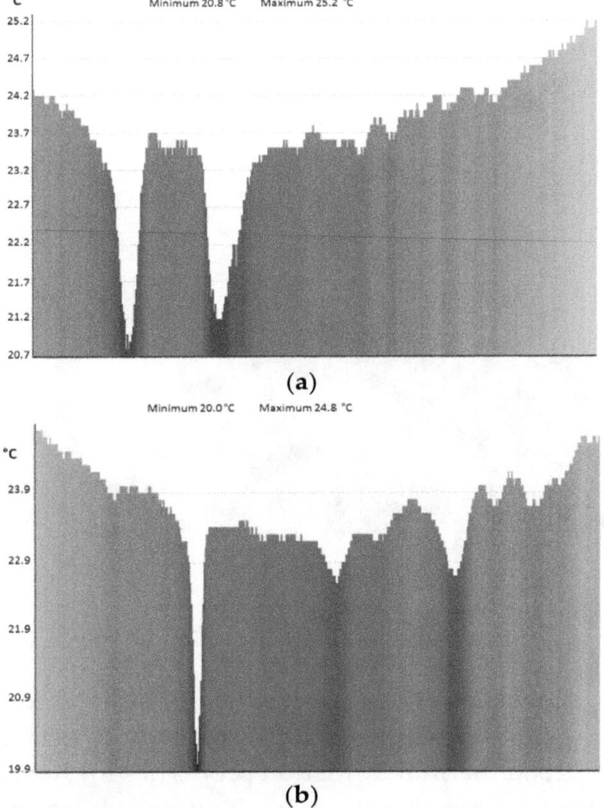

Figure 9. The temperature distribution profile of packaging surface samples B4 in line (**a**) P2, (**b**) P3.

For comparison purposes, packagings made of foamed polystyrene (XPS) were also subjected to the thermovision analysis. The image of the surface of packaging wall is shown in Figure 10 together with the profile analysis and indication of extreme values. The histogram of the temperature distribution for the entire surface area of the packaging (Figure 11) shows that about 40% of the total temperature is in the range 23.5–24.0 °C. Analysis of the temperature distribution profile on the line P1 shows the minimum temperature of 23.3 °C and maximum temperature of 24.8 °C. Whereas these values on the line P2 are, respectively, 24.7 °C and 25.7 °C (Figure 12).

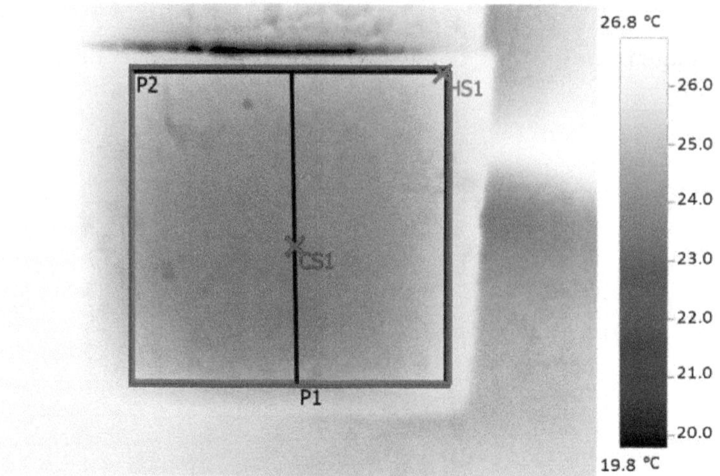

Figure 10. Infrared image of packaging surface with analysis of temperature profile and its extreme values (XPS).

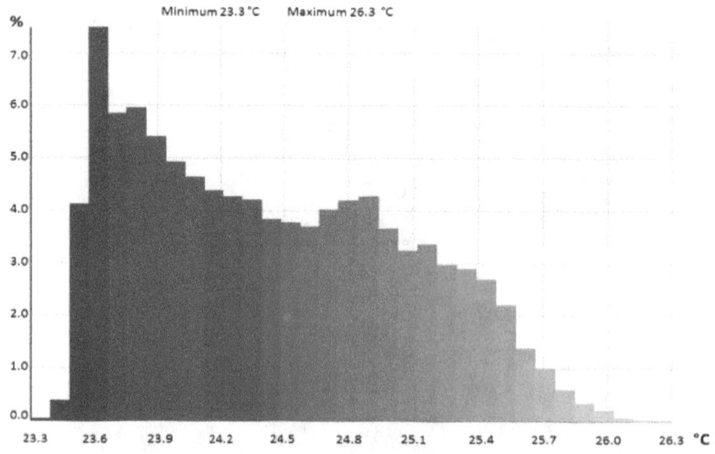

Figure 11. The histogram of temperature distribution on packaging surface (XPS).

The obtained results are consistent with the results of studies conducted previously by other authors who used lyophilized gelatin structures and lyophilized protein structures modified with agar and gelatin as packaging with good thermal insulation properties for frozen food [25,26]. The possible biodegradability of such material means that it can be a suitable replacement for foamed polystyrene used in the production of packaging.

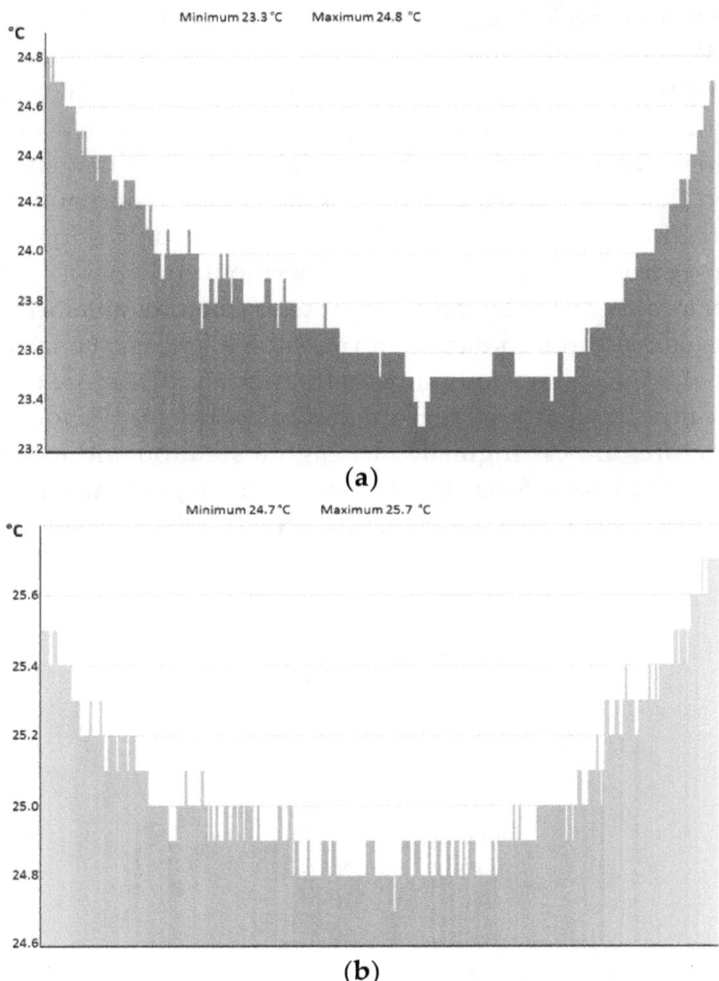

Figure 12. The temperature distribution profile of the foamed polystyrene packaging in line (**a**) P1, (**b**) P2.

The trend of using natural resources to produce packaging materials is now very common. Such materials include polysaccharides (corn and potato starch, cellulose, gums), animal proteins (casein, collagen, gelatin), vegetable (soy, gluten) and lipids (oils from fats) [34,35]. The use of proteins of animal and vegetable origin has proved to be promising in obtaining films and mixtures development of biodegradable packaging. The proteins hake films were more resistant and soluble in water, while gluten films showed greater elongation on cracking [36]. Microbial biopolymers such as gellan, bacterial cellulose, xanthan, pullulan, and curdlan are non-toxic, biocompatible, and biodegradable. They are used in the food industry as material for coating and packing purposes. The packagings prepared from these polymers are transparent and have good mechanical, moisture, and oxygen barrier properties, but weak water barrier properties. The blending of microbial gums with lipids, hydrocolloids or reinforcement agents improves the functional properties of materials [37]. However, it should continue to look for strategies that promote the improvement of the mechanical and barrier properties of these materials. Only in this way will it be possible to replace, partly or completely synthetic polymers with biodegradable polymers. These materials allow to maintain an appropriate quality of packaged raw materials. In addition, packaging made of natural materials does not require biodegradability testing.

4. Conclusions

The conducted research confirmed the possibility of using albumin foams in the form of lyophilized structures as an alternative material for frozen food packaging. In terms of thermal insulation, the proposed structures are also well suited to packaging material for frozen products as foamed polystyrene The best thermophysical properties were obtained for the sample B4 with thermal conductivity of $\lambda = 0.06$ W·(mK)$^{-1}$), thermal capacity $C = 0.29$ (MJ·(m^{3}K)$^{-1}$) and thermal diffusivity $a = 0.21$ (mm^2·s^{-1}). The density of the obtained lyophilizated protein structures was significantly lower compared to the reference material (foamed polystyrene). The obtained materials had statistically significantly lower hardness than foamed polystyrene. The analysis of thawing curves of frozen carrots with green peas showed that the product placed in a packaging made of foam polystyrene reached temperature 0 °C after the longest time (11.7 h). A slightly lower defrosting time was recorded in the product kept in packaging from material B4 (10 h). Thermovision analysis of the tested packagings showed the occurrence of local freezing. Lyophilized protein structures obtained from natural ingredients can be classified as materials made in accordance with the principles of sustainable development that meet the needs of consumers and which are environmentally friendly. These are completely biodegradable materials, suitable for composting and made from natural resources. In addition to biodegradability, lyophilized protein structures are distinguished by high insulation and low density, which provide adequate protection for stored frozen food. In addition, all the ingredients in the obtained lyophilized structures are 100% approved for contact with food, hence they can be a substitute for foamed polystyrene in the production of packaging.

Author Contributions: Conceptualization, K.K. and S.N.; methodology, K.K. and D.G.; prepared the materials, A.K.; analysed the data, M.D.; wrote the paper, K.K.; S.N., D.G.; visualization, D.G. All authors read and approved the final manuscript.

References

1. Engel, J.B.; Ambrosi, A.; Tessaro, I.C. Development of biodegradable starch-based foams incorporated with grape stalks for food packaging. *Carbohydr. Polym.* **2019**, *225*, 115–234. [CrossRef] [PubMed]
2. Malherbi, M.; Schmitz, A.C.; Grando, R.C.; Bilck, A.P.; Yamashita, F.; Tormen, L.; Fakhouri, F.M.; Velasco, J.I. Corn starch and gelatine-based films added with guabiroba pulp for application in food packaging. *Food Packag. Shelf Life* **2019**, *19*, 140–146. [CrossRef]
3. Ustawa z dnia 13 Czerwca 2013 r., O Gospodarce Opakowaniami i Odpadami Opakowaniowymi (Dz.U. z 2019 r. poz. 542). Available online: http://prawo.sejm.gov.pl/isap.nsf/DocDetails.xsp?id=WDU20130000888 (accessed on 6 December 2019).
4. Strategia Wspólnoty w dziedzinie gospodarowania odpadami. *Off. J. Eur. Commun.* **1990**. No. C 122, 18.5.1990. Available online: https://eur-lex.europa.eu/legal-content/EN/TXT/?qid=1575734282491&uri=CELEX:31990Y0518(01) (accessed on 6 December 2019).
5. Brody, A.L. Innovative food packaging solutions. *J. Food Sci.* **2008**, *73*, 9–15.
6. Wasilewska, A.; Pezała, A. Environmental aspects of sustainable packaging in FMCG industry. *Logistyka Odzysku* **2016**, *3*, 34–36.
7. Russell, D.A.M. Sustainable (food) packaging—An overview. *Food Addit. Contam.* **2014**, *31*, 396–401. [CrossRef]
8. Song, J.H.; Murphy, R.J.; Narayan, R.; Davies, G.B.H. Biodegradable and compostable alternatives to conventional plastics. *Philos. Trans. R. Soc. Lond. B. Biol. Sci.* **2009**, *364*, 2127–2139. [CrossRef]
9. Davis, G.; Song, J.H. Biodegradable packaging based on raw materials from crops and their impact on waste management. *Ind. Crops. Prod.* **2006**, *23*, 147–161. [CrossRef]
10. Chiarathanakrit, C.; Riyajan, S.A.; Kaewtatip, K. Transforming fish scale waste into an efficient filler for starch foam. *Carbohydr. Polym.* **2018**, *188*, 48–53. [CrossRef]
11. Machado, C.M.; Benelli, P.; Tessaro, I.C. Sesame cake incorporation on cassava starch foams for packaging use. *Ind. Crops. Prod.* **2017**, *102*, 115–121. [CrossRef]
12. Heydari, A.; Alemzadeh, I.; Vossoughi, M. Functional properties of biodegradable corn starch nanocomposites for food packaging applications. *Mater. Des.* **2013**, *50*, 954–961. [CrossRef]
13. Vargas-Torres, A.; Palma-Rodriguez, H.M.; Berrios, J.D.J.; Glenn, G.; Salgado-Delgado, R.; Olarte-Paredes, A.; Hernandez-Uribe, J.P. Biodegradable baked foam made with chayotextle starch mixed with plantain flour and wood fiber. *J. Appl. Polym. Sci.* **2017**, *134*, 455–465. [CrossRef]

14. Cruz-Tirado, J.P.; Siche, R.; Cabanillas, A.; Diaz-Sanchez, L.; Vejarano, R.; Tapia-Blacido, D.R. Properties of baked foams from oca (Oxalis tuberose) starch reinforced with sugarcane bagasse and asparagus peel fiber. *Procedia Eng.* **2017**, *200*, 178–185. [CrossRef]
15. Salgado, P.R.; Schmidt, V.C.; Ortiz, S.E.M.; Mauri, N.; Laurindo, J.B. Biodegradable foams based on cassava starch, sunflower proteins and cellulose fibers obtained by a baking process. *J. Food Eng.* **2008**, *85*, 435–443. [CrossRef]
16. Kaisangsri, N.; Kerdchoechuen, O.; Laohakunjit, N. Characterization of cassava starch based foam blended with plant proteins, kraft fiber and palm oil. *Carbohydr. Polym.* **2014**, *110*, 70–77. [CrossRef] [PubMed]
17. Ou, S.; Kwok, K.C.; Kang, Y. Changes in in vitro digestibility and available lysine of soy protein isolate after formation of film. *J. Food Eng.* **2004**, *64*, 301–305. [CrossRef]
18. Su, J.F.; Huang, Z.; Yuan, X.Y.; Wang, X.Y.; Li, M. Structure and properties of carboxymethyl cellulose/soy protein isolate blend edible films crosslinked by Maillard reactions. *Carbohydr. Polym.* **2010**, *79*, 145–153. [CrossRef]
19. Wiwatwongwana, F.; Pattana, S. Characterization on properties of modification gelatin films with carboxymethyl cellulose. In Proceedings of the 1st TSME International Conference on Mechanical Engineering, Ubon Ratchathani, Thailand, 20–22 October 2010; pp. 1–8.
20. Guo, J.; Ge, L.; Li, X.; Mu, C.; Li, D. Periodate oxidation of xanthan gum and its crosslinking effects on gelatin-based edible films. *Food Hydrocoll.* **2014**, *39*, 243–250. [CrossRef]
21. Sharma, B.R.; Naresh, L.; Dhuldhoya, N.C.; Merchant, S.U.; Merchant, U.C. Xanthan gum—A boon to food industry. *Food Promot. Chron.* **2006**, *1*, 27–30.
22. Sun, D. *Handbook of Frozen Food Processing and Packaging*, 2nd ed.; CRC Press: Boca Raton, FL, USA; New York, NY, USA, 2012; pp. 711–779.
23. Rozporządzenie (WE) nr 1935/2004 Parlamentu Europejskiego i Rady z dnia 27 Października 2004 r. w Sprawie Materiałów i Wyrobów Przeznaczonych do Kontaktu z Żywnością oraz Uchylające Dyrektywy 80/590/EWG i 89/109/EWG. Available online: https://eur-lex.europa.eu/legal-content/PL/TXT/?uri=celex%3A32004R1935 (accessed on 6 December 2019).
24. Grabowska, B. Frozen food packaging—Characteristics, review, norms and provisions. *Przem. Spożywczy* **2014**, *9*, 16–18.
25. Kozłowicz, K.; Góral, D.; Kluza, F.; Domin, M.; Kobus, Z.; Sagan, A.; Prazner, Ł. The porous gelatin structures as the material for packaging for frozen food. *Przem. Chem.* **2015**, *10*, 1742–1747.
26. Góral, D.; Kozłowicz, K.; Kluza, F.; Domin, M.; Blicharz-Kania, A.; Senetra, E.; Dziki, D.; Kocira, A.; Guz, T. Evaluation of thermophysical characteristics of freeze-dried protein foams as packaging material for frozen food. *Przem. Chem.* **2018**, *5*, 700–705.
27. Kozłowicz, K.; Góral, D.; Kluza, F.; Góral, M.; Andrejko, D. Experimental determination of thermophysical properties by line heat pulse method. *J. Food Meas. Charact.* **2018**, *12*, 2524–2534. [CrossRef]
28. Bornhorst, G.; Sarkar, A.; Singh, P.R. Terminal Properties of Frozen Foods. In *Engineering Properties of Foods*, 4th ed.; Rao, M.A., Rizvi, S.S.H., Datta, A.K., Ahmed, J., Eds.; CRC Press: Boca Raton, FL, USA; New York, NY, USA, 2014; pp. 247–280.
29. Marzec, A.; Jakubczak, E. Rheological properties of foams prepared for drying. *Acta Agrophys.* **2009**, *13*, 185–194.
30. Hazirah, M.A.S.P.; Isa, M.I.N.; Sarbon, N.M. Effect of xanthan gum on the physical and mechanical properties of gelatin-carboxymethyl cellulose film blends. *Food Pack. Shelf Life* **2016**, *9*, 55–63. [CrossRef]
31. Lima, M.; Carneiro, L.; Bianchini, D.; Dias, A.R.; Zavareze, R.; Prentice, C.; Moreira, A. Structure, thermal, physical, mechanical and barrier properties of chitosan films with the addition of xanthan gum. *J. Food Sci.* **2017**, *82*, 698–705. [CrossRef]
32. Kozłowicz, K.; Kluza, F.; Góral, D.; Nakonieczny, P.; Combrzyński, M. Modified gelatine structures as packaging material for frozen agricultural products. In Proceedings of the BIO Web Conference in Contemporary Research Trends in Agricultural Engineering, Kraków, Poland, 25–27 September 2017.
33. Jura, J.; Adamus, J. Thermography application for assessment of building thermal insulation. *Bud. O Zoptymalizowanym Potencjale Energetycznym* **2013**, *2*, 31–39.
34. Sanjay, M.R.; Arpitha, G.R.; Naik, L.L.; Gopalakrishna, K.; Yogesha, B. Applications of natural fibers and its composites: An overview. *Nat. Resour.* **2016**, *7*, 108–114. [CrossRef]

35. Folentarska, A.; Krystyjan, M.; Baranowska, N.M.; Ciesielski, W. Renewable raw materials as an alternative to receiving biodegradable materials. *J. Chem. Environ. Biotechnol.* **2016**, *19*, 121–124. [CrossRef]
36. Nogueira, D.; Martins, V.G. Use of different proteins to produce biodegradable films and blends. *J. Polym. Environ.* **2019**, *27*, 2027–2039. [CrossRef]
37. Alizadeh-Sani, M.; Ehsani, A.; Kia, E.M.; Khezerlou, A. Microbial gums: Introducing a novel functional component of edible coatings and packaging. *Appl. Microb. Biotechnol.* **2019**, *103*, 6853–6866. [CrossRef] [PubMed]

6

Modeling the Dependency between Extreme Prices of Selected Agricultural Products on the Derivatives Market using the Linkage Function

Zofia Gródek-Szostak [1,*], Gabriela Malik [2], Danuta Kajrunajtys [3], Anna Szeląg-Sikora [4], Jakub Sikora [4], Maciej Kuboń [4], Marcin Niemiec [5] and Joanna Kapusta-Duch [6]

1 Department of Economics and Organization of Enterprises, Cracow University of Economics, Krakow 31-510, Poland
2 Higher School of Economics and Computer Science in Krakow, Kraków 31-510, Poland
3 Department of International Management, Cracow University of Economics, Krakow 31-510, Poland
4 Institute of Agricultural Engineering and Informatics, University of Agriculture in Krakow, Krakow 30-149, Poland
5 Department of Agricultural and Environmental Chemistry, University of Agriculture in Krakow, Krakow 31-120, Poland
6 Department of Human Nutrition, Faculty of Food Technology, University of Agriculture in Krakow, Krakow 30-149, Poland
* Correspondence: grodekz@uek.krakow.pl

Abstract: The purpose of the article is to identify and estimate the dependency model for the extreme prices of agricultural products listed on the Chicago Mercantile Exchange. The article presents the results of the first stage of research covering the time interval 1975–2010. The selected products are: Corn, soybean and wheat. The analysis of the dependency between extreme price values on the selected futures was based on the estimation of five models of two-dimensional extreme value copulas, namely, the Galambos copula, the Gumbel copula, the Husler–Reiss copula, the Tawn asymmetric copula and the *t*-EV copula. The next stage of the analysis was to test whether the structure of the dependency described with the estimated copulas is a sufficient approximation of reality, and whether it is suitable for modeling empirical data. The quality of matching the estimated copulas to empirical data of return rates of agricultural products was assessed. For this purpose, the Kendall coefficient was calculated, and the methodology of the empirical combining function was used. The conducted research allowed for the determination of the conduct for this kind of phenomena as it is crucial in the process of investing in derivatives markets. The analyzed phenomena are highly dependent on e.g., financial crises, war, or market speculation but also on drought, fires, rainfall, or even crop oversupply. The conducted analysis is of key importance in terms of balancing agricultural production on a global scale. It should be emphasized that conducting market analysis of agricultural products at the Chicago Mercantile Exchange in the context of competition with the agricultural market of the European Union is of significant importance.

Keywords: agricultural product; price; modeling; management

---

1. Introduction

Sustainable agricultural production is crucial for balancing the needs of current and future populations. The benefits of modern agriculture are significant, allowing food production go hand in hand with an increase in the population [1,2]. Cooperation between business and the agricultural production circles are of particular interest at both the national and international level [3,4]. Demographic forecasts indicate that by 2050, the world's population will reach 9.7 billion. This will lead to an increase in demand for food, further aggravating environmental problems related to intensive agricultural production. For this reason, one of the main challenges is to achieve global food security and sustainable agriculture. While food security aims at ensuring sustainable and healthy food supply over time, sustainable agriculture plays a key role in maintaining resilient agro-ecosystems [5]. The above-mentioned concept of sustainable farming is fostered by ecological farming. Its development is accompanied by a high growth rate of the organic food market and a significant growth of sales of organic food. Even with strict adherence to production practices and increasing their availability, most consumers still are unaware of organic production methods. The customers' awareness of organic products does not necessarily translate into real consumption [6,7].

According to the United States Department of Agriculture (USDA), the projected global wheat trade in the 2019/2020 season is expected to increase by 6.7 million tonnes, i.e., by 4% and reach 184.6 million tonnes in the case of larger export deliveries and lower expected export prices.

The global perspective of the feed grain market in the 2019/2020 season relates to record production and consumption, as well as lower final inventories. Global production of corn is projected to increase, mostly in the USA, South Africa, Russia, Canada, India, and Brazil, compensating for slightly lower crops in China and Ukraine. The global use of corn is expected to increase by 1%, while global corn imports are expected to increase by 2%. Global soy production in the 2018/2019 season increased in relation to the May forecast by 0.7 million tonnes, reaching 355.2 million tonnes [8–11].

In the decision-making processes, as well as in many scientific disciplines, the study of the dependency of extreme values is of great importance. In many situations, their proper identification enables avoiding wrong investment decisions. Studying the dependency between extreme prices of agricultural products is of key importance in the process of analyzing the structure of dependency. This is especially true when it is important to model the dependency between maintaining the maximum and minimum values of the analyzed observations. Dependencies for extremely small or extremely large values may have different characteristics than those determined based on the entire sample. The reason for fluctuations in the prices of agricultural products are, e.g., financial crises, wars, market speculation, but also droughts, fires, rains, or even crop oversupply. For this reason, the difference in price behavior during periods of extreme change, i.e., unusual and rarely observable market events require careful monitoring because it can lead to above-average losses or profits in the process of investing on derivatives markets.

A very useful tool in the process of modeling dependencies between extreme values of time series is the functions of relations of extreme values, also called extreme value copulas. Their main advantage is the possibility of analyzing above-average losses or profits in the field of finance, insurance, but also in the case of examining futures for agricultural products.

One of the first applications of the theory of extreme value based on a two-dimensional copula appeared in the subject literature in 1964 in the study of Gumbel and Goldstein [12]. These copulas were also used in many other fields of science, e.g., in the process of analyzing the return rates of exchange rates [13], the study of capital markets characterized by high volatility [14], in the portfolio analysis and portfolio risk estimation [15] or in the field of insurance [16], and in hydrology [17].

Extreme value copulas were created as possible limit copulas in relation to the maximum i.i.d. (independent and identically distributed) distribution, i.e., a distribution in which all random variables are independent and have the same distribution of probability. Since two-dimensional copulas were used in the empirical part of this article, the basic definitions and conclusions regarding extreme value copulas presented below will be demonstrated for the two-dimensional case. More detailed information on extreme value copulas, together with an extension for a multidimensional case can be found, e.g., in a monograph by R.B. Nelsen [18], a monograph by H. Joe [19], books by J. Segers [20] and G. Gudendorf and J. Segers [21], or in the following papers: C. Genest and J. Seger [22], M. Ribatet and M. Sedki [23].

As a supplier of raw food materials, the agricultural market is of strategic importance from the point of view of food security. Therefore, it is so important to strive towards risk-reducing solutions, including models, and inference based on the statistical analyzes, as exemplified by those proposed by the authors.

In order to tackle the growing challenges of sustainable agricultural production, a better understanding of complex agricultural ecosystems is needed. This can be done using modern digital technologies that constantly monitor the physical environment, producing large amounts of data at an unprecedented rate. The analysis of the big data will enable farmers and enterprises to leverage its value and improve their efficiency. Although the analysis of large data sets leads to progress in various industries, it is not yet widely used in agriculture [24,25].

2. Materials and Methods

The source of data used for the analysis in the empirical part of the paper is the Chicago Mercantile Exchange. The choice of this exchange was motivated by its importance in the international market of agricultural products. The research sample was created using daily quotations of nominal prices of futures for three agricultural products, i.e., corn, wheat and soybean. The value of the contract is expressed in the price of a bushel, the transaction unit of a commodity in US dollars. The data includes the closing price of the contract with the shortest expiration date, so that a series of quotations could be treated as a forward price with the shortest possible execution time. The choice of products was dictated by the importance of trading on the futures market and the availability of appropriately long time series. Empirical data come from the years 1975–2010. A single time series includes 9070 observations. They have been checked for possible discontinuities and errors. In order to minimize the impact of arbitrary interference on the obtained results, no procedures have been applied to correct or supplement empirical data. The selected combining function is one, for which the distance from the empirical combining function is the smallest.

2.1. Theoretical Models of Extreme Value Copulas, the Inference Functions for Margins (IFM) Estimation Method, and the Empirical Combining Function Method

The notion of the linkage function, also often called the copula function, appeared first in subject literature in 1959 in the work of A. Sklar [26]. However, it was not until the 1990s and the monograph of H. Joe [19] and R.B. Nelsen [18] that these functions have gained immense popularity, mainly due to the presentation of their broad practical application in dependence modeling. In the general, and at the same time the simplest view, the copula is a function that allows distinguishing the component describing only the structure of dependence from the cumulative distribution function of the total distribution of a random vector. In other words, it can be said that the copula functions are functions that combine a multidimensional cumulative distribution function of a random variable with its one-dimensional limit cumulative distribution function [19–26].

Among the copula functions, one can distinguish their particularly important class, namely, the copula of extreme values [27–33]. The use of two-dimensional extreme value copulas in empirical studies has been greatly simplified by using the representation introduced into the subject literature by J. Pickands in 1981 [34]. This issue was also an essential element of research conducted in 1977 by A.A. Balkema and S.I. Resnick [35] and L. de Haan and S. I. Resnick [36]. Observations of the extremes were classified in accordance with the Extreme Value Theory (EVT), which is used to describe the behavior of limit properties of extreme values. It consists of analyzing the tails of distributions, which constitute only a small part of the entire distribution examined. Its purpose is not to describe the usual behavior of stochastic phenomena, but the unusual and rarely observed events. The Extreme Value Theory is widely used wherever modeling relationships between the behaviors of the maximum and minimum values of the analyzed observations is particularly important [37–39]. The classic approach for modeling extreme values is based on the block maxima model. This method is used for a large number of observations that have been selected from a large sample. Modeling the behavior of the extreme values of independent random variables with an identical probability distribution is in fact using the maxima or minima of the observations in fixed time blocks. These blocks are designated by means of separate time intervals of equal length, most often months, quarters or years [30].

The next part of this article will present extreme values copulas used in empirical research.

2.1.1. Theory of Extreme Copulas

Among the copula functions, one can distinguish a particularly important class, i.e., the extreme value copula. A copula is a function that allows to distinguish a component describing only the structure of dependence from a total random vector distribution. In other words, the copula functions are functions that combine a multidimensional cumulative distribution of a random variable with its one-dimensional limit distributors [40].

The definition of a copula for a two-dimensional case is as follows:

A two-dimensional copula (2-copula) is each function $C : [0, 1]^2 \rightarrow [0, 1]$ that meets the following conditions:

For each $u_1, u_2 \in [0, 1]$ there is:

$$C(u_1, 0) = C(0, u_2) = 0 \quad (1)$$

1. For each $u_1, u_2 \in [0, 1]$ there is:

$$C(u_1, 1) = u_1 \text{ and } C(1, u_2) = u_2 \quad (2)$$

2. For each $u_1, u_2, v_1, v_2 \in [0, 1]$, such that $u_1 \leq u_2$ and $v_1 \leq v_2$, there is:

$$C(u_2, v_2) - C(u_2, v_1) - C(u_1, v_2) + C(u_1, v_1) \geq 0 \quad (3)$$

For basic information in terms of modeling the price dependence of futures on agricultural products listed on the Chicago Mercantile Exchange using the copula function, let $(X_1, Y_1), (X_2, Y_2), \ldots, (X_n, Y_n)$ be independent pairs of random variables with the same distributions and common copula C. Moreover, let $C_{(n)}$ be a copula with respect to the maximum components, namely $X_n = \max\{X_i\}$, $Y_n = \max\{Y_i\}$, $i = 1, \ldots, n$. From Sklar's theorem [26], it follows that $C_{(n)}(u, v) = C^n\left(u^{\frac{1}{n}}, v^{\frac{1}{n}}\right)$ for every $u, v \in [0, 1]$ The limit of the sequence $\{C_{(n)}\}$ naturally leads to the concept of an extreme value copula. Therefore, the copula C_E is called an extreme value copula if there is a copula C, such that:

$$\forall u, v \in [0, 1] \;\; C_E(u, v) = \lim_{n \to \infty} C^n\left(u^{\frac{1}{n}}, v^{\frac{1}{n}}\right) \quad (4)$$

2.1.2. Galambos' Copula

This copula was introduced to subject literature in 1975 by J. Galambos [41]. It is an example of a symmetrical copula, described with the formula below:

$$C_\theta^{Ga}(u, v; \theta) = uv \exp\left\{\left[(-\ln u)^{-\theta} + (-\ln v)^{-\theta}\right]^{-1/\theta}\right\} \quad (5)$$

where $\theta > 0$.

The dependence function of the discussed copula is presented using the following formula:

$$A^{Ga}(t) = 1 - \left(t^{-\theta} + (1-t)^{-\theta}\right)^{-1/\theta} \quad (6)$$

2.1.3. The Gumbel Copula

This copula was introduced to literature in 1960 by E.J. Gumbel and is also called the Gumbel–Hougaard copula [42]. It is described by the following formula:

$$C_{\theta}^{Gu}(u, v; \theta) = \exp\left\{-\left[(-\ln u)^{\theta} + (-\ln v)^{\theta}\right]^{1/\theta}\right\} \tag{7}$$

where $\theta \geq 1$.

The independence function is described by the formula:

$$A^{Gu}(t) = \left(t^{\theta} + (1-t)^{\theta}\right)^{1/\theta} \tag{8}$$

It can be observed that for $\theta \to \infty$ there is an excellent dependence, while independence occurs in the case of $\theta = 1$.

2.1.4. The Husler–Reiss Copula

The name of the copula comes from its authors, who presented it for the first time in a paper in 1989 [43]. The Husler–Reiss copula is defined by the following formula:

$$C_{\theta}^{HR}(u, v; \theta) = \exp\left\{-\widetilde{u}\Phi\left[\frac{1}{\theta} + \frac{1}{2}\theta\ln\left(\frac{\widetilde{u}}{\widetilde{v}}\right)\right] - \widetilde{v}\Phi\left[\frac{1}{\theta} + \frac{1}{2}\theta\ln\left(\frac{\widetilde{v}}{\widetilde{u}}\right)\right]\right\} \tag{9}$$

wherein $\theta \geq 0, \widetilde{u} = -\ln u, \widetilde{v} = -\ln v$. On the other hand, Φ is a cumulative distribution function of a standardized normal distribution.

The dependence function of the discussed copula is determined by the formula:

$$A^{HR}(t) = t\Phi\left[\theta^{-1} + \frac{1}{2}\theta\ln\left(\frac{t}{1-t}\right)\right] + (1-t)\Phi\left[\theta^{-1} - \frac{1}{2}\theta\ln\left(\frac{t}{1-t}\right)\right] \tag{10}$$

while the parameter θ for the Husler–Reiss copula measures the degree of dependence. This means that starting from $\theta = \infty$, one is dealing with independence until the situation in which $\theta = 0$, which brings total dependence.

2.1.5. Tawn Copula

Introduced in the subject literature in 1988 by J. Tawn, this copula is an asymmetric extension of the Gumbel cupola [44]. The model of the discussed copula is presented below:

$$C_{\theta}^{Ta}(u, v; \theta) = uv\exp\left\{\theta\frac{(\ln u)(\ln v)}{\ln u + \ln v}\right\} \tag{11}$$

where $0 \leq \theta \leq 1$.

In turn, the independence function of the Tawn copula is given by the formula:

$$A^{Ta}(t) = \theta t^2 - \theta t + 1 \tag{12}$$

2.1.6. The *t*-EV Copula

This copula is known as the *t*-Student extreme copula. For a two-dimensional case, the λ degrees of freedom and the correlation coefficient $\rho \in (-1, 1)$, the *t*-EV copula is determined by the following formula:

$$C_{\lambda,\rho}^{tEV}(u, v; \rho) = \exp\left\{T_{\lambda+1}\left[-\frac{\rho}{\theta} + \frac{1}{\theta}\left(\frac{\ln u}{\ln v}\right)^{1/\lambda}\right]\ln u + T_{\lambda+1}\left[-\frac{\rho}{\theta} + \frac{1}{\theta}\left(\frac{\ln u}{\ln v}\right)^{1/\lambda}\right]\ln v\right\} \tag{13}$$

where T_λ is a cumulative t-Student distribution function with λ degrees of freedom, while the parameter θ depends on the value of the correlation coefficient ρ and the degrees of freedom λ, and is described by the equation: $\theta^2 = \frac{1-\rho^2}{\lambda+1}$ [40].

The function of t-Student extreme dependence copula is presented using the following formula:

$$A_{\lambda,\rho}^{tEV}(t) = tT_{\lambda+1}\left(\frac{\left(\frac{t}{1-t}\right)^{1/\lambda} - \rho}{\sqrt{1-\rho^2}}\sqrt{1+\lambda}\right) + (1-t)T_{\lambda+1}\left(\frac{\left(\frac{1-t}{t}\right)^{1/\lambda} - \rho}{\sqrt{1-\rho^2}}\sqrt{1+\lambda}\right) \tag{14}$$

Estimation of five models of two-dimensional extreme value copulas was carried out based on the two-stage method of maximum likelihood, i.e., the IFM method. It is the inference function method for limit distributions, also called the two-stage maximum likelihood estimation method due to the two-stage estimation process. It was proposed for the first time in the works of H. Joe and J.J. Xu [27] and H. Joe [19]. The authors suggested to divide the set of estimated parameters into two subsets, so as to first estimate the parameters associated with the limit distributions, and then find the estimator responsible for the combined distribution, i.e., the copula. The empirical combining function method is used to assess the quality of matching the parameters of the extreme value copula [35–37]. The principle of selecting an appropriate combining function boils down to selecting the best function from the finite set of candidates, which is a subset of the set of all possible combining functions. There are various ways to define the empirical combining function. In this paper, a practical formula based on the values of the combining function defined on the grid will be used [45]. Since the price series of financial instruments belong to the group of non-stationary processes, daily constant return rates were determined based on the price series to conduct statistical analysis between the prices of the agricultural products. The use of logarithmic return rates is meaningful for the properties of the data series under investigation. As one of the Box–Cox transformations, logarithmation is known to stabilize the variance of the series [38].

$$R_t = \ln\left(\frac{X_t}{X_{t-1}}\right) \tag{15}$$

where X_t is the value of the futures on the day t.

Based on the daily closing prices, the logarithmic return rates for individual agricultural products were calculated, in accordance with the formula presented above.

2.1.7. The Empirical Combining Function Method

This method is used to assess the quality of matching the parameters of the extreme value copula, based on the empirical link function [39–44]. Let the theoretical structure of the dependence be described by the combining function C_θ, dependent on the value of the parameter θ. It is then assumed that the null hypothesis H_0 means that the structure of the dependence is determined using the combining function C_θ, against the alternative hypothesis H_1, which is a negation of hypothesis H_0. The principle of selecting the appropriate combining function comes down to selecting the best function from the finite set of candidates C, which is a subset of the set of all possible combining functions. The combining function selected from the set C is that for which the distance from the empirical copula is the smallest. There are various ways to define the empirical copula. In this work a practical formula was used, based on the values of the combining function defined on the following grid:

$$L = \left\{\left(\frac{i}{m}, \frac{j}{m}\right) : i, j = 0, 1, \ldots, m\right\} \tag{16}$$

The empirical values of the combining function at the points of the L-grid are determined by the following formula:

$$C_m\left(\frac{i}{m}, \frac{j}{m}\right) = \frac{1}{m}\sum_{k=1}^{m} \mathbf{1}(R_k \le i)\mathbf{1}(S_k \le j) \tag{17}$$

where R_i, S_i are ranks of variables X and Y.

On the other hand, the distance measured between the combining functions is based on the standard L^2, and in the discrete version takes the form:

$$d_L(C_m, C) = \sqrt{\sum_{i=1}^{m}\sum_{j=1}^{m}\left(C_m\left(\frac{i}{m}, \frac{j}{m}\right) - C\left(\frac{i}{m}, \frac{j}{m}\right)\right)^2} \tag{18}$$

where:
C_m—empirical copula,
C—function of the copula from the distinguished set of copulas C.

2.1.8. Kendall's Correlation Coefficient

Empirical studies often use rank-based correlation coefficients. One of the most popular is the Kendall rank correlation coefficient. It is usually used when there are restrictions related to the use of Pearson's linear correlation coefficient. Its advantage is that, as a representative of the measure of compatibility, it does not depend on limit distributions. Although the limit distributions and the correlation matrix do not determine the form of the combined distribution, similar to the linear correlation coefficient, for any continuous edge distributions, a combined distribution with a given rank correlation coefficient can still be constructed from the entire range $[-1; 1]$.

In order to define the Kendall rank correlation coefficient, a less formal definition of compatibility should first be introduced, as follows [46].

Two different realizations (x_1, y_1) and (x_2, y_2) of the random vector (X, Y) are compatible if

$$(x_1 - x_2)(y_1 - y_2) > 0 \text{ (i.e., } x_1 > x_2 \text{ and } y_1 > y_2, \text{ or } x_1 < x_2 \text{ and } y_1 < y_2)$$

and non-compatible, if

$$(x_1 - x_2)(y_1 - y_2) < 0 \text{ (i.e., } x_1 > x_2 \text{ and } y_1 < y_2 \text{ or } x_1 < x_2 \text{ and } y_1 > y_2)$$

Let (X_1, Y_1) and (X_2, Y_2) be independent random vectors with identical distribution. Kendall's correlation coefficient τ is defined as follows:

$$\tau(X_1, Y_1) = P[(X_1 - X_2)(Y_1 - Y_2) > 0] - P[(X_1 - X_2)(Y_1 - Y_2) < 0] \tag{19}$$

Using the above terminology, it can be concluded that the Kendall's coefficient for the vector (X_1, Y_1) is the probability of compatible realizations of the random vector (X_1, Y_1) minus the likelihood of realizing the non-compatible of the vector.

However, if $\{(x_1, y_1), \ldots, (x_n, y_n)\}$ is an n-element sample of a vector (X, Y) of continuous random variables, then there are $\binom{n}{2} = \frac{n(n-1)}{2}$ different pairs (x_i, y_i) and (x_j, y_j) that are compatible or inconsistent. The Kendall factor τ for the sample version, which is designated to be distinguished from the theoretical coefficient through $\hat{\tau}$, can be calculated based on the following formula:

$$\hat{\tau} = \frac{c - d}{c + d} \tag{20}$$

where c is the number of compliant pairs, and d the number of incompatible pairs.

3. Results

Sustainability is a human-centered concept that comprises multiple aspects and objectives of different interest groups. It is not readily measurable, except as a compromise between different parts of society, of which some may try to represent future generations of mankind [47]. In order to analyze the dependence of extreme price values on selected futures listed on the Chicago Mercantile Exchange, five selected models of two-dimensional extreme values copulas defined in the previous chapter, were used. As said before, the following copulas were used in empirical research: the Gumbel copula, the Galambos copula, the Husler–Reiss copula, the *t*-EV copula and the Tawn asymmetric copula. The estimation of parameters of selected extreme values copulas was carried out based on the two-stage maximum likelihood estimation method, i.e., the IFM method for two different limit distributions, namely, for the normal distribution and the *t*-Student distribution. The selection of distributions was motivated by the results recommending their use, presented widely in the subject literature [48–52]. Results for each of the three agricultural product pairs in question are presented in Tables 1 and 2. In addition to the values of the estimated parameters and their average estimation errors given in brackets, the likelihood function (LLF) values can also be found there. In addition, the extreme t -Student dome has an additional parameter λ, responsible for the number of degrees of freedom, the estimations of which are also included in the tables.

Table 1. Results of the estimation of the parameters of extreme values copulas with the normal limit distribution.

Copula	Parameter	Maize-Soy	LLF	Corn-Wheat	LLF	Soybean-Wheat	LLF
Galambos	θ	0.897 *** (0.013)	1831	0.674 *** (0.011)	1031	0.571 *** (0.010)	693.3
Gumbel	θ	1.628 *** (0.013)	1871	1.411 *** (0.011)	1050	1.316 *** (0.010)	710.2
Hüsler–Reiss	θ	1.268 *** (0.018)	1711	1.039 *** (0.012)	981.9	0.924 *** (0.011)	660.9
Tawn	θ	0.928 *** (0.009)	1883	0.758 *** (0.013)	1033	0.633 *** (0.015)	686.4
t-EV	ρ	0.786 *** (0.012)	1885	0.672 *** (0.018)	1053	0.589 *** (0.016)	711.8
	λ	4.000 ** (1.876)		4.000 ** (1.882)		4.000 ** (1.843)	

* significance at the level of 10%, ** significance at the level of 5%, *** significance at the level of 1%. Source: [own study].

Table 2. Results of the estimation of the parameters of extreme values copulas with *t*-Student limit distribution.

Copula	Parameter	Corn–Soybean	LLF	Corn–Wheat	LLF	Soybean–Wheat	LLF
Galambos	θ	0.895 *** (0.013)	1824	0.673 *** (0.011)	1028	0.570 *** (0.010)	691.9
Gumbel	θ	1.626 *** (0.013)	1864	1.411 *** (0.011)	1047	1.315 ** (0.010)	708.9
Hüsler-Reiss	θ	1.267 *** (0.018)	1706	1.038 *** (0.012)	977.1	0.923 *** (0.011)	658.8
Tawn	θ	0.927 *** (0.009)	1875	0.757 *** (0.013)	1032	0.633 *** (0.015)	685.5
t-EV	ρ	0.785 *** (0.012)	1878	0.671 *** (0.018)	1051	0.588 *** (0.016)	710.9
	λ	4.000 ** (1.876)		4.000 ** (1.882)		4.000 ** (1.843)	

* significance at the level of 10%, ** significance at the level of 5%, *** significance at the level of 1%. Source: [own study].

The picture emerging from the analysis of the results presented in Tables 1 and 2 shows that very similar results have been obtained irrespective of the adopted limit distribution. The copula, which best describes the relationship between the extreme values of all researched agricultural product pairs, is the extreme *t*-Student copula. When comparing the obtained logarithm values of the likelihood function for individual copulas, it can also be observed that the Tawn copula may be useful for describing the relationship between extreme observations for the corn–soybean pair. A minor difference in the logarithm values of the credibility function between the t-EV copula and the Tawn copula, may suggest that the pair corn–soybean is characterized by a small degree of asymmetry, with more observations deviating in plus.

On the other hand, for pairs of agricultural products—corn–wheat and soybean–wheat—the second copula of extreme values, which may be helpful to describe the studied dependencies, with small differences in the value of the credibility function, turned out to be the Gumbel copula. Please note that in the case of an extreme values Gumbel copula, the parameter θ takes on values greater than or equal to 1; the higher the value of this parameter, the stronger the dependence of maximum losses between the analyzed observations. The results given in Tables 1 and 2 indicate a minor correlation between the extreme observations of the researched pairs of return rates of agricultural products quoted on the Chicago Mercantile Exchange. This means that unlike stock exchange indices for which, as it is well known and documented in extensive empirical studies, the occurrence of extremely negative return rates is much more likely than of extremely high rates. In the case of futures, one is not dealing with such strong dependence. The obtained results may rather point to a possible symmetry or very minor asymmetry in the tails of the return rates for the analyzed agricultural product pairs.

The next stage of the analysis was to test whether the structure of the relationship described with the estimated extreme values copulas is a sufficient approximation of reality, and whether it is suitable for modeling empirical data. In order to assess the quality of matching the estimated copulas to the empirical data on return rates of agricultural products, the methodology presented [37–44] was used. It says that one of the methods of checking the quality of matching the copula parameters is to compare the coefficients implied by the selected copula with empirical Kendall coefficients $\hat{\tau}$. Estimation of the Kendall coefficient τ was obtained for all extreme copulas using a simulation method. The results are presented in Tables 3 and 4. On the other hand, Table 5 contains the values of the Kendall correlation coefficient $\hat{\tau}$ calculated for the sample version. All estimated correlations are positive and statistically significant. Regardless of the pair of agricultural products, the strongest dependence is demonstrated by the pair corn–soy, while the weakest dependence is characterized by the pair soybean–wheat. Estimates of the Kendall coefficient for all extreme value copulas were obtained using the simulation method. The results are presented in Tables 3 and 4. Table 5, in turn, contains the values of the Kendall correlation coefficient calculated for the sample version.

Table 3. Values of Kendall's tau coefficient in the case of normal limit distribution.

Type of Copula	Corn–Soybean	Corn–Wheat	Soybean–Wheat
The Galambos' copula	0.382	0.289	0.236
The Gumbel copula	0.386	0.291	0.240
The Husler–Reiss copula	0.357	0.271	0.223
The Tawn copula	0.387	0.290	0.239
The t-EV copula	0.390	0.293	0.242

Source: [own study].

Table 4. Values of Kendall's tau coefficient in the case of t-Student distribution.

Type of Copula	Corn–Soybean	Corn–Wheat	Soybean–Wheat
The Galambos' copula	0.381	0.288	0.235
The Gumbel copula	0.385	0.291	0.239
The Husler-Reiss copula	0.356	0.270	0.222
The Tawn copula	0.386	0.289	0.239
The t-EV copula	0.388	0.293	0.241

Source: [own study].

Table 5. Sample Kendall correlation coefficients between the daily return rates of the surveyed agricultural products.

	Corn	**Soybean**	**Wheat**
Corn	1.000	0.417	0.332
Soybean		1.000	0.270
Wheat			1.000

Source: [own study].

The comparison of the empirical values of Kendall coefficients with the corresponding theoretical values demonstrates that for all three pairs of agricultural products, for both applied limit distributions, the best studied extreme dependences is described by the *t*-EV copula. In the case of the pair corn–soybean, the Tawn asymmetrical copula may also be helpful, while for the pairs corn–wheat and soybean–wheat, the Gumbel copula may also be useful. Please also note the Tawn copula, as the Kendall tau coefficient for this copula is quite satisfactory, and compared to the best-rated extreme *t*-Student copula, the differences are minimal.

An alternative method used to assess the quality of matching the parameters of the extreme value copula, is based on an empirical combining function. In order to choose the best copula function, the estimated parameters of the extreme value combining function were compared to the empirical combining function defined for the grid for $m = 303$. The selection of this particular grid size is a natural reference to the analyzed time series, which 303 sub-periods correspond to the subsequent months. The tables below demonstrate the distances of the estimated extreme value copulas from the empirical combining functions, depending on the selected limit distribution.

Analysis of the results listed in Tables 6 and 7 confirms the results obtained to assess the goodness of the adjustment of the extreme value copulas obtained with a methodology using the Kendall coefficient τ. The minor difference between the distances from the empirical joining function for the Tawn copula and the t-Student extreme copula proves that both functions may prove equally useful to describe the relationship between extreme values of all researched agricultural product pairs, regardless of the adopted limit distribution. In order to capture the dependence between extreme observations for the pair corn–soybean, the *t*-EV copula is the best, but the Gumbel copula may turn out equally important. On the other hand, for the pair corn–wheat and soybean–wheat, the Tawn copula seems the most accurate, while the *t*-EV copula ranks second. However, the differences are minor and they result from the selection of the size of the grid.

Table 6. Values of distance from the empirical combining function for selected extreme value copulas for normal limit distribution.

	Corn–Soybean	**Corn–Wheat**	**Soybean–Wheat**
The Galambos' copula	0.221	0.211	0.188
The Gumbel copula	0.209	0.197	0.174
The Husler–Reiss copula	0.285	0.252	0.219
The Tawn copula	0.218	0.180	0.163
The *t*-EV copula	0.203	0.195	0.174

Source: [own study].

Table 7. Values of distance from the empirical combining function for selected extreme value copulas for *t*-Student limit distribution.

	Corn–Soybean	**Corn–Wheat**	**Soybean–Wheat**
The Galambos' copula	0.222	0.211	0.188
The Gumbel copula	0.210	0.197	0.174
The Husler–Reiss copula	0.286	0.252	0.219
The Tawn copula	0.218	0.179	0.162
The *t*-EV copula	0.203	0.195	0.173

Source: [own study].

4. Conclusions

Sustainable agriculture is a key issue for environmentally friendly agriculture: Effective, economically viable and socially desirable. In addition, conservation of resources, environmental protection and agricultural stewardship, i.e., all requirements of sustainability, will increase and not reduce global food production. Other issues, such as the links between sustainable agriculture and the rest of the food and agri-food industry, as well as the consequences of sustainable development for rural communities and society as a whole, have not yet been finally resolved [53].

The empirical results demonstrated that the extreme values copula, which best describes the dependence between the extreme values of all researched pairs of agricultural products, irrespective of the limit distribution adopted, is the extreme *t*-Student, a.k.a. the *t*-EV copula. Moreover, in the case of the pair corn–soybean, the Tawn copula may be useful, and for corn–wheat and soybean–wheat, the Gumbel copula. The obtained results were confirmed by the assessment of the goodness of matching the parameters of the extreme value copulas. It can be observed that the results obtained in order to check the goodness of matching the estimated copula to empirical data of return rates of agricultural products reflect the results obtained in the estimation process. The final conclusion is that the return rate of the analyzed agricultural product pairs may be characterized by a minor degree of asymmetry, with the right tail being particularly heavy, which means higher probability of extreme observations than in the case of normal distribution.

The economic determinants of agricultural production have become the main reason for both producers and agri-food processors to search for exchange-based instruments to reduce the risk associated with adverse changes in product prices. A tool that is perfectly suited for this purpose is futures, which allow transferring the risk relatively cheaply to third parties. Thanks to their effectiveness, futures markets started to increase their turnover steadily. Participation of entities operating on the agricultural products market became common in countries with advanced stock exchange systems, at the same time becoming an irreplaceable link of production, processing, and trade in the agri-food industry. Analyzing agricultural markets is particularly important in terms of the need for ideation and creation of key development strategies for the agricultural sector, both at global and EU level, including the coordinated goals of the Common Agricultural Policy (CAP) [54–57].

In conclusion, our review highlights the vast possibilities of analyzing large data sets in agriculture towards optimizing management decisions based on statistical modeling. It demonstrates that the availability, techniques, and methods for analyzing large data sets, as well as the growing openness of large data sources, will encourage more academic research, public sector initiatives, and business ventures in the agricultural sector. This practice is still at an early stage of development and many barriers have to be overcome.

In practice, the results allow for rationalization of decisions of companies interested in intervention on the futures market for agricultural products, because investing there not only reduces the risk of financial losses, but also generates a high return rate.

The use of these models allows the simulation of the possible behavior of prices of agricultural products, including identification of extreme deviations. This allows programming decisions regarding possible public intervention in agricultural markets. In extraordinary situations—force majeure, weather anomalies, natural disasters—decision modeling can be used to assess the scale of damage for the agricultural market, and also to develop plans for the protection of agricultural producers within the framework of policies such as the CAP. As a consequence, they allow for securing food resources on deficit markets, e.g., with respect to geographical location.

Author Contributions: Conceptualization, Z.G.-S., D.K. and M.N.; methodology, G.M. and A.S.-S. and J.K.-D.; resources, G.M., M.K., A.S.-S. and J.S.; formal analysis, Z.G.-S., D.K. and M.K.; investigation, D.K., Z.G.-S., M.N. and J.S.; resources, M.K., G.M. and J.K.-D.; data curation, A.S.-S., D.K. and M.N.; writing—Z.G.-S., G.M., A.S.-S. and M.K.; visualization, D.K., J.S. and M.N.; funding acquisition, Z.G.-S.

References

1. Erbaugha, J.; Bierbaum, R.; Castillejac, G.; da Fonsecad, G.A.B.; Cole, S.; Hansend, B. Toward sustainable agriculture in the tropics. *World Dev.* **2019**, *121*, 158–162. [CrossRef]
2. Szeląg-Sikora, A.; Niemiec, M.; Sikora, J.; Chowaniak, M. Possibilities of designating swards of grasses and small-seed legumes from selected organic farms in Poland for feed. In Proceedings of the IX International Scientific Symposium "Farm Machinery and Processes Management in Sustainable Agriculture", Lublin, Poland, 22–24 November 2017; pp. 365–370. [CrossRef]
3. Gródek-Szostak, Z.; Szeląg-Sikora, A.; Sikora, J.; Korenko, M. Prerequisites for the cooperation between enterprises and business supportinstitutions for technological development. *Bus. Non-profit Organ. Facing Increased Compet. Grow. Cust. Demands* **2017**, *16*, 427–439.
4. Gródek-Szostak, Z.; Luc, M.; Szeląg-Sikora, A.; Niemiec, M.; Kajrunajtys, D. Economic Missions and Brokerage Events as an Instrument for Support of International Technological Cooperation between Companies of the Agricultural and Food Sector. *Infrastruct. Environ.* **2019**, 303–308. [CrossRef]
5. Rigby, D.; Cáceresb, D. Organic farming and the sustainability of agricultural systems. *Agric. Syst.* **2001**, *68*, 21–40. [CrossRef]
6. Skafa, L.; Buonocorea, E.; Dumonteta, S.; Capone, R.; Franzesea, P.P. Food security and sustainable agriculture in Lebanon: An environmental accounting framework. *J. Clean. Prod.* **2019**, *209*, 1025–1032. [CrossRef]
7. Yu, J.; Wu, J. The Sustainability of Agricultural Development in China: The Agriculture–Environment Nexus. *Sustainability* **2018**, *10*, 1776. [CrossRef]
8. USDA. *Grain: World Markets and Trade*; USDA Office of Global Analysis: Washington, DC, USA, 2019.
9. USDA. *Oilseeds: Worlds Market and Trade*; USDA Office of Global Analysis: Washington, DC, USA, 2019.
10. USDA. *World Agricultural Production*; USDA Office of Global Analysis: Washington, DC, USA, 2019.
11. EC. Short-Term Outlook for EU Agricultural Markets in 2018 and 2019. In *Agriculture and Rural Development*; Spring: Brussels, Belgium, 2019.
12. Gumbel, E.J.; Goldstein, N. Analysis of empirical bivariate extremal distribution. *J. Am. Stat. Assoc.* **1964**, *59*, 794–816. [CrossRef]
13. Starica, C. Multivariate extremes for models with constant conditional correlations. *J. Empir. Financ.* **1999**, *6*, 515–553. [CrossRef]
14. Longin, F.; Solnik, B. Extreme correlations in international equity markets. *J. Financ.* **2001**, *56*, 649–676. [CrossRef]
15. Hsu, C.P.; Huang, C.W.; Chiou, W.J.P. Effectiveness of copula—Extreme value theory in estimating value-at-risk: Empirical evidence from Asian emerging markets. *Rev. Quant. Financ. Account.* **2012**, *39*, 447–468. [CrossRef]
16. Cebrian, A.; Denuit, M.; Lambert, P. Analysis of bivariate tail dependence using extreme values copulas: An application to the SOA medical large claims database. *Belg. Actuar. J.* **2003**, *3*, 33–41.
17. Renard, B.; Lang, M. Use of a Gaussian copula for multivariate extreme value analysis: Some case studies in hydrology. *Adv. Water Resour.* **2007**, *30*, 897–912. [CrossRef]
18. Nelsen, R.B. *An Introduction to Copulas*, 2nd ed.; Springer Verlag: New York, NY, USA, 2006; pp. 97–101.
19. Joe, H. *Multivariate Models and Dependence Concepts*; Chapman & Hall: London, UK, 1997; pp. 139–168.
20. Segers, J. Nonparametric inference for bivariate extreme-value copulas. In *Topics in Extreme Values*; Ahsanullah, M., Kirmani, S.N.U.A., Eds.; Nova Science Publishers: New York, NY, USA, 2007; pp. 181–203.
21. Gudendorf, G.; Segers, J. Extreme-value copulas. In *Copula Theory and its Applications: Proceedings of the Workshop Held in Warsaw 25–26 September 2009*; Jaworski, P., Durante, F., Härdle, W., Rychlik, T., Eds.; Springer Verlag: New York, NY, USA, 2010; pp. 127–145.
22. Genest, C.; Segers, J. Rank-based inference for bivariate extreme-value copulas. *Ann. Stat.* **2009**, *37*, 2597–3097. [CrossRef]
23. Ribatet, M.; Sedki, M. Extreme value copulas and max-stable processes. *J. Société Fr. Stat.* **2012**, *153*, 138–150.
24. FAO. *The Future of Food and Agriculture: Trends and Challenges*; Food and Agriculture Organization of the United Nations: Roma, Italy, 2017.
25. Wrzaszcz, W.; Zegar, J. Challenges for Sustainable Development of Agricultural Holdings. *Econ. Environ. Stud.* **2016**, *16*, 377–402.
26. Sklar, A. Fonctions de répartition á n dimensions et leurs marges. *Publ. l'Institut Stat. l'Université Paris* **1959**, *8*, 229–231.

27. Joe, H.; Xu, J.J. The estimation method of inference function for margins for multivariate models, Department of Statistics, University of British Columbia. *Tech. Rep.* **1996**, *166*. [CrossRef]
28. Ji, Q.; Bouri, E.; Roubaud, D.; Jawad, S.; Shahzad, H. Risk spillover between energy and agricultural commodity markets: A dependence-switching CoVaR-copula model. *Energy Econ.* **2018**, *75*, 14–27. [CrossRef]
29. Lia, R.L.; Wang, D.H.; Tu, J.Q.; Li, S.P. Correlation between agricultural markets in dynamic perspective—Evidence from China and the US futures markets. *Phys. A Stat. Mech. Appl.* **2016**, *464*, 83–92. [CrossRef]
30. Hill, J.; Schneeweis, T.; Yau, J. International trading/non-trading time effects on risk estimation in futures markets. *J. Futures Mark.* **1990**, *10*, 407–423. [CrossRef]
31. Yang, K.; Tian, F.; Chen, L.; Li, S. Realized volatility forecast of agricultural futures using the HAR models with bagging and combination approaches. *Int. Rev. Econ. Financ.* **2017**, *49*, 276–291. [CrossRef]
32. Cheng, B.; Nikitopoulos, C.S.; Schlögl, E. Pricing of long-dated commodity derivatives: Do stochastic interest rates matter? *J. Bank. Financ.* **2018**, *95*, 148–166. [CrossRef]
33. Greiner, R.; Puig, J.; Huchery, C.; Collier, N.; Garnett, S.T. Scenario modelling to support industry strategic planning and decision making. *Environ. Model. Softw.* **2014**, *55*, 120–131. [CrossRef]
34. Pickands, J. Multivariate extreme value distributions. *Bull. Int. Stat. Inst.* **1981**, *2*, 859–878.
35. Balkema, A.A.; Resnick, S.I. Max-infinite divisibility. *J. Appl. Probab.* **1977**, *14*, 309–319. [CrossRef]
36. De Haan, L.; Resnick, S.I. Limit theory for multivariate sample extremes. *Z. Wahrscheinlichkeitstheorie Verwandte Geb.* **1977**, *40*, 317–337. [CrossRef]
37. Davison, A.C.; Smith, R.L. Models for Exceedances over High Thresholds. *J. R. Stat. Soc. Ser. B (Methodological)* **1990**, *52*, 393–442. [CrossRef]
38. Katz, R.W.; Parlange, M.B.; Naveau, P. Statistics of extremes in hydrology. *Adv. Water Resour.* **2002**, *25*, 1287–1304. [CrossRef]
39. McNeil, A.J. *Extreme Value Theory for Risk Managers, Internal Modelling and CAD II*; RISK Books: London, UK, 1999.
40. Demarta, S.; McNeil, A.J. The t copula and related copulas. *Int. Stat. Rev.* **2005**, *73*, 111–129. [CrossRef]
41. Galambos, J. Order statistics of samples from multivariate distributions. *J. Am. Stat. Assoc.* **1975**, *9*, 674–680.
42. Gumbel, E.J. Bivariate exponential distributions. *J. Am. Stat. Assoc.* **1960**, *55*, 698–707. [CrossRef]
43. Husler, J.; Reiss, R.D. Maxima of normal random vectors: Between independence and complete dependence. *Stat. Probab. Lett.* **1989**, *7*, 283–286. [CrossRef]
44. Tawn, J.A. Bivariate extreme value theory: Models and estimation. *Biometrika* **1988**, *75*, 397–415. [CrossRef]
45. Heilpern, S. *Funkcje Łączące, Wydawnictwo Akademii Ekonomicznej im*; Oskara Langego we Wrocławiu: Wroclaw, Poland, 2007.
46. Hsieh, J.J. Estimation of Kendall's tau from censored data. *Comput. Stat. Data Anal.* **2010**, *54*, 1613–1621. [CrossRef]
47. Fisher, J.; Rucki, K. Re-conceptualizing the Science of Sustainability: A Dynamical Systems Approach to Understanding the Nexus of Conflict, Development and the Environment. *Sustain. Dev.* **2017**, *25*, 267–275. [CrossRef]
48. Durrleman, V.; Nikeghbali, A.; Roncalli, T. *Which Copula Is the Right One?* Working Paper; Groupe de Recherche Opérationnelle, Crédit Lyonnais: Lyon, France, 2000.
49. Kamnitui, N.; Genest, C.; Jaworski, P.; Trutschnig, W. On the size of the class of bivariate extreme-value copulas with a fixed value of Spearman's rho or Kendall's tau. *J. Math. Anal. Appl.* **2019**, *472*, 920–936. [CrossRef]
50. Li, F.; Zhou, J.; Liu, C.h. Statistical modelling of extreme storms using copulas: A comparison study. *Coast. Eng.* **2018**, *142*, 52–61. [CrossRef]
51. Gudendorf, G.; Segers, J. Nonparametric estimation of multivariate extreme-value copulas. *J. Stat. Plan. Inference* **2012**, *142*, 3073–3085. [CrossRef]
52. Sriboonchitta, S.; Nguyen, H.T.; Wiboonpongse, A.; Liu, J. Modeling volatility and dependency of agricultural price and production indices of Thailand: Static versus time-varying copulas. *Int. J. Approx. Reason.* **2013**, *54*, 793–808. [CrossRef]
53. Fousekis, P.; Tzaferi, D. Price returns and trading volume changes in agricultural futures markets: An empirical analysis with quantile regressions. *J. Econ. Asymmetries* **2019**, *19*, e00116. [CrossRef]

54. Sreekumar, S.; Sharma, K.C.; Bhakar, R. Gumbel copula based multi interval ramp product for power system flexibility enhancement. *Int. J. Electr. Power Energy Syst.* **2019**, *112*, 417–427. [CrossRef]
55. Keya, N.; Anowar, S.; Eluru, N. Joint model of freight mode choice and shipment size: A copula-based random regret minimization framework. *Transp. Res. Part E Logist. Transp. Rev.* **2019**, *125*, 97–115. [CrossRef]
56. Niemiec, M.; Komorowska, M.; Szeląg-Sikora, A.; Sikora, J.; Kuboń, M.; Gródek-Szostak, Z.; Kapusta-Duch, J. Risk Assessment for Social Practices in Small Vegetable farms in Poland as a Tool for the Optimization of Quality Management Systems. *Sustainability* **2019**, *11*, 3913. [CrossRef]
57. Kapusta-Duch, J.; Szeląg-Sikora, A.; Sikora, J.; Niemiec, M.; Gródek-Szostak, Z.; Kuboń, M.; Leszczyńska, T.; Borczak, B. Health-Promoting Properties of Fresh and Processed Purple Cauliflower. *Sustainability* **2019**, *11*, 4008. [CrossRef]

7

Survivability of Probiotic Bacteria in Model Systems of Non-Fermented and Fermented Coconut and Hemp Milks

Agnieszka Szparaga [1], Sylwester Tabor [2], Sławomir Kocira [3,*], Ewa Czerwińska [4], Maciej Kuboń [2], Bartosz Płóciennik [1] and Pavol Findura [5]

1 Department of Agrobiotechnology, Koszalin University of Technology, Racławicka 15–17, 75-620 Koszalin, Poland; agnieszka.szparaga@tu.koszalin.pl (A.S.); b.plociennik2@gmail.com (B.P.)
2 Department of Production Engineering, Logistics and Applied Computer Science, Agricultural University Krakow, Balicka 116B, 30-149 Krakow, Poland; sylwester.tabor@urk.edu.pl (S.T.); maciej.kubon@ur.krakow.pl (M.K.)
3 Department of Machinery Exploitation and Management of Production Processes, University of Life Sciences in Lublin, Akademicka 13, 20-950 Lublin, Poland
4 Department of Biomedical Engineering, Koszalin University of Technology, Śniadeckich 2, 75-453 Koszalin, Poland; ewa.czerwinska@tu.koszalin.pl
5 Department of Machines and Production Biosystems, Slovak University of Agriculture in Nitra, Tr. A. Hlinku 2, 949 76 Nitra, Slovakia; pavol.findura@uniag.sk
* Correspondence: slawomir.kocira@up.lublin.pl

Abstract: This study aimed at determining the survivability of probiotic bacteria cultures in model non-dairy beverages subjected or not to the fermentation and storage processes, representing milk substitutes. The experimental material included milks produced from desiccated coconut and non-dehulled seeds of hemp (*Cannabis sativa* L.). The plant milks were subjected to chemical and microbiological evaluation immediately after preparation as well as on day 7, 14, and 21 of their cold storage. Study results proved that the produced and modified plant non-dairy beverages could be the matrix for probiotic bacteria. The fermentation process contributed to increased survivability of *Lactobacillus casei* subsp. *rhamnosus* in both coconut and hemp milk. During 21-day storage of inoculated milk substitutes, the best survivability of *Lactobacillus casei* was determined in the fermented coconut milk. On day 21 of cold storage, the number of viable *Lactobacillus casei* cells in the fermented coconut and hemp milks ensured meeting the therapeutic criterion. Due to their nutritional composition and cell count of bacteria having a beneficial effect on the human body, the analyzed groceries—offering an alternative to milk—represent a category of novel food products and their manufacture will contribute to the sustainable development of food production and to food security assurance.

Keywords: probiotic; non-dairy beverages; survivability; fermentation; bacteria; coconut; hemp; sustainable food production

---

1. Introduction

Sustainable food production should be considered through the perspective of a better understanding of food security. In recent years, many researchers and policy makers have focused only on the physical availability of food, owing to the sufficient agricultural production [1,2]. This has partly been driven by widespread claims that we need to boost the global food production to feed the world in 2050 [3]. However according to the FAO (Food and Agriculture Organization of the United Nations) definition [4]: "food security exists when all people, at all times, have physical and economic access to sufficient, safe and nutritious food that meets their dietary needs and food preferences for an active and healthy life". Hence, it needs to be emphasized that the sustainable food production is inevitably related to the food security in its three aspects: food security, food safety, and food quality, without which the development of the food industry sector would not be possible [5,6]. Due to the current dynamic development of sciences related to food and human nutrition, a correlation has been confirmed between the health status and nutritional patterns. A well-balanced diet is the key factor in diseases prevention and treatment. The growing nutritional interests and awareness of consumers have prompted many producers to manufacture functional food [7–9]. A functional food definition covers certain strains of microorganisms being constituents of food of plant and animal origin that contain physiologically active compounds. These compounds are beneficial for human health and help minimizing the risk of chronic diseases development [10]. One of the multiple examples of functional food products are these containing microorganisms endogenous to the human gastrointestinal tract and exhibiting a positive effect on human health [11]. So far, the greatest part of probiotic products has been offered by fermented beverages made of animal milk. Currently, research is underway into other products that may be matrix for probiotic bacteria [12,13]. Consumers avoiding milk because of allergies or lactose intolerance, and consumers following a vegan diet can replace milk with other plant-based substitutes. Beverages derived from soybeans have for many years been the predominant equivalents of milk. Today, coconut, almonds, hemp, and various cereals (e.g., oats, buckwheat, and rice) are also used to produce plant-based beverages [14,15]. A drawback of these products is however their specific taste that does not suit to everyone. A solution to this problem is offered by lactic acid fermentation, which imparts a characteristic, pleasant after-taste to these products and contributes to the improvement of the digestibility [13,16]. Numerous attempts have recently been undertaken to ferment vegan beverages serving as milk substitutes using various strains of probiotic bacteria, which was expected to additionally increase their health value [16,17]. However, most of the study results reported in literature concern the feasibility of producing fermented soybean milk [13–16]. This is related to the fact, that manufacture of high quality plant-based beverages containing probiotic bacteria poses a serious challenge [18,19]. According to Yuliana et al. [20], the production of coconut-based beverages is difficult because of the suppressed growth and survivability of these probiotic microorganisms, compared to dairy beverages. Difficulties in the manufacture and fortification of hemp milk were encountered by Batkiene et al. [21]. The first ones were related to the stability of produced emulsions, whereas the latter ones to the survivability of probiotic bacteria during storage. Worthy of notice is that the production of hemp-based products has increased in recent years due to the confirmed nutritional value and low allergenicity of seeds of this plant [22]. This has been feasible owing to new varieties characterized by a low concentration of a psychoactive compound delta-9-tetrahydrocannabinol (THC) [23] and to cultivations with the use of elite category sowing material [24].

The current definition of a probiotic means those microbial strains that positively affect consumer health when taken in the right amount [25]. Accordingly to FAO/WHO guidelines, the count of probiotic bacteria cannot be less than the value corresponding to 10^6 cfu per 1 mL of a product through the entire period of its storage till the end of its shelf life. This value has been deemed the therapeutic minimum [26–28].

The main problems associated with the fermentation of plant beverages are related to the sensory quality of the final product and to the resistance of probiotic microorganisms. Producers encounter difficulties with the physical stability caused by milk coagulation (it occurs at the beginning or in the course of storage). The appearance of these products resembles that of low-fat yoghurt [29–32]. Additional problems concern the survivability of probiotic bacteria, which is dependent on multiple factors, including e.g., presence of other microorganisms in the product, time and conditions of strains culture and product storage, product processing technology or pH value [13,26,33,34].

Considering the above, the major objective of this study was to determine the survivability of probiotic bacterial cultures in model fermented and non-fermented stored plant beverages being milk substitutes, because today the plant-based alternative milks provide a huge perspective for the sustainable development of the healthy food market and should therefore be widely scrutinized. Evaluation of the effect of production and processing techniques, and also of fortification techniques, of plant-based beverages may serve to develop a nutritionally complete beverage with a high overall acceptability and health values.

2. Materials and Methods

2.1. Plant Material and Beverages Production

The experimental material included non-dairy beverages produced from the following plant raw materials: desiccated coconut (Bakalland, Warszawa, Poland) and non-dehulled seeds of hemp (*Cannabis sativa* L.; Sante, Warszawa, Poland; Figure 1). Seeds and desiccated coconut were ground in a WZ-1 laboratory mill (Sadkiewicz Instruments, Bydgoszcz, Poland). Chemical composition of analyzed material is presented in Table 1.

Figure 1. Raw materials used for non-dairy beverages production: (**a**) hemp seeds and (**b**) desiccated coconut.

Table 1. Contents of protein, lipids, and sugars of desiccated coconut and non-dehulled hemp seeds (nutritional information available on respective product labels).

Product	Protein	Lipids	Sugars
		(%)	
Desiccated coconut	5.6	63.2	5.9
Non-dehulled seeds of hemp	25.2	36.1	5.4

Milk substitutes to be analyzed were produced according to the schemes presented in block diagrams in Figures 2 and 3.

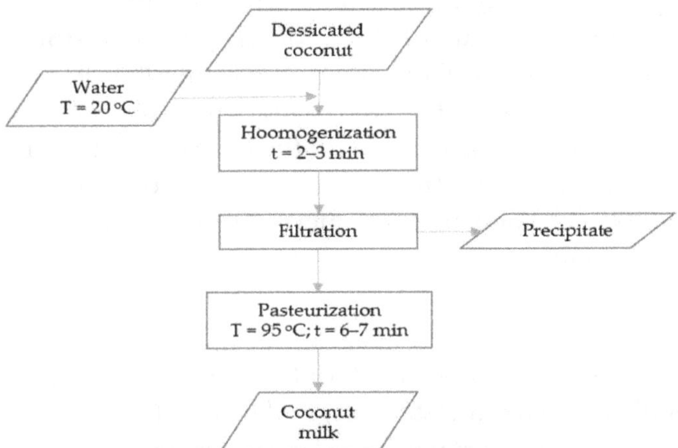

Figure 2. Technological scheme of coconut milk production (own study based on Blasco [35]).

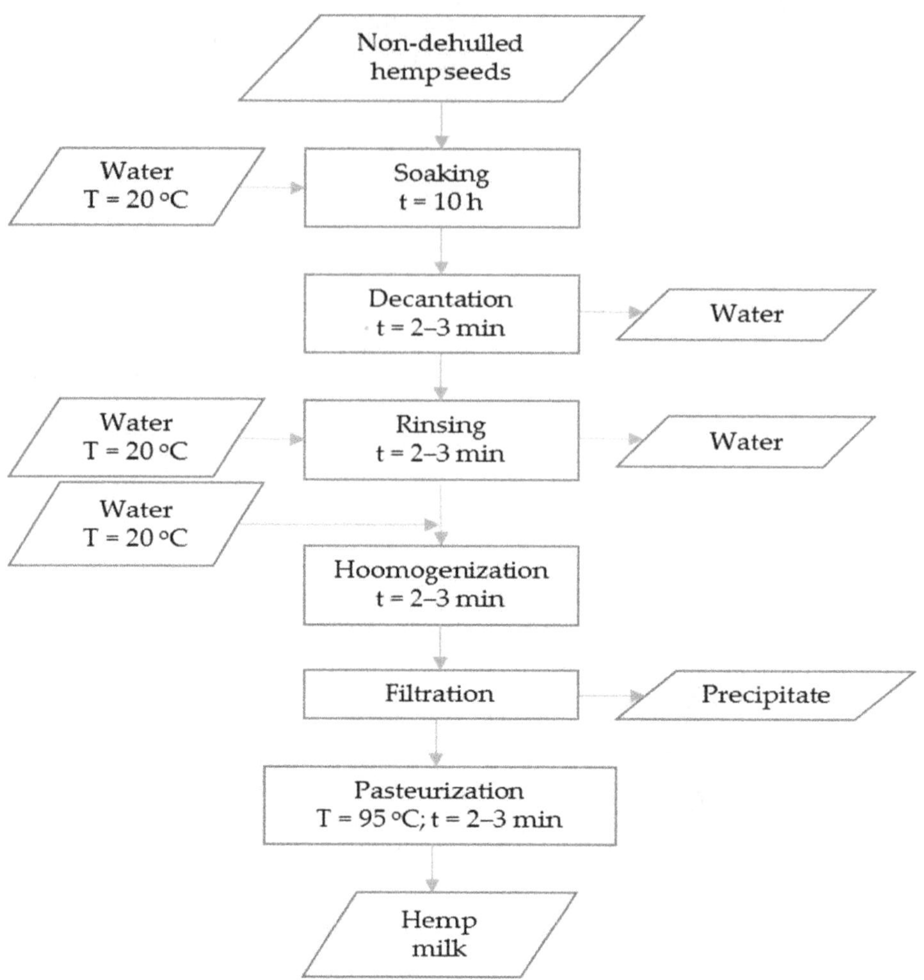

Figure 3. Technological scheme of hemp milk production (own study based on Szakuła [36]).

Immediately after production, the plant-based milks were subjected to the physicochemical analysis to determine content of their protein, lipids, and carbohydrates, and their acidity. Their microbiological status was assessed as well.

Models of non-dairy beverages intended for the determination of counts of viable bacterial cells during storage were divided into four groups, each containing three samples—two groups included beverages with probiotic addition and two groups—beverages without the probiotic. Both, the samples supplemented and not supplemented with a probiotic monoculture were fermented in a laboratory incubator (Binder BD 260) at a temperature of 37 °C for 6 h, and then stored at a temperature of 4 °C. The remaining samples were cold stored. Duration of the fermentation process was chosen based on results of a study conducted by Zielińska et al. [37], who demonstrated that intensive proliferation of *Lactobacillus casei* subsp. *rhamnosus* cells proceeded till the 6th hour of the process. Optimal parameters of the fermentation process, established by Zielińska et al. [38], allow producing plant-based fermented beverages with sensory quality acceptable by consumers [39]. Based these findings, investigations have not assumed own sensory evaluation.

2.2. Probiotic Microorganisms

The study was conducted with probiotic bacterial strain *Lactobacillus casei* subsp. *rhamnosus* LCR 3013 in the form of a lyophilizate (Serowar, Szczecin, Poland). Before analyses, the strain was stored at a temperature of −18 °C to preserve its properties. The bacterial culture was activated by transferring 0.1 g of the lyophilized strain to 5 mL of an MRS broth (Merck, Warszawa, Poland). Next, the suspension was incubated at a temperature of 37 °C for 24 h. The resultant culture was centrifuged at 10,000× *g* for 5 min. Centrifugation was repeated after rinsing the resultant precipitate with a physiological saline solution. Cell biomass suspended in a physiological saline solution (5 mL) and having the optical density of 1° McF (Densimat, bioMérieux, Grassina FI, Italy) was added to 100 mL of plant milk (5% *v/v*), to achieve bacterial cell count of approximately 10 log (cfu/mL) [21].

2.3. Chemical Evaluation of Non-Dairy Beverages

The produced non-dairy beverages were analyzed for content of: protein with the Kjeldahl's method [40] (Büchi Distilation Unit, K314) and lipids with the Gerber's method [41]. Their active acidity (pH) was measured as well using a VOLTCRAFT KBM-110 m (Conrad Electronic SE, Hirschau, Germany) with a pH electrode [42].

Additional analyses were conducted to determine the content of reducing sugars. They were made with the colorimetric method using 3,5-dinitrosalicilic acid (DNS) [43]. DNS acid (1950 μL) were added to 50 μL of the analyzed non-dairy beverages, and the mixture was incubated in a water bath at a temperature of 99 °C for 10 min. After cooling the mixture, 900 μL of DNS acid were added to 100 μL of mixture sample. Absorbance was measured at $\lambda = 540$ nm (UV-vis spectrophotometer, VWR UV-6300PC, USA) and results of these measurements were converted based on the standard curve into equivalents of glucose (g/L) contained in non-dairy beverages.

2.4. Microbiological Status of Plant Beverages

The produced plant beverages were pasteurized and afterwards subjected to a microbiological analysis based on the pour-plate Koch's method [44] and the sterile serial dilutions method (from 10^{-1} to 10^{-8}). For this purpose, 1 mL of each of the two subsequent dilutions (10^{-6} and 10^{-8}) and 9 mL of each of the appropriately selected medium were transferred onto Petri dishes and left to solidify [16]. Nutrient agar (BTL, Łódź, Poland) was used to isolate mesophilic and psychrophilic bacteria, whereas Sabouraud agar enriched with chloramphenicol (BTL, Łódź, Poland) was used for fungi and yeast isolation from the beverages. Samples were incubated at a temperature of 30 °C for 48 h (mesophiles) and at 20 °C for 72 h (psychrophiles), and at 20 °C for 5 days (fungi and yeast) [44]. Total bacterial count (TBC) was determined as well. Bacterial cultures were inoculated with the pour-plate method in three replications for each sample. Plates with inoculates were incubated at a temperature of 37 °C for 48 h under anaerobic conditions using anaerostats with anaerocult A inserts (Merck, Darmstadt, Germany). Nutrient broth agar (BTL, Łódź, Poland) was used for inoculations. After completed incubation, results were converted into the number of colony forming units per 1 mL of product (cfu/mL). Dilutions of 10^{-6} and 10^{-8} were used for analyses and for TBC determination in each sample. The above analyses were carried out for plant beverages without probiotic strain addition.

2.5. Microbiological Analyses of Counts of Viable Bacterial Cells During Storage of Fermented and Non-Fermented Non-Dairy Beverages

Analyses of the survivability of the probiotic strain *Lactobacillus casei* subsp. *rhamnosus* in fermented and non-fermented models of coconut and hemp beverages (with added starter monoculture of probiotic bacteria) were conducted immediately after their preparation, after their fermentation as well as on day 7, 14, and 21 of their storage at a temperature of 4 °C. The maximal cold storage time assumed in the study was selected based on results of a research conducted by Gustaw et al. [45] into the survivability of *Lactobacillus casei* strain in fermented beverages with the addition of selected protein preparations.

Having been diluted in sterile water, the analyzed samples were transferred onto sterile Petri dishes (1 mL of sample from each dilution of 10^{-6} and 10^{-8}), to which 9 mL of the selective MRS Agar medium (by de Man, Rogosa and Sharpe) [46] (BTL, Poland) were added afterwards. Next, the samples were incubated at a temperature of 30 °C for 72 h. The total count of viable lactic acid bacteria per 1 mL of the sample was computed according to the following formula:

$$N = n_c \cdot d_r, \quad (1)$$

where: N—number of viable bacterial cells (cfu/mL), n_C—number of bacterial colonies, and d_r—dilution rate.

2.6. Statistical Analysis

All experiments were carried out in three replications. Results were expressed as arithmetic means. The Shapiro–Wilk test was used to evaluate the normal distribution of data. Results were analyzed using a one-way analysis of variance (ANOVA). The significance of differences between mean values was estimated based on Tukey confidence intervals, at a $p < 0.05$. Values followed by different small letters are significantly different at $p < 0.05$ (effect of storage). Values followed by different big letters are significantly different at $p < 0.05$ (effect of treatment). The standard deviation (±SD) value was determined for all reported mean values. The statistical analysis was performed using Statistica 13.3 software (StatSoft, Kraków, Poland).

3. Results and Discussion

3.1. Evaluation of the Produced Non-Dairy Beverages

One of the key technological aspects in probiotic food production is to maintain optimal conditions that would ensure the proper growth and viability of potentially probiotic bacteria during fermentation and storage [47]. This may be accomplished through, i.e., the appropriate choice of a carrier and product supplementation with nutrients [48]. Hence, raw materials of plant origin need to be analyzed for the content of nutrients indispensable for probiotic bacteria metabolism, and for the effect of environment on their survivability [27,49].

In order to identify factors that affect probiotics survivability, a study with non-dairy beverages produced from desiccated coconut and hemp seeds under laboratory conditions was conducted. It needs to be emphasized that the nutritional value of non-dairy beverages is largely determined by their protein content [28,50]. In the study, protein content was determined at 3.23 g/100g in coconut beverage and at 6.96 g/100 g in hemp beverage (Table 2).

Table 2. Contents of protein, lipids, and glucose, and active acidity of coconut and hemp milks.

Non-Dairy Beverage	Protein	Lipids	Reducing Sugars	Active Acidity
	(% ± SD)		(g glucose/L ± SD)	(pH ± SD)
Coconut milk	3.23 ± 0.28	21.08 ± 0.41	34.53 ± 0.39	6.15 ± 0.15
Hemp milk	6.96 ± 0.19	18.02 ± 0.54	30.21 ± 0.33	6.81 ± 0.11

Differences in protein content were determined depending on the plant raw material used for beverages production. Discrepancies were also noted when comparing protein content determined in the study and these declared by selected producers of plant beverages intended for the European market [13], i.e., obtained results were higher than protein content declared by producers of coconut and hemp beverages. These differences might be due to the various quality of raw materials used for beverages production and to treatments applied in the production process (e.g., heat treatment) that contribute to a decrease in total protein concentration.

Hoffman and Kostyra [15] evaluated plant-based milk substitutes in terms of their nutritional value and demonstrated that only the beverage made of soybean seeds equaled milk in this respect. The other plant materials had significantly lower content of protein, i.e., two-fold lower—quinoa, and three-fold lower—a mixed beverage made of soybean, rice, and oats. Beverages made of coconuts and almonds were characterized by trace amounts of protein.

This indicates that although coconut-based milk substitutes provide very low amounts of protein, they may offer an alternative to consumers who seek for gluten-free food.

Results of study confirm findings reported by Sethi et al. [13], who demonstrated that plant beverages are inexpensive substitutes of milk, especially for consumers allergic to milk, but are not comparable nor equal with it in terms of their nutritional value, as hemp milk (living harvest) provides barely 2 g and coconut milk less than 1 g of protein in 240 mL of the product.

Coconut beverages are food products characterized by a high content of lipids, including significant content of the following saturated fatty acids: lauric (50%) and myristic (6%–7%). The unsaturated fatty acids of coconut include oleic acid (monounsaturated) and linolic acid (polyunsaturated) [51]. In turn, the hemp beverage contains approximately 80% of essential unsaturated fatty acids (EFAs), including linolic acid (56%) and α-linolenic acid (19%). According to dieticians, the optimal ratio of these acids should reach 3:1, as is the case with the hemp beverage [52]. Figure 4 presents content of lipids in the analyzed raw beverages made of coconut and hemp seeds. As it results from our study, the coconut beverage contained 21% and the hemp beverage contained 18% of lipids.

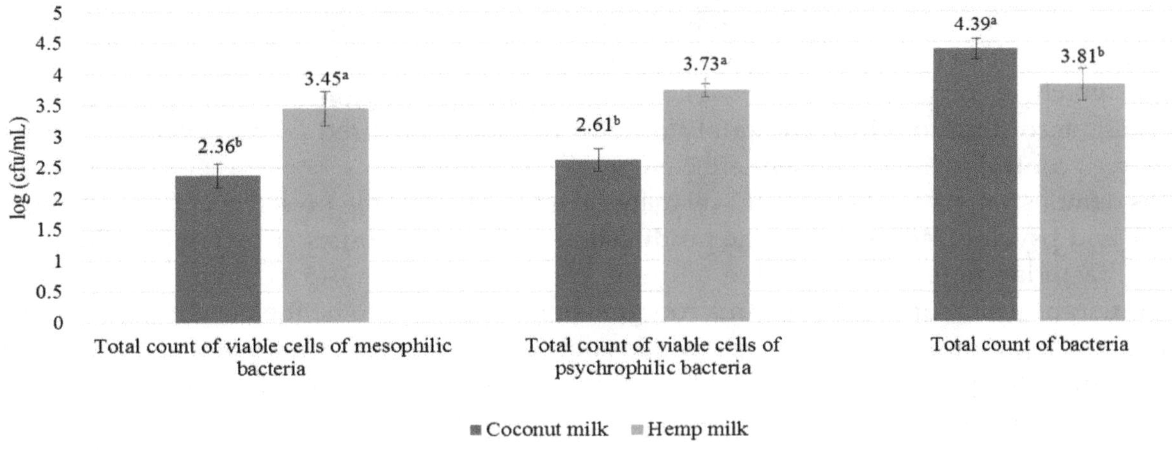

Figure 4. Total count of microorganisms isolated from pasteurized non-diary beverages (a, b—statistically significant differences, $p \leq 0.05$).

Due to the composition of the plant milks, their fatty acid profile differed significantly from that of milk. Milk contains approximately 1.2% of saturated fatty acids, whereas their content in its substitutes does not exceed 0.7%. Similar conclusions were formulated by Belewu and Belewu [53], who determined lipids content in the produced coconut milk at 24.10%. In turn, Sethi et al. [13] demonstrated that the analyzed plant-derived milks were characterized by a similar concentration of lipids reaching 6 g in hemp milk (living harvest) and 5 g in coconut milk per 240 mL of the product.

Content and types of carbohydrates in fermented beverages have a significant impact on the development and activity of health-promoting bacteria. Reducing sugars present in the medium may be good sources of carbon necessary for probiotics metabolism [54]. This was confirmed by Jurkowski and Błaszczyk [55], who demonstrated monosaccharides to be indispensable substrates during the fermentation process by lactic acid bacteria. Hence, control of their content seems necessary in the production of probiotic foods. In the study, the content of reducing sugars was expressed as glucose concentration in the produced non-dairy beverages (Figure 4), and reached 34.52 g glucose/L in coconut milk and 30.21 g glucose/L in hemp milk. Data obtained demonstrate that the coconut milk is a better raw material for the production of probiotic non-dairy beverages because it is richer in compounds necessary for fermentation (e.g., glucose). This is in line with results reported by Quasem et al. [17] who analyzed a sesame beverage as a bacteria matrix. Differences in the content of reducing sugars are one of the factors, which determine the possibility of using plant-based milk as a natural medium for lactic acid bacteria development. According to Sethi et al. [13], however, the choice of a matrix for these beneficial microorganisms may also be driven by the total content of carbohydrates. In hemp milk, their concentration reaches barely 1 g, whereas in coconut milk it is high and reaches 7 g/240 mL of product.

The content of acids is a factor that determines food freshness, but at the same time it largely affects its color and taste. Acids in beverages and other food products, contribute to control microflora growth [56]. In our study, the active acidity of coconut beverage (before fermentation) was determined at pH 6.15 (Figure 4). This value is similar to results reported by other authors. For instance, when investigating the effect of strawberry beverage supplementation with a soybean protein isolate, Dłużewska et al. [57] observed changes in the real acidity, values of which ranged from pH 5.82 to pH 6.61. In turn, acidity of a coconut beverage analyzed by Belewu and Belewu [53] reached pH 6.23. A negligibly higher active acidity (pH 6.81) was determined in our study for the hemp seed beverage. Many scientists have reported on the decreased acidity of the medium in the case of fermented products, which appeared to result from the accumulation of lactic acid caused by the activity of microorganisms.

3.2. *Microbiological Purity of Non-Dairy Beverages*

From the perspective of health safety, pasteurization is an indispensable process during the manufacture of plant-based beverages. It contributes to eradication of pathogenic microorganisms and to inactivation of certain spore-forms. Therefore, in the present study, we evaluated its effectiveness (Figure 4), which is consistent with provisions of Commission Regulations (EC) 2073/2005 and 229/2019 [58], according to which the safety of food is mainly ensured by a preventive approach including, e.g., control of heat treatment effectiveness.

It was demonstrated that the type of raw material influenced on the presence of bacteria, both the mesophilic and psychrophilic ones, in the produced non-dairy beverages. The determined count of mesophilic bacteria ranged from 2.36 log (cfu/mL; coconut milk) to 3.45 log (cfu/mL; hemp milk). A more numerous group of the isolated microorganisms turned out to be the psychrophilic bacteria, with counts ranging from 2.6 to 3.7 log (cfu/mL). In contrast, no fungi or yeast were detected. The total bacterial count in the produced non-dairy beverages reached 3.81 log (cfu/mL) in hemp milk and 4.39 cfu/mL in coconut milk. It was, therefore, concluded that the short-term temperature increase to 95 °C did not contribute to the complete neutralization of the microflora of the non-dairy beverages. Thus, it seems necessary to develop some other method that would be more effective in ensuring the appropriate microbiological purity of plant-based beverages. Lee et al. [59] evaluated the effect of increased hydrostatic pressure coupled with high temperature on counts of viable, spore-forming cells of pathogenic microorganisms. They reported a significant decrease in the number of active resting spores to a negligible level (below 1 cfu/mL) upon the coupled use of pressure of 207 MPa and temperature of 90 °C.

3.3. *Evaluation of the Effect of Fermentation Process on Chemical and Microbiological Properties of Non-Dairy Beverages*

Immediately after the addition of the inoculum from *Lactobacillus casei* subsp. *rhamnosus* lyophilizate, the total count of viable lactic acid bacteria (LAB) cells reached 11.72 log (cfu/mL) in coconut beverage and 8.41 log (cfu/mL) in hemp beverage. The fermentation process (37 °C/6 h) caused an increase in bacteria count in the samples to 13.26 and 10.92 log (cfu/mL), respectively (Figure 5). Initially, the difference in the count of viable LAB cells could be due to the viability of probiotics themselves in the food matrix, which may be affected by pH (initial pH values were at 6.12 for coconut milk and 6.79 for hemp milk), oxygen level, and presence of competing microorganisms (bacterial cells and their resting spores undamaged during pasteurization: 4.39 log (cfu/mL) in coconut milk and 3.81 log (cfu/mL) in hemp milk) [60–62]. Therefore, LABs resistance to inconvenient conditions appears to be an important technological trait, which enables selecting strains for untypical food matrix like, e.g., non-dairy beverages [63]. Initial differences in the count of probiotic bacterial cells in both analyzed types of plant milks could be due to the fact that viable, metabolically active cells may rapidly lose their capability for growth, and this dormancy state of a part of the population may occur especially when the cells are exposed to unbeneficial factors. For explicit confirmation of these assumptions, fluorescent techniques should be employed that allow monitoring subtle changes in the dynamics of proliferation and decay of microorganisms that may be in the viable but non-culturable (VBNC) or active but non-culturable (ABNC) state, i.e., in the state of bacteria transition to the dormancy state under unfavorable colonization conditions [63–66].

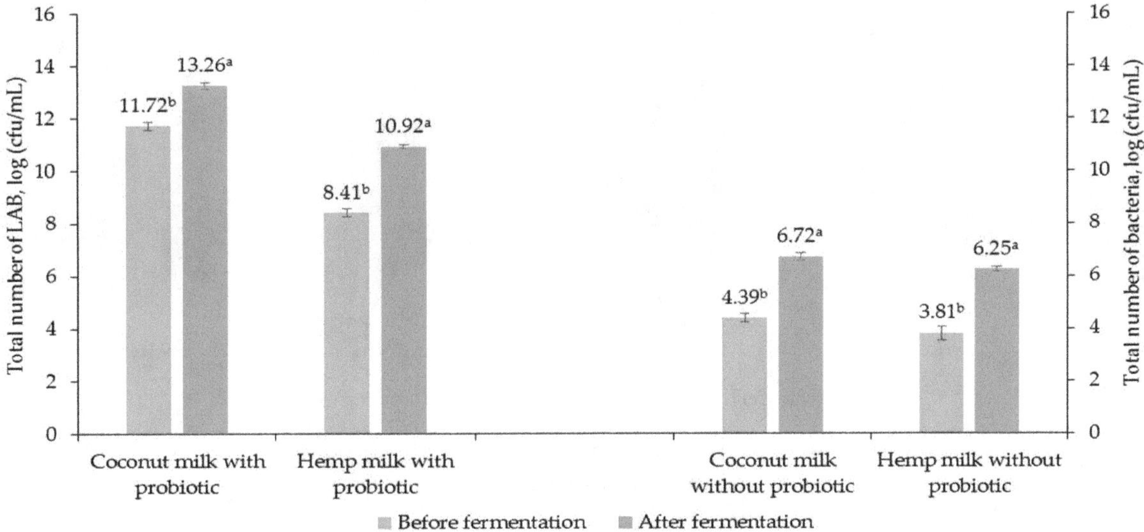

Figure 5. Total count of viable bacterial cells in the analyzed non-dairy beverages before and after fermentation (a, b—statistically significant differences, $p \leq 0.05$).

A similar dependency was demonstrated in the study conducted by Zaręba [67], who noted that 4-h fermentation process of soybean milk contributed to the proliferation of *Lactobacillus* species bacteria by approximately 0.5 log (cfu/mL). A study conducted by Bartkiene et al. [21] also showed that *L. casei* cell count in fermented hemp milk reached 8.78 log (cfu/mL) and was higher compared to the count determined before the fermentation process (8 log (cfu/mL)).

The analyzed models of non-dairy beverages differed in terms of the total bacteria count. TBC was also determined in the control samples (without the probiotic). Before fermentation it reached 4.39 log (cfu/mL) in coconut milk and 3.81 log (cfu/mL) in hemp milk; whereas after fermentation for the respective values were at 6.72 and 6.25 log (cfu/mL; Figure 5). Presumably, these were resting spores that had survived pasteurization and whose proliferation was promoted by fermentation temperature (37 °C) being optimal for their growth. These speculations may be confirmed by results reported earlier by Czaczyk et al. [68], who noticed the greatest growth of *Bacillus* ssp. bacilli under these conditions. Similar observations were made by Huy et al. [69]. The activity of microorganisms during incubation is also affected by the type and amount of nutrients available in the medium. However, considering the "Microbiological Limits for Assessment of Microbiological Quality of Ready-to-eat Foods" [70], the criterion related to the microbiological quality did not exceed the maximum value of 7 log (cfu/mL), set by the International Commission for Microbiological Specification of Food.

Monosaccharides, including mainly glucose, present in plant beverages represent a good source of carbon to bacteria. The physicochemical analysis conducted in the study allowed concluding that coconut milk (34.53 g glucose/L) was a better source of these compounds compared to hemp beverage (30.21 g glucose/L). The content of reducing sugars decreased significantly after the fermentation process (Figure 6).

The non-dairy beverages with probiotic addition were characterized by a greater reduction in glucose content after fermentation, i.e., to 23.05 g glucose/L in coconut milk and to 19.79 g glucose/L in hemp milk. In turn, milks without the probiotic bacteria were characterized by noticeably higher glucose content, which decreased due to the activity of undesirable microorganisms.

Figure 7 presents results of measurements of active acidity of the non-dairy beverages before and after fermentation.

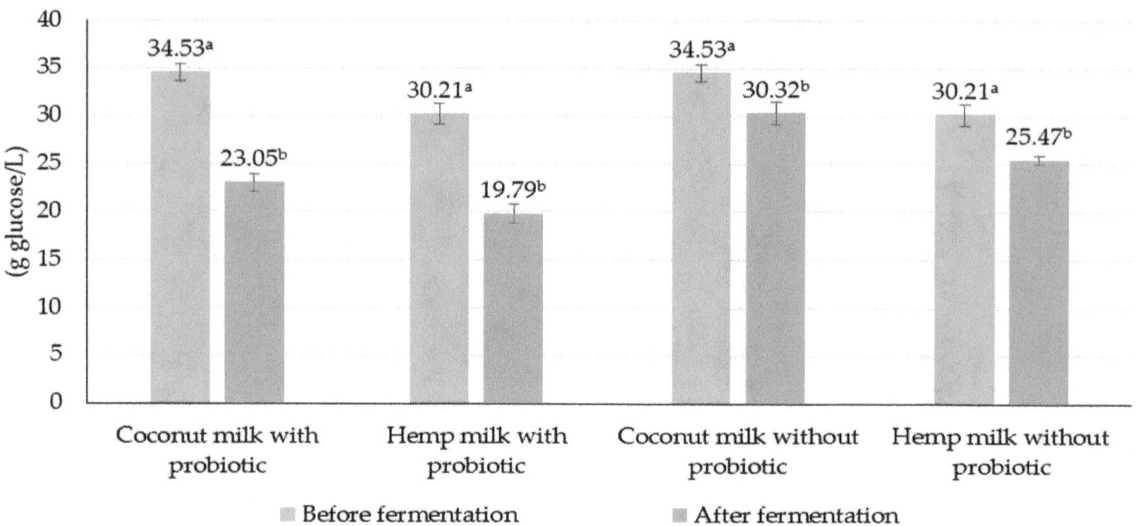

Figure 6. Reducing sugars concentration in the analyzed non-dairy beverages before and after fermentation (a, b—statistically significant differences, $p \leq 0.05$).

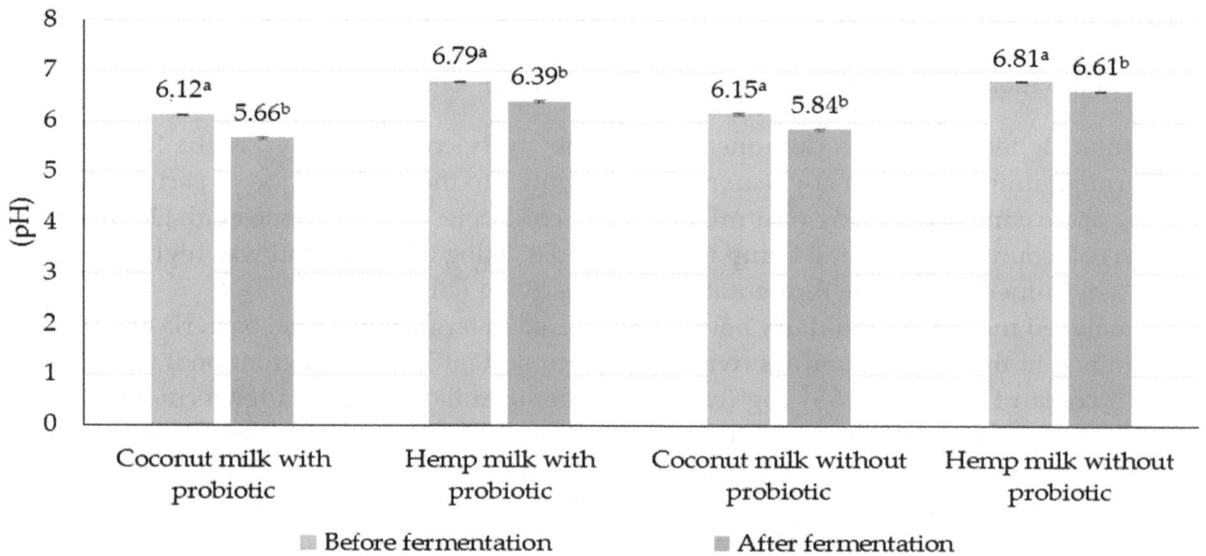

Figure 7. Active acidity of the analyzed non-dairy beverages before and after fermentation (a, b—statistically significant differences, $p \leq 0.05$).

The greatest differences in the active acidity, before and after fermentation, were demonstrated in the fermented probiotic beverage made of coconut, i.e., pH 6.12 and pH 5.66, respectively. In turn, the smallest decrease in the pH value (by 0.21) was noted in the fermented hemp seed milk without the probiotic. It may, therefore, be concluded that the acidity level is determined by the accumulation of organic acids caused by monosaccharides metabolism. So negligible pH changes may be due to the fact that *Lactobacillus casei* subsp. *rhamnosus* strain used in the study is a representative of facultatively heterofermentative bacteria. Apart from lactic acid, these bacteria are capable of producing CO_2, acetic acid (aerobic conditions), acetic aldehyde, and/or ethanol (anaerobic conditions) [71–73]. During the heterofermentation process, glucose degradation proceeds accordingly to the pentose phosphate pathway, and the capability of lactic acid bacteria for this fermentation results from a lack of certain enzymes like, e.g., triphosphate isomerase and aldolase [74].

3.4. Assessment of the Quality of the Produced Non-Dairy Beverages During Cold Storage

Changes of the active acidity of the non-dairy beverages during 21-day cold storage were presented in Table 3. As expected, storage time had a significant effect on the acidity level of milk substitutes, causing vast differences in pH values of the fermented samples of coconut and hemp milks between day 1 and 21 of storage.

Table 3. Changes in the active acidity of fermented and non-fermented non-dairy beverages during 21-day storage.

Treatment	Storage Time			
	Day 1	Day 7	Day 14	Day 21
	Fermented			
CP	5.61 ± 0.02 ^{aF}	5.11 ± 0.02 ^{bG}	4.93 ± 0.02 ^{cB}	4.81 ± 0.02 ^{dB}
HP	6.47 ± 0.01 ^{aC}	6.08 ± 0.02 ^{bA}	5.91 ± 0.02 ^{cA}	5.78 ± 0.02 ^{dA}
C	5.84 ± 0.01 ^{aE}	4.85 ± 0.01 ^{bH}	4.08 ± 0.01 ^{cG}	3.58 ± 0.02 ^{dG}
H	6.62 ± 0.02 ^{aB}	5.49 ± 0.02 ^{bD}	4.39 ± 0.02 ^{cF}	3.41 ± 0.01 ^{dH}
	Non-Fermented			
CP	6.15 ± 0.02 ^{aD}	5.31 ± 0.02 ^{bE}	4.88 ± 0.02 ^{cC}	4.21 ± 0.01 ^{dC}
HP	6.81 ± 0.02 ^{aA}	5.87 ± 0.02 ^{bB}	4.65 ± 0.01 ^{cD}	3.95 ± 0.02 ^{dE}
C	6.15 ± 0.01 ^{aD}	5.25 ± 0.02 ^{bF}	4.53 ± 0.02 ^{cE}	4.01 ± 0.02 ^{dD}
H	6.81 ± 0.02 ^{aA}	5.71 ± 0.01 ^{bC}	4.35 ± 0.02 ^{cF}	3.70 ± 0.02 ^{dF}

CP—Coconut beverage with probiotic; HP—Hemp beverage with a probiotic; C—Coconut beverage without probiotic; H—Hemp beverage without probiotic. Means in the rows, followed by different small letters (a–d) are significantly different at $p < 0.05$ (effect of storage). Means in the columns, followed by different big letters (A–F) are significantly different at $p < 0.05$ (effect of treatment).

Usually, the active acidity (pH) of fermented plant-based beverages should not be lower than 4.0 throughout the storage period [75]. Results presented in Table 3 indicate that the pH value remained above 4.0 within fourteen days of cold storage of the analyzed milk substitutes. However, a pH decline was recorded at the end of the storage period in the case of the samples not inoculated with the probiotic monoculture. This is in agreement with results reported by Paseephol and Sherkat [76] and by Colakoglu and Gursoy [77]. Guo et al. [78] reported that the pH value of fermented buffalo milk containing *Lactobacillus casei* decreased from 5.02 (day 1) to 4.00 (day 30 of storage) [78]. In turn, Akalin et al. [79] demonstrated that the probiotic bacteria decreased the pH value of various yoghurts from 4.51 to 4.40 after 28 days of their cold storage [79].

Data obtained in the study indicate that pH values of all beverage samples decreased during cold storage. This dependency may be explained by the persistent metabolic activity of the probiotic monoculture, which was also noticed by Bonczar et al. [80] during cold storage of fermented beverages. When comparing pH values over the storage period, these researchers observed decrease in all samples. Similar conclusions were drawn by Bartkiene et al. [21], who analyzed hemp milk in a 15-day storage model. The pH value of the fermented hemp milk with the addition of a probiotic culture of *L. casei* decreased slightly from 5.15 in the first day to 4.77 in the last day of cold storage.

The study demonstrated also a decrease in protein content of the produced non-dairy beverages along with storage time (Table 4).

The analysis of the nutritional value of non-fermented milk substitutes demonstrated that, after 21 days of storage, the total protein content was higher in the non-dairy beverages supplemented with probiotic bacteria.

A similar tendency was noted during storage of fermented non-dairy beverages (i.e., decreased protein concentration). However, the conducted analyses showed a higher total protein content in the fermented than in the non-fermented beverages, and lower in the non-dairy beverages fortified with the probiotic monoculture.

Table 4. Protein content of fermented and non-fermented non-dairy beverages during 21-day storage.

Treatment	Storage Time			
	Day 1	Day 7	Day 14	Day 21
	Fermented			
CP	3.02 ± 0.21 ^{aB}	2.95 ± 0.23 ^{aB}	2.84 ± 0.15 ^{aB}	2.67 ± 0.26 ^{aB}
HP	6.68 ± 0.26 ^{aA}	6.41 ± 0.28 ^{aA}	6.33 ± 0.21 ^{aA}	6.19 ± 0.24 ^{aA}
C	3.02 ± 0.19 ^{aB}	2.96 ± 0.17 ^{aB}	2.87 ± 0.18 ^{aB}	2.71 ± 0.18 ^{aB}
H	6.68 ± 0.29 ^{aA}	6.45 ± 0.25 ^{aA}	6.34 ± 0.23 ^{aA}	6.22 ± 0.23 ^{aA}
	Non-Fermented			
CP	3.23 ± 0.22 ^{aB}	3.17 ± 0.19 ^{aB}	3.10 ± 0.21 ^{aB}	3.01 ± 0.24 ^{aB}
HP	6.96 ± 0.27 ^{aA}	6.91 ± 0.25 ^{aA}	6.82 ± 0.23 ^{aA}	6.68 ± 0.26 ^{aA}
C	3.23 ± 0.18 ^{aB}	3.15 ± 0.20 ^{aB}	3.09 ± 0.20 ^{aB}	2.98 ± 0.18 ^{aB}
H	6.96 ± 0.27 ^{aA}	6.89 ± 0.22 ^{aA}	6.78 ± 0.21 ^{aA}	6.63 ± 0.25 ^{aA}

CP—Coconut beverage with probiotic; HP—Hemp beverage with a probiotic; C—Coconut beverage without probiotic; H—Hemp beverage without probiotic. Means in the rows, followed by different small letters (a) are significantly different at $p < 0.05$ (effect of storage). Means in the columns, followed by different big letters (A,B) are significantly different at $p < 0.05$ (effect of treatment).

According to Bernat et al. [81], the significant decrease in protein content is due to the fact that during the fermentation and storage of beverages, the bacterial starters could hydrolyze proteins to synthesize amino acids necessary for their nutrition. Investigations conducted by the aforementioned authors demonstrated also that fermented plant-based substitutes of milk had by approximately 17% lower content of β-glucan than their non-fermented counterparts, while this compound is capable of proteins crosslinking in food [82].

Changes observed in lipids content of the analyzed non-dairy beverages resembled these of protein (Table 5). In the entire period of cold storage, lipids concentration decreased negligibly in all milks. It may, thus, be concluded that the study demonstrated stability of this component in the produced and modified plant-based beverages.

Table 5. Lipids content in fermented and non-fermented non-dairy beverages during 21-day storage.

Treatment	Storage Time			
	Day 1	Day 7	Day 14	Day 21
	Fermented			
CP	21.06 ± 0.06 ^{aA}	20.98 ± 0.06 ^{abA}	20.87 ± 0.06 ^{bcA}	20.75 ± 0.06 ^{cA}
HP	18.01 ± 0.06 ^{aB}	17.95 ± 0.06 ^{abB}	17.81 ± 0.06 ^{bcB}	17.66 ± 0.06 ^{cB}
C	21.06 ± 0.07 ^{aA}	20.96 ± 0.07 ^{aA}	20.89 ± 0.07 ^{abA}	20.73 ± 0.06 ^{bA}
H	18.01 ± 0.06 ^{aB}	17.89 ± 0.05 ^{abB}	17.78 ± 0.06 ^{bB}	17.62 ± 0.06 ^{cB}
	Non-Fermented			
CP	21.08 ± 0.02 ^{aA}	21.01 ± 0.06 ^{aA}	20.94 ± 0.06 ^{aA}	20.84 ± 0.06 ^{bA}
HP	18.02 ± 0.02 ^{aB}	17.97 ± 0.06 ^{aB}	17.92 ± 0.06 ^{aB}	17.69 ± 0.06 ^{bB}
C	21.08 ± 0.02 ^{aA}	21.02 ± 0.06 ^{aA}	20.93 ± 0.06 ^{abA}	20.79 ± 0.06 ^{bA}
H	18.02 ± 0.02 ^{aB}	17.95 ± 0.06 ^{aB}	17.89 ± 0.06 ^{aB}	17.66 ± 0.06 ^{bB}

CP—Coconut beverage with probiotic; HP—Hemp beverage with a probiotic; C—Coconut beverage without probiotic; H—Hemp beverage without probiotic. Means in the rows, followed by different small letters (a–c) are significantly different at $p < 0.05$ (effect of storage). Means in the columns, followed by different big letters (A,B) are significantly different at $p < 0.05$ (effect of treatment).

As reported by Bernat et al. [82], who analyzed the microstructure of oat milk during storage, the similar concentration of lipids may be due to their embedding in a polysaccharide network. Stability of such system is additionally associated with the cross-linking properties of β-glucans [75]. Furthermore, almost all lipid droplets are retained in the polysaccharide-protein matrix, which is responsible for the physical stability of plant milk. It has been demonstrated that certain proteins may be attached to lipid globules, thereby ensuring protection of emulsions against destabilization processes [82].

Changes in concentrations of individual nutrients in plant-based beverages during cold storage may also be due to the fermentation process, which contributes to decreased content of carbohydrates and also of some non-digestible poly- and oligo-saccharides, to the improvement of protein quality, to the facilitated synthesis of selected amino acids, and to the improved availability of vitamins. In addition, it ensures optimal pH conditions for the enzymatic degradation of many compounds being important growth factors for the potentially probiotic bacteria [83].

The *Lactobacillus* strains used in the study are complex microorganisms that need carbohydrates, amino acids, B-group vitamins, nucleic acids, and minerals for their proper growth [84]. For this reason, it can be concluded that the fermentation of plant-based beverages may offer an inexpensive method for the synthesis of substrates in the product that would promote the growth of beneficial microorganisms [85].

3.5. *Quantitative Analysis of Viable Bacterial Cells During Storage of Fermented and Non-Fermented Non-Dairy Beverages*

An important aspect determining the quality of health-promoting fermented beverages is the analysis of changes in the number of viable lactic acid bacteria in 1 mL of a product over the entire period of its shelf life [86]. Al-Otaibi [87] emphasized that 6 log (cfu/mL) of viable probiotic cells should be consumed every day to ensure health benefits to consumers. Both types of the fermented non-dairy beverages met this requirement regarding viability of *L. casei* till the end of the storage period. On the first day of storage (after fermentation), the number of active bacterial LAB cells ranged from 10.92 log (cfu/mL) in hemp milk to 13.26 log (cfu/mL) in coconut milk (Table 6).

Table 6. Changes in the number of *Lactobacillus casei* subsp. *rhamnosus* during 21-day storage (4 °C) of fermented and non-fermented non-dairy beverages.

Treatment	Storage Time			
	Day 1	Day 7	Day 14	Day 21
		Fermented		
CP	13.26 ± 0.14^{aA}	11.46 ± 0.18^{bA}	11.26 ± 0.17^{bA}	9.41 ± 0.24^{cA}
HP	10.92 ± 0.10^{aC}	10.31 ± 0.25^{aB}	8.28 ± 0.33^{bB}	7.35 ± 0.26^{cB}
		Non-Fermented		
CP	11.72 ± 0.04^{aB}	6.81 ± 0.13^{bC}	5.42 ± 0.15^{cC}	3.12 ± 0.13^{dC}
HP	8.41 ± 0.18^{aD}	6.35 ± 0.16^{bC}	4.53 ± 0.08^{cD}	3.54 ± 0.20^{dC}

CP—Coconut beverage with probiotic; HP—Hemp beverage with a probiotic. Means in the rows, followed by different small letters (a–d) are significantly different at $p < 0.05$ (effect of storage). Means in the columns, followed by different big letters (A–D) are significantly different at $p < 0.05$ (effect of treatment).

Opposite observations were made in the case of non-fermented models of plant-based beverages. The number of LAB cells isolated from these samples on the first day of storage reached 11.72 log (cfu/mL) in coconut milk and 8.41 log (cfu/mL) in hemp milk. These values decreased to 3.12 log (cfu/mL) and 3.54 log (cfu/mL), respectively, after 21 days of cold storage (Table 6).

Bakirci and Kavaz [88] reported that the total counts of *Lactobacillus acidophilus*, *Bifidobacterium* ssp., and *Streptococcus thermophilus* decreased during cold storage of banana yoghurts, but remained at the required level (above 6 log (cfu/mL)) until day 14. In addition, a few other authors demonstrated that lactic acid bacteria (*L. debrueckii* spp. *bulgaricus* and *S. thermophilus*) survived well in yoghurt throughout its shelf life [79,89]. Olson and Aryana [90] showed that the number of *Lactobacillus acidophilus* strain cells decreased in natural yoghurt from 6.84 to 4.43 log (cfu/mL) over an 8-week storage period. Results of own study pointed to a higher survivability of potentially probiotic bacteria, compared to that reported by Mousavi et al. [91] for a probiotic juice from pomegranate. These authors observed reductions in counts of *Lactobacillus delbrueckii* and *Lactobacillus plantarum* by three logarithmic cycles after 14 days of cold storage. High survival rates of *Lactobacillus casei* under cold storage conditions were also demonstrated by Pereira et al. [92], who investigated fermentation and survivability of this probiotic in a juice from cashew apple.

The reduced count of *L. casei* subsp. *rhamnosus* during cold storage may be due to the production by these microorganisms of agents exhibiting anti-microbial activity like e.g., organic acid, bacteriocins, and hydrogen peroxide [93]. The concentration of hydrogen peroxide has an important impact, because *L. casei* subsp. *rhamnosus* do not produce the catalase enzyme. Accumulation of metabolism products during storage of non-dairy beverages may lead to the transition of *L. casei* subsp. *rhamnosus* at VBNC (viable but non-culturable). At this stage, the cells have an "unsatisfactory" physiological state, which means that they are alive but do not divide, and as a result do not have the ability to grow and reproduction [63]. Al-Otaibi [87] noticed that the number of bifidobacteria in eight commercial fermented dairy products decreased significantly since the day of manufacture till the end of cold storage (5 °C). This author reported also that the count of viable bacteria maintained at 10^6 cfu/mL till the end of the storage period in only two of the analyzed products. Reduction in the number of viable health-promoting bacteria can also be caused by decreased acidity, presence of post-production acid [94] sensitivity to oxygen [95] and metabolites, i.e., hydrogen peroxide and ethanol, and also to bacteriocins produced by lactic acid bacteria [96].

The determination of the number of active *L. casei* cells in the analyzed non-dairy beverages is difficult due to the presence of other microorganisms [97]. For this reason, simultaneous analyses were carried out for control samples (without probiotic). In their case, the total bacterial count was observed to decrease. In contrast, interesting seem to be changes in the viability of microorganisms (other than LABs) in the non-fermented control beverages, in which acidity approximating the neutral pH contributed to the development of undesirable microorganisms. However, the total bacterial count in the beverages not fortified with the probiotic monoculture, fitted within the range from 3.82 log (cfu/mL) to 3.93 log (cfu/mL) in fermented non-dairy beverages and from 5.92 to 6.67 log (cfu/mL) the non-fermented ones (Table 7). Thus, the criterion related to the microbiological quality did not exceed the maximum value of 7 log (cfu/mL).

Table 7. Changes in the total number of bacteria during 21-day storage (4 °C) of fermented and non-fermented non-dairy beverages.

Treatment	Storage Time			
	Day 1	Day 7	Day 14	Day 21
	Fermented			
C	6.72 ± 0.17 aA	6.23 ± 0.30 aA	5.54 ± 0.24 bA	3.82 ± 0.18 cB
H	6.25 ± 0.13 aB	5.69 ± 0.29 aA	4.37 ± 0.20 bB	3.93 ± 0.27 bB
	Non-Fermented			
C	4.39 ± 0.17 bC	4.41 ± 0.15 bB	4.79 ± 0.19 bB	5.67 ± 0.18 aA
H	3.81 ± 0.14 bD	3.86 ± 0.08 bB	5.76 ± 0.18 aA	5.92 ± 0.16 aA

C—Coconut beverage without probiotic; H—Hemp beverage without probiotic. Means in the rows, followed by different small letters (a–c) are significantly different at $p < 0.05$ (effect of storage). Means in the columns, followed by different big letters (A–D) are significantly different at $p < 0.05$ (effect of treatment).

Results of the present study indicate that cell viability was maintained at a satisfactory level throughout the storage period of fermented non-dairy beverages fortified with the probiotic monoculture. According to Pereira et al. [98], sugars, proteins, and lipids are only some of the factors that may affect the growth of probiotic bacteria and their survival rates in food products. Hence, the relative stability observed in content of nutrients in the plant-based beverages contributed indirectly to ensuring the therapeutic minimum of the analyzed products throughout their storage period.

4. Conclusions

The conducted study proved that the produced and modified plant-based beverages could serve as a food matrix for probiotic bacteria. The growth and survivability of probiotic bacteria in food products was determined by many factors including e.g., storage conditions, medium acidity, and sensitivity of oxygen and metabolites. The fermentation process contributed to the increased survival rates of *Lactobacillus casei* subsp. *rhamnosus* in both coconut and hemp milk. During 21-day storage of inoculated milk substitutes, the highest survivability of *Lactobacillus casei* subsp. *rhamnosus* was demonstrated in the fermented coconut milk (9.41 log (cfu/mL)). On day 21 of cold storage, the number of viable *Lactobacillus casei* subsp. *rhamnosus* cells in fermented coconut and hemp milk ensured meeting the therapeutic minimum (>6 log (cfu/mL)). Due to their nutrients composition and number of bacterial cells exhibiting a positive effect on a human body, the analyzed non-dairy beverages, offering an alternative to milk, represent a category of novel food products, and their manufacture will contribute to the sustainable development of food production and to the assurance of food safety.

Author Contributions: A.S. and S.K. conceived and designed the research. A.S., S.K., B.P. and E.C. performed the experiments. A.S., S.T., S.K., E.C., M.K., and B.P. prepared the materials. A.S., S.T., S.K., M.K. and P.F. analyzed the data. A.S., E.C. and S.K. wrote the paper. A.S., E.C., S.T., S.K., M.K. and P.F. revised the manuscript. All authors read and approved the final manuscript.

References

1. Dam, L.R.; Boafo, Y.A.; Degefa, S.; Gasparatos, A.; Saito, O. Assessing the food security outcomes of industrial crop expansion in smallholder settings: Insights from cotton production in Northern Ghana and sugarcane production in Central Ethiopia. *Sustain. Sci.* **2017**, 677–693. [CrossRef]
2. Szparaga, A.; Kocira, S. Generalized logistic functions in modelling emergence of *Brassica napus* L. *PLoS ONE* **2018**, *13*, e0201980. [CrossRef] [PubMed]
3. Alexandratos, N.; Bruinsma, J. World Agriculture: Towards 2015/2030: The 2012 Revision. Available online: http://www.fao.org/3/a-ap106e.pdf (accessed on 24 July 2019).
4. FAO. Food security information for action programme. In *Food Security Concepts and Frameworks*; FAO: Rome, Italy, 2008.
5. Blay-Palmer, A.; Sonnino, R.; Custot, J. A food politics of the possible? Growing sustainable food systems through networks of knowledge. *Agric. Hum. Values* **2016**, *33*, 27–43. [CrossRef]
6. Szparaga, A.; Stachnik, M.; Czerwińska, E.; Kocira, S.; Dymkowska-Malesa, M.; Jakubowski, M. Multi-objective optimization of osmotic dehydration of plums in sucrose solution and its storage using utopian solution methodology. *J. Food Eng.* **2019**, *245*, 104–111. [CrossRef]
7. Saluk-Juszczak, J.; Kołodziejczyk, J.; Babicz, K. Functional food—A role of nutraceuticals in cardiovascular disease prevention. *Kosmos* **2010**, *59*, 527–538.
8. Martirosyan, D.; Singh, J. A new definition of functional food by FFC: What makes a new definition unique? *Funct. Food Health Dis. J.* **2015**, *5*, 209–223. [CrossRef]
9. Dymkowska-Malesa, M.; Szparaga, A.; Czerwińska, E. Evaluation of polychlorinated biphenyls content in chosen vegetablesfrom Warmia and Mazury region. *Rocz. Ochr. Srodowiska* **2014**, *16*, 290–299.
10. Mullin, G.; Delzenne, N.M. Functional foods and dietary supplements in 2017: Food for thought. *Curr. Opin. Clin. Nutr. Metab. Care* **2017**, *20*, 453–455. [CrossRef]
11. Douglas, L.C.; Sanders, M.E. Probiotics and prebiotics in dietetics practice. *J. Am. Diet. Assoc.* **2008**, *108*, 510–521. [CrossRef]

12. Kapka-Skrzypczak, L.; Niedźwiecka, J.; Wojtyła, A.; Kruszewski, M. Probiotics and prebiotics as a bioactive component of functional food. *Pediatr. Endocrinol. Diabetes Metab.* **2012**, *18*, 79–83.
13. Sethi, S.; Tyagi, S.K.; Anurag, R.K. Plant-based milk alternatives an emerging segment of functional beverages: A review. *J. Food Sci. Technol.* **2016**, *53*, 3408–3423. [CrossRef] [PubMed]
14. Kandylis, P.; Pissaridi, K.; Bekatorou, A.; Kanellaki, M.; Koutinas, A. Dairy and non-dairy probiotic beverages. *Curr. Opin. Food Sci.* **2016**, *7*, 58–63. [CrossRef]
15. Hoffman, M.; Kostyra, E. Sensory quality and nutritional value of vegan substitutes of milk. *PTPS* **2015**, *1*, 52–57.
16. Zielińska, D. *Lactobacillus* Strain Survival Study in Fermeted Soy Beverage. *Zywnosc-Nauka Technol. Jakosc* **2006**, *49*, 121–128.
17. Quasem, J.M.; Mazahreh, A.S.; Abu-Alruz, K. Development of Vegetable Based Milk from Decorticated Sesame (Sesamum Indicum). *Am. J. Appl. Sci.* **2009**, *6*, 888–896. [CrossRef]
18. Yeung, P.S.M.; Sanders, M.E.; Kitts, C.L.; Cano, R.; Tong, P.S. Species specific identification on commercial probiotic strains. *J. Dairy Sci.* **2002**, *8*, 1039–1051. [CrossRef]
19. Hadadji, M.; Bensoltane, A. Growth and lactic acid production by Bifidobacterium longum and Lactobacillus acidophilus in goat's milk. *Afr. J. Biotechnol.* **2006**, *5*, 505–509.
20. Yuliana, N.; Rangga, A. Rakhmiati Manufacture of fermented coco milk-drink containing lactic acid bacteria cultures. *Afr. J. Food Sci.* **2010**, *4*, 558–562.
21. Bartkiene, E.; Zokaityte, E.; Lele, V.; Sakiene, V.; Zavistanaviciute, P.; Klupsaite, D.; Bendoraitiene, J.; Navikaite-Snipaitiene, V.; Ruzauskas, M. Technology and characterisation of whole hemp seed beverages prepared from ultrasonicated and fermented whole seed paste. *Int. J. Food Sci. Technol.* **2019**, 1–14. [CrossRef]
22. Wang, Q.; Jiang, J.; Xiong, Y.L. High pressure homogenization combined with pH shift treatment: A process to produce physically and oxidatively stable hemp milk. *Food Res. Int.* **2018**, *106*, 487–494. [CrossRef] [PubMed]
23. Scholz-Ahrens, K.E.; Ahrens, F.; Barth, C.A. Nutritional and health attributes of milk and milk imitations. *Eur. J. Nutr.* **2019**, 1–16. [CrossRef] [PubMed]
24. Kaniewski, R.; Pniewska, I.; Kubacki, A.; Strzelczyk, M.; Chudy, M.; Oleszak, G. Konopie siewne (*Cannabis sativa* L.)-cenna roślina przydatna i lecznicza. *Postępy Fitoter.* **2017**, *2*, 139–144. [CrossRef]
25. Mojka, K. Probiotics, prebiotics and synbiotics–characteristics and functions. *Probl. Hig. Epidemiol.* **2014**, *95*, 541–549.
26. Zaręba, D.; Ziarno, M.; Obiedziński, M. Viability of yoghurt bacteria and probiotic strains in models of fermented and non-fermented milk. *Med. Weter.* **2008**, *64*, 1007–1011.
27. Jeske, S.; Zannini, E.; Arendt, E.K. Past, present and future: The strength of plant-based dairy substitutes based on gluten-free raw materials. *Food Res. Int.* **2018**, *110*, 42–51. [CrossRef] [PubMed]
28. Akin, Z.; Ozcan, T. Functional properties of fermented milk produced with plant proteins. *LWT* **2017**, *86*, 25–30. [CrossRef]
29. Mårtensson, O.; Öste, R.; Holst, O. Lactic Acid Bacteria in an Oat-based Non-dairy Milk Substitute: Fermentation Characteristics and Exopolysaccharide Formation. *LWT-Food Sci. Technol.* **2000**, *33*, 525–530. [CrossRef]
30. Cruz, N.; Capellas, M.; Jaramillo, D.; Trujillo, A.-J.; Guamis, B.; Ferragut, V. Soymilk treated by ultra high-pressure homogenization: Acid coagulation properties and characteristics of a soy-yogurt product. *Food Hydrocoll.* **2009**, *23*, 490–496. [CrossRef]
31. Chaiwanon, P.; Puwastien, P.; Nitithamyong, A.; Sirichakwal, P.P. Calcium Fortification in Soybean Milk and In Vitro Bioavailability. *J. Food Compos. Anal.* **2000**, *13*, 319–327. [CrossRef]
32. Patel, H.; Pandiella, S.; Wang, R.; Webb, C. Influence of malt, wheat, and barley extracts on the bile tolerance of selected strains of lactobacilli. *Food Microbiol.* **2004**, *21*, 83–89. [CrossRef]
33. Vera-Pingitore, E.; Jimenez, M.E.; DallAgnol, A.; Belfiore, C.; Fontana, C.; Fontana, P.; Von Wright, A.; Vignolo, G.; Plumed-Ferrer, C. Screening and characterization of potential probiotic and starter bacteria for plant fermentations. *LWT* **2016**, *71*, 288–294. [CrossRef]
34. Navarro, R.R.; Faronilo, K.M.L.; Eom, S.H.; Jeon, K.H. Isolation, characterization, and identification of probiotic lactic acid bacterium from sabeng, a Philippine fermented drink. *J. ISSAAS* **2018**, *24*, 127–136.
35. Blasco, M. *Milk Power. Mleko roślinne 80 przepisów*; Wydawnictwo Burda: Warszawa, Poland, 2014.
36. Szakuła, M. *Konopie w kuchni*; Smiling Spoon: Warszawa, Poland, 2015.

37. Zielińska, D.; Kołożyn-Krajewska, D.; Goryl, A. Survival models of potentially probiotic *Lactobacillus casei* KN291 bacteria in a fermented soy beverage. *Zywnosc-Nauka Technol. Jakosc* **2008**, *5*, 126–134.
38. Zielińska, D.; Uzarowicz, U. Development of ripening and storage conditions of probiotic soy beverage. *Zywnosc-Nauka Technol. Jakosc* **2007**, *5*, 186–193.
39. Zielińska, D. Selecting Suitable Bacterial Strains of Lactobacillus and Identifying Soya Drink Fermentation Conditions. *Zywnosc-Nauka Technol. Jakosc* **2005**, *2*, 289–297.
40. AOAC Official Method. *18 Fat Content of Raw and Pasteurized Whole Milk Gerber Method by Weight (Method I) First Action*; AOAC International: Rockville, MD, USA, 2000.
41. AOAC. *Official Methods of Analysis of AOAC International*, 17th ed.; Horwitz, W., Ed.; AOAC International: Gaithersburg, MD, USA, 2000.
42. AOAC. *Official Method of Analysis of the Association of Official Analytical Chemists*; AOAC International: Arlington, VA, USA, 1997.
43. Miller, G.L. Use of Dinitrosalicylic Acid Reagent for Determination of Reducing Sugar. *Anal. Chem.* **1959**, *31*, 426–428. [CrossRef]
44. Czerwińska, E.; Piotrowski, W. Potential Sources of Milk Contamination Influencing its Quality for Consumption. *Rocz. Ochr. Srodowiska* **2011**, *13*, 635–652.
45. Gustaw, W.; Kozioł, J.; Waśko, A.; Skrzypczak, K.; Michalak-Majewska, M.; Nastaj, M. Physicochemical Properties and Survival of Lactobacillus casei in Fermented Milk Beverages Produced With Addition of Selected Milk Protein Preparations. *Zywnosc-Nauka Technol. Jakosc* **2015**, *6*, 129–139. [CrossRef]
46. De Man, J.C.; Rogosa, M.; Sharpe, M.E. A MEDIUM FOR THE CULTIVATION OF LACTOBACILLI. *J. Appl. Bacteriol.* **1960**, *23*, 130–135. [CrossRef]
47. Champagnem, C.P.; Rastall, R.A. Some technological challenges in the addition of probiotic bacteria to foods. In *Prebiotics and Probiotics Science and Technology*, 2nd ed.; Charalampopoulos, D., Rastall, R.A., Eds.; Springer: London, UK, 2009; pp. 763–806.
48. Yamaguishi, C.T.; Spier, M.R.; Lindner, J.D.D.; Soccol, V.T.; Soccol, C.R. Current Market Trends and Future Directions. In *Probiotics*; Springer: Berlin, Germany, 2011; Volume 21, pp. 299–319. [CrossRef]
49. Szydłowska, A.; Kołożyn-Krajewska, D. Applying potencially probiotic bacterial strains to pumpkin pulp fermentation. *Zywnosc-Nauka Technol. Jakosc* **2010**, *73*, 109–119. [CrossRef]
50. Sobczyk, M.; Ziarno, M. Characteristics of Microbial Quality and Nutrient Content (Protein, Fat, Mineral Compounds) Some Petals Muesli. *ZPPNR* **2013**, *575*, 119–129.
51. Jessa, J.; Hozyasz, K.K. Health value of coconut products. *Pediatr. Pol.* **2015**, *90*, 415–423. [CrossRef]
52. Aizpurua-Olaizola, O.; Omar, J.; Navarro, P.; Olivares, M.; Etxebarria, N.; Usobiaga, A. Identification and quantification of cannabinoids in *Cannabis sativa* L. plants by high performance liquid chromatography-mass spectrometry. *Anal. Bioanal. Chem.* **2014**, *406*, 7549–7560. [CrossRef] [PubMed]
53. Belewu, M.A.; Belewu, K.Y. Comparative Physico-Chemical Evulation of Tiger-nut, Soybean and Coconut Milk Sources. *Int. J. Agric. Biol.* **2007**, *9*, 785–787.
54. Yeo, S.-K.; Liong, M.-T.; Yeo, S.; Liong, M. Effect of prebiotics on viability and growth characteristics of probiotics in soymilk. *J. Sci. Food Agric.* **2010**, *90*, 267–275. [CrossRef] [PubMed]
55. Jurkowski, M.; Błaszczyk, M. Physiology and Biochemistry of Lactic Acid Bacteria. *Kosmos* **2012**, *61*, 493–504.
56. Nogala-Kałucka, M. *Analiza żywności. Wybrane metody oznaczeń jakościowych i ilościowych składników żywności. Wyd*; Uniwersytetu Przyrodniczego w Poznaniu: Poznań, Poland, 2016.
57. Dłużewska, E.; Nizler, M.; Maszewska, M. Technological Aspects of Obtaining Soy Beverages. *Acta Sci. Technol. Aliment* **2004**, *3*, 93–102.
58. Rozporządzenie Komisji (WE) 2073/2005 i 2019/229. Available online: https://eur-lex.europa.eu/legal-content/PL/TXT/PDF/?uri=CELEX:32019R0229&from=ES (accessed on 26 August 2019).
59. Lee, S.-Y.; Dougherty, R.H.; Kang, D.-H. Inhibitory Effects of High Pressure and Heat on Alicyclobacillus acidoterrestris Spores in Apple Juice. *Appl. Environ. Microbiol.* **2002**, *68*, 4158–4161. [CrossRef]
60. Ouwehand, A.C.; Salminen, S.J. The Health Effects of Cultured Milk Products with Viable and Non-viable Bacteria. *Int. Dairy J.* **1998**, *8*, 749–758. [CrossRef]
61. Shah, N.P. Functional foods from probiotics and prebiotics. *Food Technol.* **2001**, *55*, 46–53.
62. Gupta, S.; Abu-Ghannam, N. Probiotic Fermentation of Plant Based Products: Possibilities and Opportunities. *Crit. Rev. Food Sci. Nutr.* **2012**, *52*, 183–199. [CrossRef] [PubMed]

63. Olszewska, M.; Łaniewska-Trokenheim, Ł. Study of the nonculturable state of lactic acid bacteria cells under adverse growth conditions. *Med. Weter.* **2011**, *67*, 105–109.
64. Warmińska-Radyko, I.; Olszewska, M.; Mikś-Krajnik, M. Effect of temperature and sodium chloride on the growth and metabolism of Lactococcus strains in long-term incubation of milk. *Milchwissenschaft* **2010**, *65*, 32–35.
65. Joux, F.; LeBaron, P. Use of fluorescent probes to assess physiological functions of bacteria at single-cell level. *Microbes Infect.* **2000**, *2*, 1523–1535. [CrossRef]
66. Lahtinen, S.J.; Gueimonde, M.; Ouwehand, A.C.; Reinikainen, J.P.; Salminen, S.J. Probiotic Bacteria May Become Dormant during Storage. *Appl. Environ. Microbiol.* **2005**, *71*, 1662–1663. [CrossRef] [PubMed]
67. Zaręba, D. Fatty Acid Profile of Soya Milk Fermented by Various Bacteria Strains of Lactic Acid Fermentation. *Zywnosc-Nauka Technol. Jakosc* **2009**, *67*, 59–71.
68. Czaczyk, K.; Marciniak, A.; Białas, W.; Mueller, A.; Myszka, K. The Effect of Environmental Factors Influencing Lipopeptide Biosurfactants Biosynthesis by *Bacillus* Spp. *Zywnosc-Nauka Technol. Jakosc* **2007**, *50*, 140–149.
69. Huy, D.N.A.; Haom, P.A.; Hungm, P.V. Screening and identification of *Bacillus* sp. Isolated from traditional Vietnamese soybean-fermented products for high fibrinolytic enzyme production. *Int. Food Res. J.* **2016**, *23*, 326–331.
70. Food and Environmental Hygiene Department FEHD. Available online: https://www.cfs.gov.hk/english/whatsnew/whatsnew_act/files/MBGL_RTE%20food_e.pdf (accessed on 2 September 2019).
71. Wee, Y.J.; Yun, J.S.; Kim, D.; Ryu, H.W. Batch and repeated batch production of L(+)-lactic acid by Enterococcus faecalis RKY1 using wood hydrolyzate and corn steep liquor. *J. Ind. Microbiol. Biotechnol.* **2006**, *33*, 431–435. [CrossRef]
72. Libudzisz, Z.; Kowal, K.; Żakowska, Z. *Mikrobiologia Techniczna*; Wydawnictwo Naukowe PWN: Warszawa, Polnd, 2007.
73. Gajewska, J.; Błaszczyk, M.K. Probiotyczne bakterie fermentacji mlekowej (LAB). *Postępy Mikrobiol.* **2012**, *51*, 55–65.
74. Makarova, K.; Slesarev, A.; Wolf, Y.; Sorokin, A.; Mirkin, B.; Koonin, E.; Pavlov, A.; Pavlova, N.; Karamychev, V.; Polouchine, N.; et al. Comparative genomics of the lactic acid bacteria. *Proc. Natl. Acad. Sci.* **2006**, *103*, 15611–15616. [CrossRef] [PubMed]
75. Gupta, S.; Cox, S.; Abu-Ghannam, N. Process optimization for the development of a functional beverage based on lactic acid fermentation of oats. *Biochem. Eng. J.* **2010**, *52*, 199–204. [CrossRef]
76. Paseephol, T.; Sherkat, F. Probiotic stability of yoghurts containing Jerusalem artichoke inulins during refrigerated storage. *J. Funct. Foods* **2009**, *1*, 311–318. [CrossRef]
77. Colakoglu, H.; Gursoy, O. Effect of lactic adjunct cultures on conjugated linoleic acid (CLA) concentration of yoghurt drink. *J. Food Agric. Environ.* **2011**, *9*, 60–64.
78. Guo, Z.; Wang, J.; Yan, L.; Chen, W.; Liu, X.-M.; Zhang, H.-P. In vitro comparison of probiotic properties of *Lactobacillus casei* Zhang, a potential new probiotic, with selected probiotic strains. *LWT* **2009**, *42*, 1640–1646. [CrossRef]
79. Fenderya, S.; Akbulut, N.; Akalın, A.S. Viability and activity of bifidobacteria in yoghurt containing fructooligosaccharide during refrigerated storage. *Int. J. Food Sci. Technol.* **2004**, *39*, 613–621. [CrossRef]
80. Bonczar, G. The effects of certain factors on the properties of yoghurt made from ewe's milk. *Food Chem.* **2002**, *79*, 85–91. [CrossRef]
81. Bernat, N.; Cháfer, M.; Chiralt, A.; González-Martínez, C. Vegetable milks and their fermented derivative products. *Int. J. Food Stud.* **2014**, *3*, 93–124. [CrossRef]
82. Bernat, N.; Cháfer, M.; González-Martínez, C.; Rodríguez-García, J.; Chiralt, A. Optimisation of oat milk formulation to obtain fermented derivatives by using probiotic *Lactobacillus reuteri* microorganisms. *Food Sci. USA Technol. Int.* **2015**, *21*, 145–157. [CrossRef]
83. Blandino, A.; Al-Aseeri, M.; Pandiella, S.; Cantero, D.; Webb, C. Cereal-based fermented foods and beverages. *Food Res. Int.* **2003**, *36*, 527–543. [CrossRef]
84. Gomes, A.M.; Malcata, F.; Gomes, A.M.; Malcata, F. Bifidobacterium spp. and Lactobacillus acidophilus: Biological, biochemical, technological and therapeutical properties relevant for use as probiotics. *Trends Food Sci. Technol.* **1999**, *10*, 139–157. [CrossRef]
85. Rivera-Espinoza, Y.; Gallardo-Navarro, Y. Non-dairy probiotic products. *Food Microbiol.* **2010**, *27*, 1–11. [CrossRef] [PubMed]

86. Skrzyplonek, K.; Jasińska, M. Quality of Fermented Probiotic Beverages Made From Frozen Acid Whey and Milk During Refrigerated Storage. *Zywnosc-Nauka Technol. Jakosc* **2016**, *104*, 32–44. [CrossRef]
87. At-Otaibi, M.M. Evaluation of some probiotic fermented milk products from Al-Ahsa markets, Saudi Arabia. *J. Food Technol.* **2009**, *4*, 1–8. [CrossRef]
88. Bakirci, I.; Kavaz, A. An investigation of some properties of banana yogurts made with commercial ABT-2 starter culture during storage. *Int. J. Dairy Technol.* **2008**, *61*, 270–276. [CrossRef]
89. Pescuma, M.; Hébert, E.M.; Mozzi, F.; De Valdez, G.F. Functional fermented whey-based beverage using lactic acid bacteria. *Int. J. Food Microbiol.* **2010**, *141*, 73–81. [CrossRef] [PubMed]
90. Olson, D.; Aryana, K. An excessively high Lactobacillus acidophilus inoculation level in yogurt lowers product quality during storage. *LWT* **2008**, *41*, 911–918. [CrossRef]
91. Mousavi, Z.E.; Mousavi, S.M.; Razavi, S.H.; Emam-Djomeh, Z.; Kiani, H. Fermentation of pomegranate juice by probiotic lactic acid bacteria. *World J. Microbiol. Biotechnol.* **2011**, *27*, 123–128. [CrossRef]
92. Pereira, A.L.F.; Maciel, T.C.; Rodrigues, S. Probiotic beverage from cashew apple juice fermented with Lactobacillus casei. *Food Res. Int.* **2011**, *44*, 1276–1283. [CrossRef]
93. Salas, M.L.; Mounier, J.; Valence, F.; Coton, M.; Thierry, A.; Coton, E. Antifungal Microbial Agents for Food Biopreservation—A Review. *Microorganisms* **2017**, *5*, 37. [CrossRef]
94. Wang, Y.-C.; Yu, R.-C.; Chou, C.-C. Growth and survival of bifidobacteria and lactic acid bacteria during the fermentation and storage of cultured soymilk drinks. *Food Microbiol.* **2002**, *19*, 501–508. [CrossRef]
95. Frank, J.F.; Marth, E.M. Fermentations. In *Fundamentals of Dairy Chemistry*; Wong, N.P., Jenness, R., Keeney, M., Marth, E.H., Eds.; Springer: Boston, MA, USA, 1988; pp. 655–738. [CrossRef]
96. Medina, L.M.; Jordano, R. Survival of constitutive microflora in commercially fermented milk containing *Bifidobacteria* during refrigerated storage. *J. Food Prot.* **1994**, *56*, 731–733. [CrossRef] [PubMed]
97. Kosikowska, M.; Jakubczyk, E. Metody oznaczania bakterii probiotycznych w produktach mlecznych. *Przeg. Mlecz.* **2007**, *57*, 12–17.
98. Pereira, A.L.F.; Almeida, F.D.L.; de Jesus, A.L.T.; da Costa, J.M.C.; Rodrigues, S. Storage Stability and Acceptance of Probiotic Beverage from Cashew Apple Juice. *Food Bioprocess Technol.* **2013**, *6*, 3155–3165. [CrossRef]

8

Numerical Modeling of the Shape of Agricultural Products on the Example of Cucumber Fruits

Andrzej Anders *, Dariusz Choszcz, Piotr Markowski, Adam Józef Lipiński, Zdzisław Kaliniewicz and Elwira Ślesicka

Department of Heavy Duty Machines and Research Methodology, University of Warmia and Mazury in Olsztyn, Olsztyn 10-957, Poland; choszczd@uwm.edu.pl (D.C.); pitermar@uwm.edu.pl (P.M.); adam.lipinski@uwm.edu.pl (A.J.L.); arne@uwm.edu.pl (Z.K.); elwira.slesicka@uwm.edu.pl (E.Ś.)

* Correspondence: andrzej.anders@uwm.edu.pl

Abstract: The aim of the study was to build numerical models of cucumbers cv. *Śremski* with the use of a 3D scanner and to analyze selected geometric parameters of cucumber fruits based on the developed models. The basic dimensions of cucumber fruits–length, width and thickness—were measured with an electronic caliper with an accuracy of $d = 0.01$ mm, and the surface area and volume of fruits were determined by 3D scanning. Cucumber fruits were scanned with an accuracy of $d = 0.13$ mm. Six models approximating the shape of cucumber fruits were developed with the use of six geometric figures and their combinations to calculate the surface area and volume of the analyzed agricultural products were identified. The surface area and volume of cucumber fruits calculated by 3D scanning and mathematical formulas were compared. The surface area calculated with the model combining two truncated cones and two hemispheres with different diameters, joined base-to-base, was characterized by the smallest relative error of 3%. Fruit volume should be determined with the use of mathematical formulas derived for a model composed of an ellipsoid and a spheroid. The proposed geometric models can be used in research and design.

Keywords: 3D scanner; geometric model; reverse engineering; fruit; cucumber

---

1. Introduction

Advanced measurement techniques and software supporting complex simulations of selected technological processes are required to introduce new products and technologies on the market and to improve product quality. Models of agricultural products should account for the designed technological processes and should accurately reflect the products' shape [1]. A 3D model that accurately describes a product's geometric and physical parameters can be used in the design process. A traditional approach to modeling relies on the assumption that agri-food products are homogeneous and isotropic, and the modeled objects are assigned regular shapes (e.g., cylinder, sphere, cone, etc.) Computer-Aided Design (CAD) and Computational Fluid Dynamics (CFD) software can be applied to simulate complex processes that occur during the processing of agri-food products [2]. The development of a model that closely approximates the shape of the original agricultural product and can be used in computer simulations poses the key challenge in the research and design of food processing equipment. Numerical modeling based on traditional methods is a laborious and difficult task, in particular when the studied objects have irregular shape [3]. In the process of measuring fruits and seeds, many researchers rely solely on image analysis tools and measuring devices such as calipers and micrometers [4,5]. In the literature, traditional methods have been used to determine the geometric parameters of soybeans (*Glycine max* L. Merr.) [6], sunflower seeds (*Helianthus annuus* L.) [7], oilseed rape seeds (*Brassica napus* L.) [8,9], mustard seeds (*Sinapis alba*) [10] and flax seeds (*Linnum usitatissimum* L.) [11]. In small objects such as seeds, only basic dimensions can be measured with a caliper or a micrometer. In larger products such as fruits and vegetables, the analyzed parameters can be measured with a caliper or a micrometer at any point on the object's surface.

In the literature, traditional and advanced measuring techniques have been deployed to accurately render the shape of the analyzed products. Erdogdu et al. [12] relied on a machine vision system designed by Luzuriaga et al. [13] to determine the geometric parameters of shrimp cross-sections and to develop mathematical models of the thermal processing of shrimp. Crocombe et al. [14] analyzed the surface of meat pieces by laser scanning to develop a numerical model and simulate meat refrigeration time. Jancsok et al. [15] used a machine vision system to build numerical models of pears cv. Konferencja. Borsa et al. [16] performed computed tomography scans and calculated the radiation dose absorbed by the examined food products. Sabliov et al. [17] proposed an image analysis method for measuring the volume and surface area of axially symmetric agricultural products. Zapotoczny [18] developed a test stand for measuring the geometric parameters of cucumber fruits with the use of digital image analysis. The cited author registered changes in the shape and size of greenhouse-grown cucumbers during storage. Scheerlinck et al. [19] relied on a machine vision system to develop a 3D model of strawberries and a thermal system for disinfecting fruit surfaces. Du and Sun [20] and Zheng et al. [21] developed an image analysis technique for measuring the surface area and volume of beef loin and beef joints. Kim et al. [22] generated 3D geometric models of food products with a complex shape with the use of computed tomography. Goni et al. [23] modeled the geometric properties of the studied objects with the involvement of magnetic resonance imaging. Siripon et al. [24] analyzed chicken half-carcasses with a 3D scanner (Atos, GOM, Germany) and used the results to simulate cooking processes. Mieszkalski [25,26] developed computer models of carrots, apples cv. Jonagored and chicken eggs. The shape of biological objects was described with Bézier curves. The resulting mathematical models were used to generate 3D figures that accurately rendered the shape and basic dimensions of the studied products. Balcerzak et al. [27] modeled the geometric parameters of corn and oat kernels in the 3ds Max environment. Images of kernel cross-sections were used to acquire geometric data, generate meshes and determine nodal coordinates. Ho Q. T. and others used multiscale modeling in food engineering. Multiscale models support evaluations of the phenomena occurring inside agricultural raw materials on a micro and macro scale. The authors relied on X-ray tomography to generate multiscale models [28]. The volume of agricultural raw materials can also be determined by water displacement. However, this method cannot be applied to materials that easily absorb water [29].

The majority of methods require complex and expensive measuring devices and software. A thorough knowledge of various imaging techniques is required to model irregularly shaped objects. Models that accurately render the shape of the analyzed products can be developed with the use of a 3D scanner. This technique is considerably simpler, but it is not yet widely used. 3D models can be used to analyze the shape of whole products or their fragments [30,31].

The dimensions and basic geographic parameters of agricultural materials have been long determined with the use of simple measuring devices, including analog and digital calipers, micrometers and dial indicators. The main limitation of conventional measuring techniques is that they investigate only characteristic points in the examined objects, and the measured values can be used to calculate selected parameters, such as surface area and volume, with mathematical formulas [29]. In contrast, indirect methods rely on the acquisition of images of the investigated object and digital image analysis. The advances made in digital technology and computing power have contributed to the widespread popularity of indirect measuring methods. Indirect measurements produce linear dimensions as well as images of the analyzed surfaces. The main advantage of indirect methods is that measurements are rapid, whereas the main limitation stems from the fact that measurements are performed along the contours of the acquired image, which are projected onto a plane [9]. A relatively new method has been proposed for registering the shape of a sample as a cloud of points. The location of every point in the modeled sample is determined with the use of 3D scanners, which register the position of the laser beam, a structured light source. The points registered by a 3D scanner support the development of a numerical model, which can be used in metrological analyses. The development of a numerical model with the described method is time-consuming, but the results can be stored in computer memory [32,33].

The presented methods for measuring the geometric properties of objects produce highly similar results, provided that the required precision thresholds are met. However, the time and conditions of measurement can vary. Approximation formulas are widely applied to calculate volume and area. The main problem is the selection of the optimal model for determining the above parameters with the required accuracy. The aim of this study was to compare selected geometric parameters of cucumber fruits acquired from 3D models and models based on basic geometric figures and direct caliper measurements.

2. Materials and Methods

The experiment was performed on cucumber fruits cv. *Śremski* stored indoors at a constant temperature of 18.1 °C and 60% humidity. Cucumbers were purchased from the Pozorty Production and Experimental Station in Olsztyn. Fifty whole cucumber fruits without visible signs of damage were randomly selected for the experiment. Cucumbers were purchased on five occasions in the second half of August 2018, and 10 cucumbers were purchased each time. The length, width and thickness of cucumber fruits were measured with an electronic caliper with an accuracy of d = 0.01 mm. Each fruit was additionally measured with an electronic caliper at the points presented in Figure 1. Cucumbers were scanned with the Nextengine 3D scanner with a resolution of 15 points per mm^2. Scanning precision was 0.13 mm. Cucumbers were mounted on a turntable. Individual images were combined in the ScanStudio HD PRO program [34]. The developed numerical models were used to determine the surface area and volume of cucumber fruits. The above parameters were measured in the MeshLab program [35].

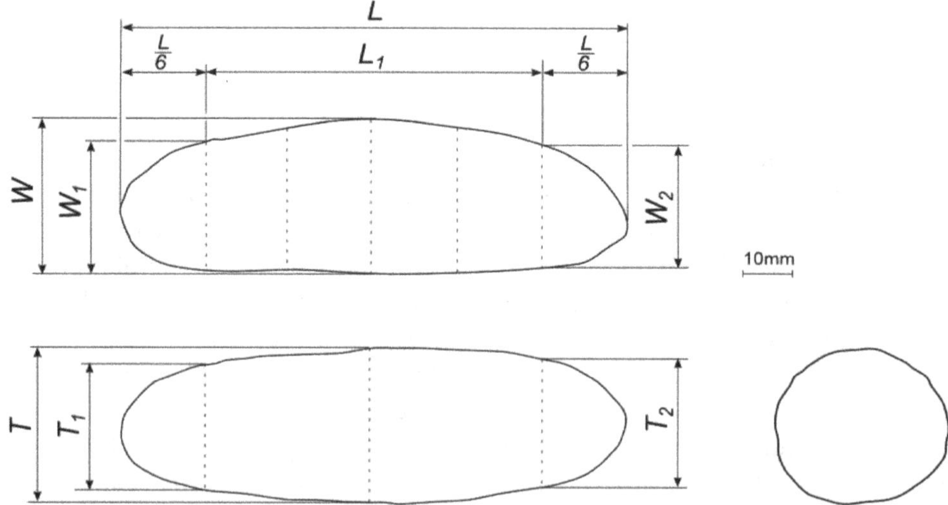

Figure 1. Shape of a selected cucumber fruit: L—length (mm), W—width (mm), T—thickness (mm), L_1—length of the middle section (mm), W_1, W_2—width of the terminal section (mm), T_1, T_2—thickness of the terminal section (mm).

The measured dimensions were used to build six geometric models whose shape resembled the shape of cucumber fruits. The surface area and volume of fruits were calculated from the developed models. Geometric models were built based on basic geometric figures, including an ellipsoid, cylinder, hemisphere, truncated cone and a combination of selected figures. The analyzed geometric models are presented in Figure 2.

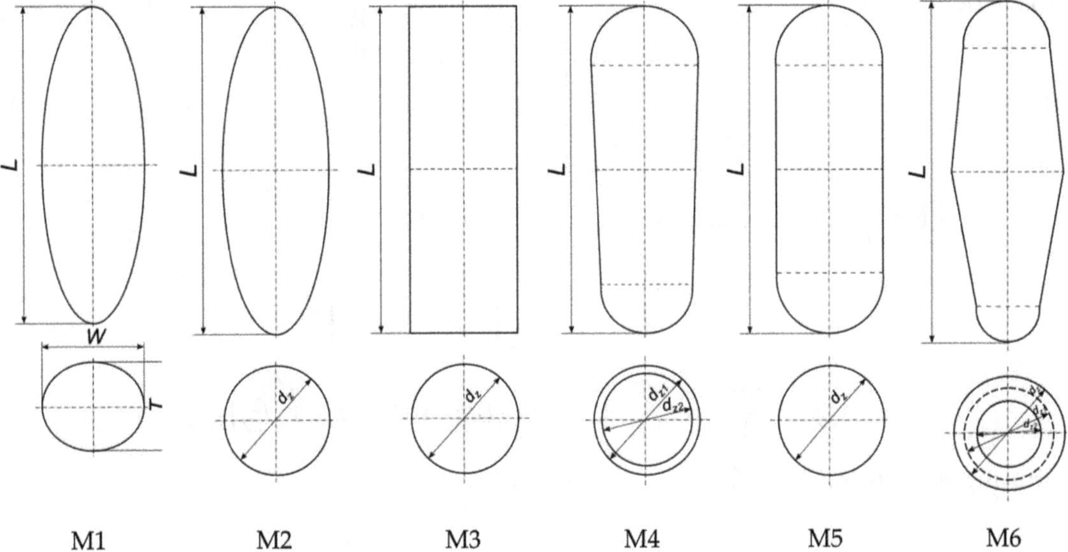

Figure 2. Models of cucumber fruits: M1—ellipsoid, M2—spheroid, M3—cylinder, M4—truncated cone and two hemispheres, M5—cylinder and two hemispheres, M6—two truncated cones and two hemispheres.

Mathematical formulas were derived for every geometric model and were used to calculate the surface area and volume of cucumbers [36,37]:

ellipsoid model (M1):

$$A_{M1} = 2 \cdot \pi \cdot \left(\left(\frac{L}{2}\right)^2 + \frac{\frac{T}{2} \cdot \left(\frac{L}{2}\right)^2}{\sqrt{\left(\frac{W}{2}\right)^2 - \left(\frac{L}{2}\right)^2}} \cdot F(\Theta, m) + \frac{T}{2} \cdot \sqrt{\left(\frac{W}{2}\right)^2 - \left(\frac{L}{2}\right)^2} \cdot E(\Theta, m) \right) \tag{1}$$

where:

$$m = \frac{\left(\frac{L}{2}\right)^2 \cdot \left(\left(\frac{T}{2}\right)^2 - \left(\frac{L}{2}\right)^2\right)}{\left(\frac{T}{2}\right)^2 \cdot \left(\left(\frac{W}{2}\right)^2 - \left(\frac{L}{2}\right)^2\right)} = \frac{L^2 \cdot T^2 - L^4}{T^2 \cdot W^2 - L^2 \cdot T^2} \tag{2}$$

$$\Theta = \arcsin\left(\sqrt{\frac{\sqrt{W^2 - L^2}}{|W|}} \right) \tag{3}$$

and where $F(\Theta,m)$ and $E(\Theta,m)$ are incomplete elliptic integrals of the first and second kind [37].

$$V_{M1} = \frac{\pi \cdot T \cdot W \cdot L}{6} \tag{4}$$

spheroid model (M2), when: $\frac{L}{2} > \frac{d_z}{2}$, then:

$$A_{M2} = 2 \cdot \pi \cdot \left(\frac{d_z}{2}\right)^2 \cdot \left(1 + \frac{\frac{L}{2}}{\frac{d_z}{2} \cdot e} \cdot \arcsin(e)\right) = \frac{4 \cdot \pi \cdot d_z^2 + \pi \cdot L \cdot d_z \cdot e \cdot \arcsin(e)}{8} \tag{5}$$

where:

$$e = \sqrt{1 - \frac{d_z^2 \cdot L^2}{16}} \tag{6}$$

$$V_{M2} = \frac{\pi \cdot d_z^2 \cdot L}{6} \tag{7}$$

cylinder model (M3):

$$A_{M3} = \pi \cdot d_z \cdot L + 2 \cdot \pi \cdot \left(\frac{d_z}{2}\right)^2 \tag{8}$$

$$V_{M3} = \frac{\pi \cdot d_z^2 \cdot L}{4} \tag{9}$$

model combining a truncated cone and two hemispheres (M4)

$$A_{M4} = \frac{\pi}{2} \cdot \left(d_{z1}^2 + d_{z2}^2\right) + \pi \cdot \sqrt{\left(\frac{d_{z1}}{2}\right)^2 + L_1^2} \cdot \left(\frac{d_{z1}}{2} + \frac{d_{z2}}{2}\right) \tag{10}$$

$$V_{M4} = \frac{\pi}{12} \cdot \left(d_{z1}^3 + d_{z2}^3 + L_1 \cdot \left(d_{z1}^2 + d_{z1} \cdot d_{z2} + d_{z2}^2\right)\right) \tag{11}$$

model combining a cylinder and two hemispheres (M5)

$$A_{M5} = \pi \cdot d_w \cdot \left(\frac{d_w}{2} + \frac{d_w}{2} + L_1\right) \tag{12}$$

$$V_{M5} = \pi \cdot d_w^2 \cdot \left(\frac{d_w}{6} + \frac{L_1}{4}\right) \tag{13}$$

model combining two truncated cones and two hemispheres (M6)

$$A_{M6} = \frac{(\pi \cdot d_{z1} + \pi \cdot d_z) \cdot \sqrt{d_{z1}^2 + L_1^2} + 2 \cdot \pi \cdot d_{z1}^2 + (\pi \cdot d_{z2} + \pi \cdot d_z) \cdot \sqrt{d_{z2}^2 + L_1^2} + 2 \cdot \pi \cdot d_{z2}^2}{4} \tag{14}$$

$$V_{M6} = \frac{2 \cdot \pi \cdot d_{z2}^3 + \pi \cdot L_1 \cdot d_{z2}^2 + \pi \cdot L_1 \cdot d_z \cdot d_{z2} + 2 \cdot \pi \cdot d_{z1}^3 + \pi \cdot L_1 \cdot d_{z1}^2 + \pi \cdot L_1 \cdot d_z \cdot d_{z1} + 2 \cdot \pi \cdot L_1 \cdot d_z^2}{24} \tag{15}$$

In models M2, M3, M4, M5 and M6, geometric mean diameter was calculated with the following formulas:

$$d_w = \frac{W_1 + W_2 + T_1 + T_2}{4} \tag{16}$$

$$d_z = \frac{W + T}{2} \tag{17}$$

$$d_{z1} = \frac{W_1 + T_1}{2} \tag{18}$$

$$d_{z2} = \frac{W_2 + T_2}{2} \tag{19}$$

every cucumber fruit was weighed on the Radwag WAA 100/C/2 electronic scale to the nearest 0.001 g. The significance of differences between the mean values of the measured parameters was determined in the Kruskal-Wallis test with multiple comparisons of mean ranks. The aim of the analysis was to identify homogeneous groups. The results were processed statistically in the Statistica 13.3 PL program at a significance level of $\alpha = 0.05$.

3. Results and Discussion

Cucumber fruits (*Cucumis sativus* L.) cv. *Śremski* are botanical berries with a more or less elongated shape, varied size, smooth or spiny skin. Cucumbers are filled with seeds, and their color ranges from dark green to yellow. At harvest maturity, cucumbers are cylindrical in shape, without a neck, with a gently tapering end at the flower base and a small seed chamber. The smallest of the examined cucumbers weighed 43.05 g, and the largest123.70 g. The surface area of cucumbers determined in the 3D scanner ranged from 74.84 cm^2 do 145.38 cm^2, with an average of 111.25 cm^2. Based on the generated 3D images, the volume of cucumbers was determined in the range of 46.65 cm^3 to 127.38 cm^3, with an average of 77.26 cm^3 (Table 1). Exemplary 3D models of cucumber fruits are presented in Figures 3 and 4.

Table 1. Geometric parameters of cucumber fruits.

Variable1	Mean	Range	Standard Deviation
L (mm)	113.14	39.10	9.94
W (mm)	37.23	13.04	3.28
T (mm)	35.47	14.82	3.31
A^{3D} (mm^2)	111.25	70.54	16.12
V^{3D} (mm^3)	77.26	80.73	18.89

^{3D}-3D scan

Figure 3. 3D model of a cucumber fruit with a texture overlay.

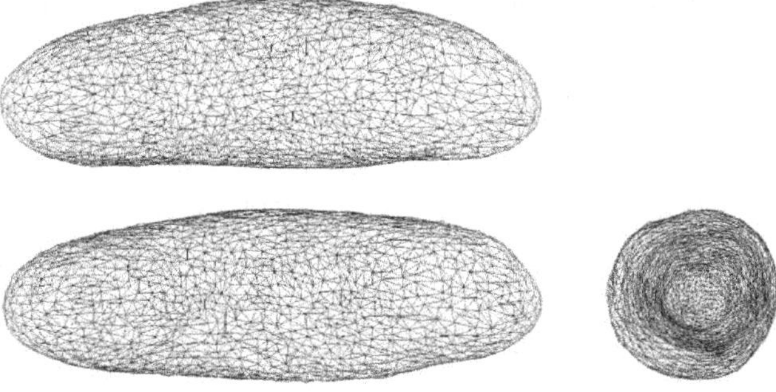

Figure 4. 3D model of a cucumber fruit represented by a triangle mesh.

The mean dimensions, surface area and volume of the analyzed cucumber fruits are presented in Table 1.

The significance of differences between the mean surface area and mean volume of cucumbers was determined in the Kruskal-Wallis nonparametric test. The significance of differences between the parameters acquired by 3D scanning and the parameters calculated with mathematical formulas is presented in Tables 2 and 3. The mean surface area of cucumber fruits calculated from the 3D model did not differ significantly from the mean surface area calculated from the spheroid model (M2—formula 5) and the model combining two truncated cones and two hemispheres with different diameters (M6—formula 14).

The mean volume of cucumber fruits calculated from the 3D model did not differ significantly from the mean volume calculated from the ellipsoid model (M1—formula 4), spheroid model (M2—formula 7) and the geometric model combining two truncated cones and two hemispheres with different diameters (M6—formula 15).

Table 2. The significance of differences between the mean surface area of cucumber fruits.

Surface Area *A* (Kruskal-Wallis Test) H(6, N = 350) = 132.2065; *p* = 0.000				
Probability of Multiple Comparisons				
Measurement Method	**Number of Observations N**	**Rank Sum**	**Mean Rank**	**Mean**
3D	50	9288.50	185.77	111.25 [bc]
M1	50	6939.50	138.79	101.71 [a]
M2	50	7863.00	157.26	105.93 [ab]
M3	50	15,564.00	311.28	150.45 [d]
M4	50	6181.00	123.62	100.17[a]
M5	50	5737.00	114.74	98.57 [a]
M6	50	9852.00	197.04	114.06 [c]

Values marked with the same letters in columns do not differ significantly; [a,b,c,d] ($p \leq 0.05$).

Table 3. The significance of differences between the mean volume of cucumber fruits.

Volume *V* (Kruskal-Wallis Test) H(6, N = 350) = 124.2550; *p* = 0.000				
Probability of Multiple Comparisons				
Measurement Method	**Number of Observations N**	**Rank Sum**	**Mean Rank**	**Mean**
3D	50	8301.00	166.02	77.26 [a]
M1	50	8910.00	178.20	79.21 [a]
M2	50	8982.00	179.64	79.29 [a]
M3	50	15,085.00	301.70	118.93 [c]
M4	50	5492.00	109.84	65.85 [b]
M5	50	5262.00	105.24	65.16 [b]
M6	50	9393.00	187.86	81.27 [a]

Values marked with the same letters in columns do not differ significantly; [a,b,c] ($p \leq 0.05$).

The distribution of surface area values computed from the 3D model and the proposed geometric models is presented in Figure 5. The distribution of volume values computed from the same models is presented in Figure 6.

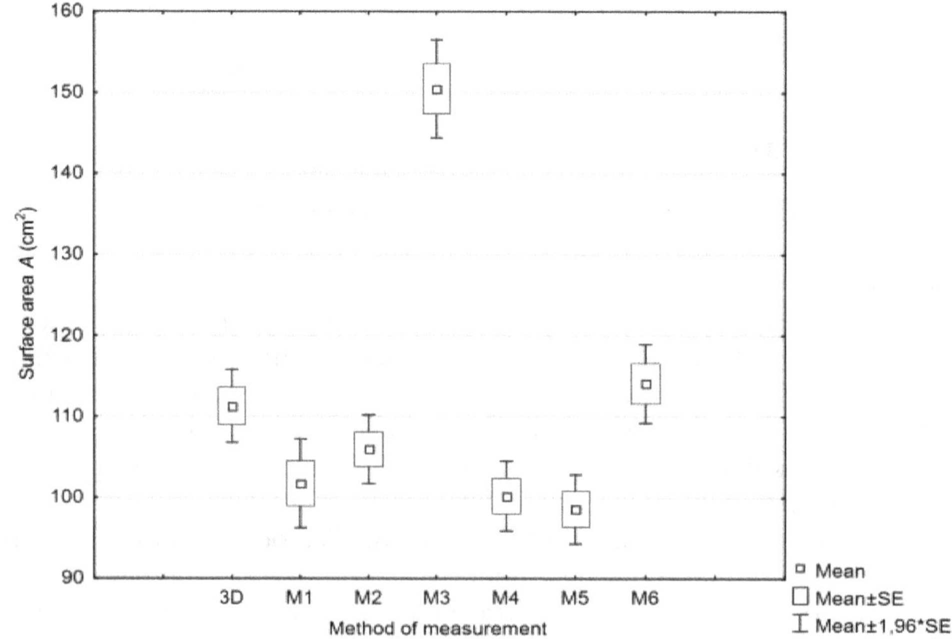

Figure 5. Parameters of normal distribution of cucumber surface area.

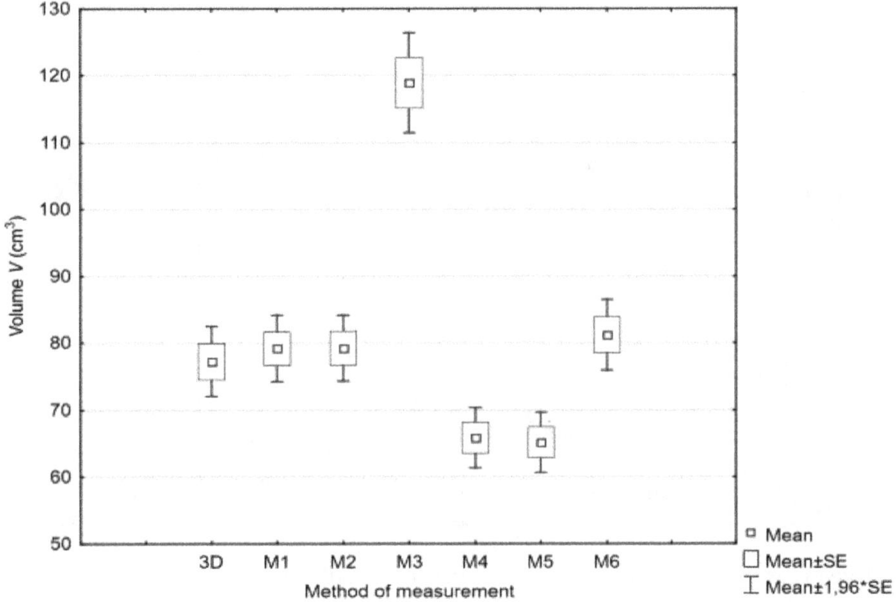

Figure 6. Parameters of normal distribution of cucumber volume.

If we assume that fruit dimensions acquired from 3D scans are burdened by a small error, these parameters can be used as a reference to compare the results of caliper measurements and to describe the shape of cucumber fruits with selected geometric figures. The relative error between the values acquired from 3D scans and direct measurements was regarded as the error of the method. The data presented in Figure 7 indicate that the error in direct measurements of cucumber surface area was smallest for the model combining two truncated cones and two hemispheres with different diameters (M6) where it did not exceed 3%. The error was estimated at 5% when model M2 and formula 4 were used. The data presented in Figure 8 indicate that the error in direct measurements of cucumber volume was smallest for the ellipsoid model (M1), the spheroid model (M2), and the model combining two truncated cones and two hemispheres with different diameters (M6). The error did not exceed 6% when ellipsoids were used, and it was estimated at 6% when model M6 was used.

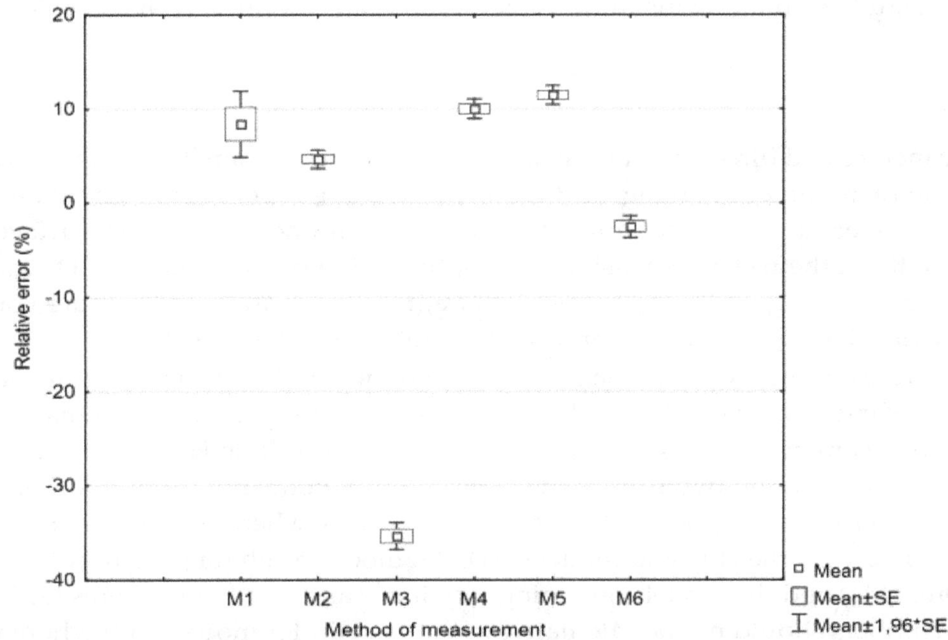

Figure 7. Relative error of cucumber surface area determined with geometric models and the 3D model.

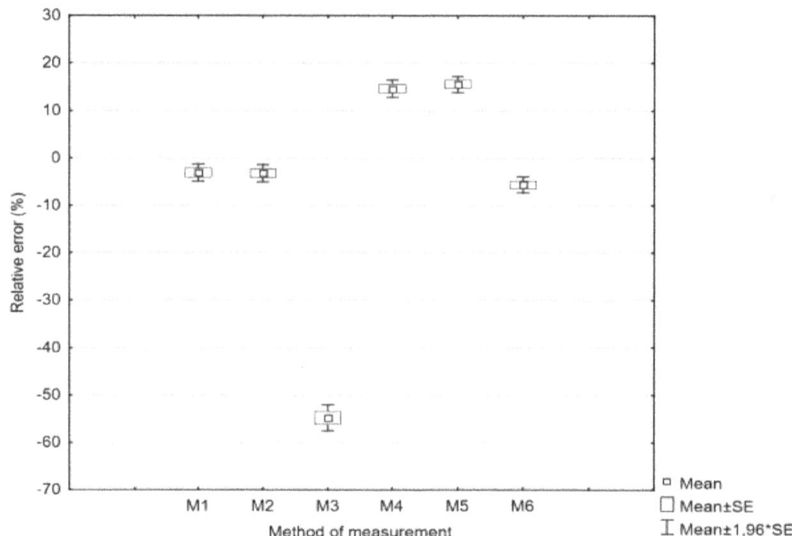

Figure 8. Relative error of cucumber volume determined with geometric models and the 3D model.

The results of this study were compared with the findings of other authors. Zapotoczny (2002) investigated the geometrical parameters of greenhouse-grown cucumbers under laboratory conditions with the use of image analysis methods. The cited author analyzed 2D images of 27 greenhouse-grown cucumbers and determined their average length at 163.17 mm, average width at 32.00 mm, and average projected area at 51.20 cm^2. The multiscale modeling approach deployed by Ho et al. (2013) supports the description of the phenomena occurring inside agricultural raw materials. Multiscale models consist of interconnected sub-models that describe the behavior of raw material in different spatial scales. This approach supports the prediction of processes and phenomena occurring inside raw materials. However, multiscale modeling is relatively complex, and not widely used. Rahmi and Ferruh (2009) described the applicability of 3D models for processing agricultural raw materials and for food production. The presented models were generated based on 3D scans of selected materials, including chicken egg, pear fruit, strawberry fruit, banana and apple. The authors modeled the cooling process in pear fruit and compared the results with experimental findings. Cucumber fruits have never been analyzed in studies on modeling and determination of geometrical parameters of agricultural raw materials.

4. Conclusions

1. Geometric models and direct measurements of the geometric parameters of agricultural products facilitate the planning of spraying, sorting and packaging operations. These methods enable small-scale farmers to easily determine the geometric parameters (volume, surface area) of raw materials without the use of expensive and sophisticated devices such as 3D scanners. Direct measurements of the geometric parameters of agricultural raw materials are consistent with sustainable development principles and can be applied on a large scale.
2. Models where the relative error of measurement does not exceed 5% are recommended when the surface area of cucumbers is calculated with an electronic caliper and mathematical formulas of the presented geometric models. The above condition was fulfilled by the spheroid model (M2) and the model combining two truncated cones and two hemispheres with different diameters (M6). Relative error was higher in the range of 8% to 12% when the surface area of cucumbers was determined with the ellipsoid model (M1), the model combining a truncated cone and two hemispheres (M4) and the model combining a cylinder and two hemispheres (M5). The surface area of cucumbers should not be calculated with the cylinder model (M3) where relative error reached 35%.

3. The volume of fruits can be calculated with the use of the ellipsoid model (M1), the spheroid model (M2) and, similarly to surface area measurements, the model combining two truncated cones and two hemispheres with different diameters (M6). The relative error of the above geometric models did not exceed 5.5%. Relative error was higher in the range of 14% to 16% when cucumber volume was determined with the model combining a truncated cone and two hemispheres (M4) and the model combining a cylinder and two hemispheres (M5). The relative error of the cylinder model (M3) was determined at 54%.
4. The significance of differences between the mean values of surface area was determined in the Kruskal-Wallis test, and no significant differences were observed in models M1, M2, M4 and M5. However, models M1, M4 and M5 cannot be used to determine the surface area of cucumber fruit due to high mean relative error at 8.37%, 9.98% and 11.44%, respectively.
5. In the literature, the mathematical formula for calculating the volume of an ellipsoid (M1) is often used to determine the volume of agricultural products with an ellipsoidal shape. Relative error is estimated at 3% when the volume of ellipsoidal fruits is calculated with the above mathematical formula.
6. In the group of the evaluated methods for determining the geometric parameters of agricultural materials, 3D scanning is the most informative approach. Numerical models support the determination of a full range of geometric parameters (dimensions, area, volume) of entire objects and their fragments. The shape of the analyzed object is stored in computer memory as a cloud of points, and can be used to measure volume without the involvement of displacement methods where the sample is immersed in liquid. Numerical models can also be archived and used for future research.
7. The measurable result of the study was the development of models supporting the determination of the geometric parameters (surface area, volume) of agricultural materials based on their basic dimensions (length, width, thickness). In most cases, the proposed models support the determination of the above geometric parameters with a relative error below 5% within a short period time. Therefore, they can be used in the research and design of new cucumber processing equipment.
8. Further research should focus on the development of models of agricultural raw materials that facilitate the determination of geometric parameters for planning and performing of production processes in agriculture.

Author Contributions: A.A. developed the concept and design of the study; A.A., P.M. and Z.K. conducted the experiments; A.A., Z.K., P.M. and E.Ś. contributed to the literature study; D.C., P.M., Z.K., A.A. and A.J.L. analyzed the data and made final calculations; A.A., Z.K. and P.M. wrote the paper; A.A., P.M. and Z.K. critically revised it.

References

1. Datta, A.K.; Halder, A. Status of food process modeling and where do we go from here (synthesis of the outcome from brainstorming). *Compr. Rev. Food Sci. Food Saf.* **2008**, *7*, 117–120. [CrossRef]
2. Verboven, P.; De Baerdemaeker, J.; Nicolai, B.M. Using computational fluid dynamics to optimize thermal processes. In *Improving the Thermal Processing of Foods*; Richardson, P., Ed.; CRC Press: Boca Raton, FL, USA, 2004; pp. 82–102.
3. Goni, S.M.; Purlis, E.; Salvadori, V.O. Three-dimensional reconstruction of irregular foodstuffs. *J. Food Eng.* **2007**, *82*, 536–547. [CrossRef]
4. Frączek, J.; Wróbel, M. Methodic aspects of seed shape assessment. *Inżynieria Rolnicza* **2006**, *12*, 155–163. (In Polish)
5. Szwedziak, K.; Rut, J. Assessment of pollutants of the grain corn with the help of computer analysis of the image. *Postępy Techniki Przetwórstwa Spożywczego* **2008**, *1*, 14–15. (In Polish)
6. Deshpande, S.D.; Bal, S.; Ojha, T.P. Physical properties of soybean. *J. Agric. Eng. Res.* **1993**, *56*, 89–98. [CrossRef]

7. Gupta, R.K.; Das, S.K. Physical properties of Sunflower seeds. *J. Agric. Eng. Res.* **1997**, *66*, 1–8. [CrossRef]
8. Cahsir, S.; Marakoglu, T.; Ogut, H.; Ozturk, O. Physical properties of rapeseed (*Brassica napus oleifera* L.). *J. Food Eng.* **2005**, *69*, 61–66.
9. Tańska, M.; Rotkiewicz, D.; Kozirok, W.; Konopka, I. Measurement of the geometrical features and surface color of rapeseeds using digital image analysis. *Food Res. Int.* **2005**, *38*, 741–750. [CrossRef]
10. Jadwisieńczak, K.; Kaliniewicz, Z. Analysis of the mustard seeds cleaning process. Part 1. Physical properties of seeds. *Inżynieria Rolnicza* **2011**, *9*, 57–64. (In Polish)
11. Coskuner, Y.; Karababa, E. Some physical properties of flaxseed (*Linum usitatissimum* L.). *J. Food Eng.* **2007**, *78*, 1067–1073. [CrossRef]
12. Erdogdu, F.; Balaban, M.O.; Chau, K.V. Modeling of heat conduction in elliptical cross-section: II, Adaptation to thermal processing of shrimp. *J. Food Eng.* **1998**, *38*, 241–258. [CrossRef]
13. Luzuriaga, D.A.; Balaban, M.O.; Yeralan, S. Analysis of visual quality attributes of white shrimp by machine vision. *J. Food Sci.* **1997**, *62*, 113–118. [CrossRef]
14. Crocombe, J.P.; Lovatt, S.J.; Clarke, R.D. Evaluation of chilling time shape factors through the use of three-dimensional surface modeling. In Proceedings of the 20th International Congress of Refrigeration 1999: IIR/IIF, Sydney, Australia, 19–24 September 1999; p. 353.
15. Jancsok, P.T.; Clijmans, L.; Nicolai, B.M.; De Baerdemaeker, J. Investigation of the effect of shape on the acoustic response of 'conference' pears by finite element modeling. *Postharvest Biol. Technol.* **2001**, *23*, 1–12. [CrossRef]
16. Borsa, J.; Chu, R.; Sun, J.; Linton, N.; Hunter, C. Use of CT scans and treatment planning software for validation of the dose component of food irradiation protocols. *Radiat. Phys. Chem.* **2002**, *63*, 271–275. [CrossRef]
17. Sabliov, C.M.; Bolder, D.; Keener, K.M.; Farkas, B.E. Image processing method to determine surface area and volume of axi-symmetric agricultural products. *Int. J. Food Prop.* **2002**, *5*, 641–653. [CrossRef]
18. Zapotoczny, P. Measuring geometrical parameters of cucumbers fruits using computer image analysis. *Problemy Inżynierii Rolniczej* **2002**, *4*, 57–64. (In Polish)
19. Scheerlinck, N.; Marquenie, D.; Jancsok, P.T.; Verboven, P.; Moles, C.G.; Banga, J.R.; Nicolai, B.M. A model-based approach to develop periodic thermal treatments for surface decontamination of strawberries. *Postharvest Biol. Technol.* **2004**, *34*, 39–52. [CrossRef]
20. Du, C.; Sun, D.W. Estimating the surface area and volume of ellipsoidal ham using computer vision. *J. Food Eng.* **2006**, *73*, 260–268. [CrossRef]
21. Zheng, C.; Sun, D.W.; Du, C.J. Estimating shrinkage of large cooked beef joints during air-blast cooling by computer vision. *J. Food Eng.* **2006**, *72*, 56–62. [CrossRef]
22. Kim, J.; Moreira, R.G.; Huang, Y.; Castell-Perez, M.E. 3-D dose distributions for optimum radiation treatment planning of complex foods. *J. Food Eng.* **2007**, *79*, 312–321. [CrossRef]
23. Goni, S.M.; Purlis, E.; Salvadori, V.O. Geometry modeling of food materials from magnetic resonance imaging. *J. Food Eng.* **2008**, *88*, 561–567. [CrossRef]
24. Siripon, K.; Tansakul, A.; Mittal, G.S. Heat transfer modeling of chicken cooking in hot water. *Food Res. Int.* **2007**, *40*, 923–930. [CrossRef]
25. Mieszkalski, L. Computer-aiding of mathematical modeling of the carrot (*Daucus carota* L.) root shape. *Ann. Wars. Univ. Life Sci. SGGW Agric.* **2013**, *61*, 17–23.
26. Mieszkalski, L. Bezier curves in modeling the shapes of biological objects. *Ann. Wars. Univ. Life Sci. SGGW Agric.* **2014**, *64*, 117–128.
27. Balcerzak, K.; Weres, J.; Górna, K.; Idziaszek, P. Modeling of agri-food products on the basis of solid geometry with examples in AutoDesk 3ds Max and finite element mesh generation. *J. Res. Appl. Agric. Eng.* **2015**, *60*, 5–8.
28. Ho, Q.T.; Carmeliet, J.; Datta, A.K.; Defraeye, T.; Delele, M.A.; Herremans, E.; Opara, L.; Ramon, H.; Tijskens, E.; Sman, R.; et al. Multiscale modeling in food engineering. *J. Food Eng.* **2013**, *114*, 279–291. [CrossRef]
29. Anders, A. *Determination of the Geometric Parameters of Seeds with Different Methods*; Wydawnictwo Uniwersytetu Warmińsko-Mazurskiego w Olsztynie: Olsztyn, Poland, 2019; ISBN 978-83-8100-163-2. (In Polish)
30. Rahmi, U.; Ferruh, E. Potential use of 3-dimensional scanners for food process modeling. *J. Food Eng.* **2009**, *93*, 337–343.

31. Anders, A.; Markowski, P.; Kaliniewicz, Z. Numerical modelling of agricultural products on the example of bean and yellow lupine seeds. *Int. Agrophys.* **2015**, *29*, 397–403. [CrossRef]
32. Anders, A.; Markowski, P.; Kaliniewicz, Z. Evaluation of geometric and physical properties of chosen pear cultivars based on numerical models obtained by a 3D scanner. *Zeszyty Problemowe Postępów Nauk Rolniczych* **2014**, *577*, 3–12. (In Polish)
33. Anders, A.; Markowski, P.; Kaliniewicz, Z. The application of a 3D scanner for the evaluation of geometric properties of *Cannabis sativa* L. seeds. *Acta Agrophys.* **2014**, *21*, 391–402. (In Polish)
34. NextEngine User Manual. 2010. Available online: http://www.nextengine.com (accessed on 1 July 2018).
35. MeshLab Visual Computing Lab—ISTI—CNR. 2013. Available online: http://meshlab.sourceforge.net (accessed on 1 July 2018).
36. Gastón, A.L.; Abalone, R.M.; Giner, S.A. Wheat drying kinetics. Diffusivities for sphere and ellipsoid by finite elements. *J. Food Eng.* **2002**, *52*, 313–322. [CrossRef]
37. Bronsztejn, I.N.; Siemiendiajew, K.A.; Musiol, G.; Muhling, H. *Nowoczesne Kompendium Matematyki*; PWN: Warszawa, Poland, 2009. (In Polish)

9

Modeling and Prediction of the Uniformity of Spray Liquid Coverage from Flat Fan Spray Nozzles

Paweł A. Kluza [1], Izabela Kuna-Broniowska [1,*] and Stanisław Parafiniuk [2]

1 Department of Applied Mathematics and Computer Science, University of Life Sciences in Lublin, 20-612 Lublin, Poland; pawel.kluza@up.lublin.pl

2 Department of Machinery Exploitation and Management of Production Processes, University of Life Sciences in Lublin, 20-612 Lublin, Poland; stanislaw.parafiniuk@up.lublin.pl

* Correspondence: izabela.kuna@up.lublin.pl

Abstract: The effectiveness and quality of agricultural spraying largely depends on the technical efficiency of the nozzles installed in agricultural sprayers. The uniform spraying of plants results in a decrease in the amount of pesticides used in agricultural production and affects environmental safety. Both newly developed sprayers and those currently in use need quality control as well as an assessment of the performance of the spraying process, especially its uniformity. However, the models applied presently do not ensure accurate estimates or predictions of the spray liquid coverage uniformity of the treated surface. Generally, the distribution of the atomized liquid quantity is symmetrical and leptokurtic, which means that it does not fit well to the commonly used standard distribution. Therefore, there is a need to develop and design new tools for the evaluation, modeling, and prediction of such a process. The research problem studied in the present work was to find a new model for the distribution of atomized liquid quantity that could provide capabilities better than have been available so far to assess and predict the spraying process results. The research problem was solved through the formulation of a new function for the probability density distribution of sprayed liquid accumulation on the surface of the preset dimension size. The development of the new model was based on the results from a series of water atomization tests with an appropriate measurement device design based on the widely applied flat fan nozzles (AZ-MM type).

Keywords: flat fan nozzle; liquid coverage; coefficient of variation (CV); crop yields

---

1. Introduction

Spraying liquids is a technically and technologically important process in all areas of economy and everyday life. Particular importance is ascribed to the correctness of the process and the application thereof in agricultural production and the food industry, taking into account the effect of the sprayed liquids on the environment. A fundamental issue in these processes is their quality and efficiency. They determine the requirements for the structure and the use of technical equipment and the parameters of spraying liquids in practical terms. Hence, the uniformity and predictability of the distribution of the sprayed liquid are the basic characteristics.

The quality of agricultural spraying largely depends on the uniformity of the distribution of the sprayed liquid over the spraying surface. In field crops, currently slot nozzles are the most commonly used, which wear out during exploitation. A worn nozzle causes the amount of liquid flowing out to be greater and, thus, the uniformity of liquid distribution is disturbed. The current Directive 2009/128EC of the European Parliament and of the Control concerning the sustainable use of pesticides in agriculture contains guidelines regarding, among others, the testing of spray nozzles installed in agricultural sprayers [1]. In field sprayers, the even parameter is the even distribution of sprayed liquid on the sprayed surface. Specialized devices with a grooved table are used to measure the uniformity of spray distribution [2,3].

The stream of the sprayed liquid has the shape of a cone with characteristics depending on the design of the nozzle and the geometry of its working slot, which is subject to wear and can therefore change its shape due to mechanical damage or limescale deposition. Another consequence of nozzle wear is also the change in flow rate.

Field spraying is an example of the use of the liquid spraying process. One of the key issues in this process is the uniformity of the liquid amount distribution achieved from the spray boom.

The coefficient of variation is a measure of the process uniformity expressed (usually in percentages) as the ratio of the standard deviation to the arithmetic mean of the sample. This parameter is used most commonly for the comparison of the variation of a trait in two different distributions.

In the case of field spraying, its value should not exceed 10%, as specified by the EN ISO 16119-2 part 2 standard [4].

When the level defined in the standard is not exceeded, the spray uniformity is regarded to be correct. Its level is mainly determined by the shape of the distribution of the amount of liquid sprayed from a single nozzle, based on which spray boom is constructed. Distribution data from a single nozzle is replicated to simulate a complete virtual boom.

Manufacturers strive to improve the sprayer structure to ensure the best uniformity of surface coverage.

The field sprayer boom that will be equipped in such nozzles should provide the greatest uniformity of the treated surface spray, i.e., it should be characterized by the minimum value of the coefficient of variation for the sprayed liquid. To date, there has been no model ensuring a relatively low value of the coefficient of variation of liquid coverage from a virtual field boom containing all the same new nozzles and, hence, a very high uniformity of spraying. Similarly, there is no model that would satisfactorily determine the amounts of accumulated liquid allowing the achievement of the permissible value of the coefficient of variation defined by the standard.

Therefore, there is a need to propose a new model that will substantially increase the level of liquid spray uniformity from the spray boom, which will result in the improvement of the spray quality.

An analysis of liquid sprayed from a single nozzle and a sprayer in terms of the distribution characteristics has been carried out [5].

To ensure the most optimal spray, in [6–8], the impact of the sprayer and nozzle parameters on the spray distribution and on the coefficient of variation, which is a measure of the uniformity of treated surface coverage, was analyzed. Various types of nozzles have been tested in laboratory conditions to analyze this parameter [9].

The simulation of a virtual field boom illustrating the work of the sprayer with the analysis of spray uniformity has been carried out [10].

A review of various distribution patterns and spray coverage achieved by some nozzles has been presented [11].

Mathematical models describing the physical motion of a sprayed liquid particle [12] and presenting the wear of flat fan nozzles [13] have been developed. In [14], as well as [15], the impact of an external air stream simulating the wind with varied velocity on the distribution of sprayed liquid has been analyzed. It has compared the patterns of liquid distributions and the values of the coefficient of variation after the spraying process in experiments conducted on tables with 5 cm and 10 cm groove spacing [16].

The scope of the research in the present work is to find a new model for the distribution of atomized liquid quantity thanks to which better capabilities than have been available so far to assess and predict the spraying process results could be provided. Achieving this aim will help to increase environmental efficiency in agriculture.

2. Materials and Methods

2.1. *Description of the Experiment*

To obtain data for the development and verification of the new model, the experiments were conducted in the Laboratory of Techniques for Application of Agrochemicals, Department of Machinery Exploitation and Management of Production Processes, University of Life Sciences in Lublin.

It was assumed that mounting new nozzles of good quality to the spray boom would ensure the most uniform spray distribution.

Typically, the working spray boom was located at a height of 0.5 m above the sprayed surface. The liquid outflow pressure was set at 0.3 MPa. The process time was preset to one minute. The spray angle of the flat fan nozzles was set at 110°. In this case, the width of the area sprayed with a single sprayer nozzle was approx. 1.5 m, and the surface coverage of all 25 nozzles mounted on the virtual boom was approx. 12.5 m.

The tests and measurements were carried out on grooved tables with a width corresponding to that of the virtual field boom. Two types of flat fan nozzles were included in the study. The results obtained from AZ-MM spray nozzles were used for the development of the model. The goodness of the fit of the model to the data was evaluated based on results obtained from RS-MM spray nozzles.

Fifty new nozzles were used in the experiment.

The tests were carried out using the blue nozzles, characterized by a liquid flow rate of 1.2 liters per minute.

Water was used as the test liquid, following the standard practice for this type of experiment conducted in laboratory conditions.

The experimental measurement stand shown in Figure 1 facilitated the analysis of the distribution of the atomized liquid sprayed from the nozzle. The results obtained in this design were used for the construction of a virtual field boom. The sprayed liquid reaches the table and falls along the grooves into 50 containers with the same volume arranged serially at equal distances from each other. Each measurement determines the amount of the liquid accumulated in the consecutive containers and is expressed in milliliters. The capacity of a single container is 250 ml. All experiments were carried out on the measurement stand with 5 cm groove spacing.

Figure 1. Measurement table (Department of Machinery Exploitation and Management of Production Processes, University of Life Sciences in Lublin).

With the parameters determined during the experimental process, the spray spectrum for a single nozzle covered approximately 30 containers located in the center of the measurement table. Hence, 0 mL value was in the 10 containers located on the left-hand side and 0 mL value was in the 10 containers located on the right side of the analyzed spectrum area.

The data used in the experiment were the results provided by all 50 available nozzles (25 of each type used for the development and verification of the model) included in the measurement stand. This yielded 50 different data sets. Each of these sprays duplicated 25 times simulated the data generated by the virtual spray boom as described below.

The width of the sprayed area was scaled to an appropriate number of the consecutively numbered containers located at an identical distance as on the measurement table. Distribution data from a single nozzle was replicated to simulate a complete virtual boom.

I. The nozzle was located on the extreme left of the virtual field boom;

II. Next, another identical nozzle was placed on the boom on the right side of the first one at a distance of 0.5 m (i.e., at a distance covered by 10 containers to the right);

III. By mounting successive identical nozzles on the boom, the 25th container was reached (Figure 2). The total range of measurements comprised 270 containers. The amounts of the liquid collected in each container served for the calculation of the coefficient of variation.

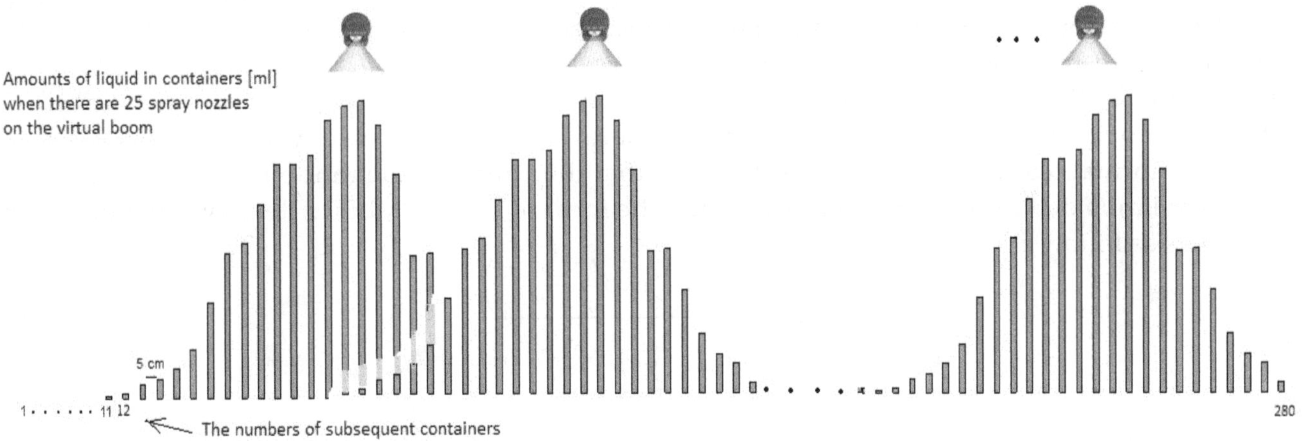

Figure 2. Data collection scheme (final stage 3).

In 50 measurement series performed in the experiment, data from 50 virtual field sprayer booms, created from single nozzle distribution, were collected. Each boom was assigned the same number as that of each consecutive nozzle with a specified degree of wear.

Within the adopted scope of work, we carried out a comprehensive statistical analysis of the distribution of the liquid amounts determined for each simulated virtual field boom and for the real data from each single nozzle.

2.2. *Distribution of Droplets after the Spraying Process*

The first attempts to match the distribution of droplets to the real sprays provided by flat fan nozzles were undertaken [17,18]. The researchers considered a triangular pattern. The effect of changes in the sprayer nozzle angle on the symmetry and shift of spray distribution was emphasized [17]. The density function of the triangular distribution is described by an equation that combines the nozzle height, the set angle, and the distance between water-sensitive papers:

$$f(x) = x\left(H - d_i ctg\frac{\propto}{2}\right) \tag{1}$$

where:

x is the distance between water-sensitive papers [m];
H is the distance between the spray nozzle and the sprayed surface [m];
d_i is the distance between the ith water-sensitive paper and the nozzle [m];

α is the preset angle of the nozzle spray [°].

Another probability distribution that was analyzed in the description of spray was beta distribution [18]. It is defined by two shape parameters denoted as α and β, which are the exponents of a random variable in the formula and determine the shape of the distribution. The density function of this distribution is as follows:

$$f(x) = \frac{\Gamma(\alpha+\beta)}{\Gamma(\alpha)\Gamma(\beta)} x^{\alpha-1}(1-x)^{\beta-1} \tag{2}$$

where:

x is a random variable specifying the distance between two water-sensitive papers [m];

$\alpha, \beta > 0$ are shape parameters estimated from the experimental data;

$\Gamma(z) = \int_0^{+\infty} t^{z-1}e^{-t}dt$ is the special gamma function for $z > 0$.

Joint research was conducted on this distribution and found that it was sometimes more suitable for the description of spray than the normal distribution [5].

This most popular distribution mentioned above was fitted and applied in comprehensive investigations of spray [19]. The issue was divided by the researchers into two stages. The first stage consisted of the selection of an appropriate model and density function, whereas the second stage was focused on the estimation of parameters determining the spray quality, i.e., pressure and nozzle height and size. Given the basic fact that the function of spray distribution density is a symmetrical curve with an approximate bell shape, the distribution of sprayed liquid was effectively fitted to the doubly truncated normal distribution [19]. This indicates that the values of the random variable had upper and lower limits determined by a certain value. The form of the probability density function $f(x)$ in this case is as follows:

$$f(x) = \frac{\frac{1}{\sigma\sqrt{2\pi}} \exp\left(-\frac{(x-\mu)^2}{2\sigma^2}\right)}{F\left(\frac{b-\mu}{\sigma}\right) - F\left(\frac{a-\mu}{\sigma}\right)} \tag{3}$$

where:

x is a random variable specifying the distance between two water-sensitive papers [m];

u is the expected value of the random variable;

σ is the standard deviation for normal distribution;

$F(x)$ is the cumulative distribution function that complies with $f(x)$;

$a,\ b$ are the lower and upper cut-off points of normal distribution, respectively.

After the development of the model presented above, a relationship was discovered between its parameters and factors that influence spraying [19]. By the application of the multiple regression equation, the researchers were able to express the standard deviation and spray width with the coefficients of determination of 0.98 and 0.976, respectively:

$$\begin{aligned} \sigma &= -78 + 27.1\ln(H) + 6.15\ln(P) + 1.72\ \ln(Q_2) \\ W &= -401 + 1.39\ln(H) + 36.5\ \ln(P) + 8.94\ \ln(Q_2) \end{aligned} \tag{4}$$

where:

σ is the standard deviation [cm];

W is the spray width [cm];

H is the nozzle height [cm];

P is the spray pressure [105 Pa];

Q_2 is the standard flow rate at a pressure of 0.2 MPa [l/min].

Another model [20] is the so-called mean distribution model. It consists of the calculation of the mean values for the parameters of the distribution of liquid amounts collected from several single nozzles. This approach generates highly precise results, but at the cost of a substantially greater number of measurements. Additionally, since different types of nozzles are often used, this will very likely be a random model.

Figure 3 shows a comparison of the consistency of the data obtained from triangular, beta, and normal distributions with the measurement results generated by a Teejet 110 04 VS nozzle at a pressure of 0.2 MPa and a spray height of 0.5 m [21].

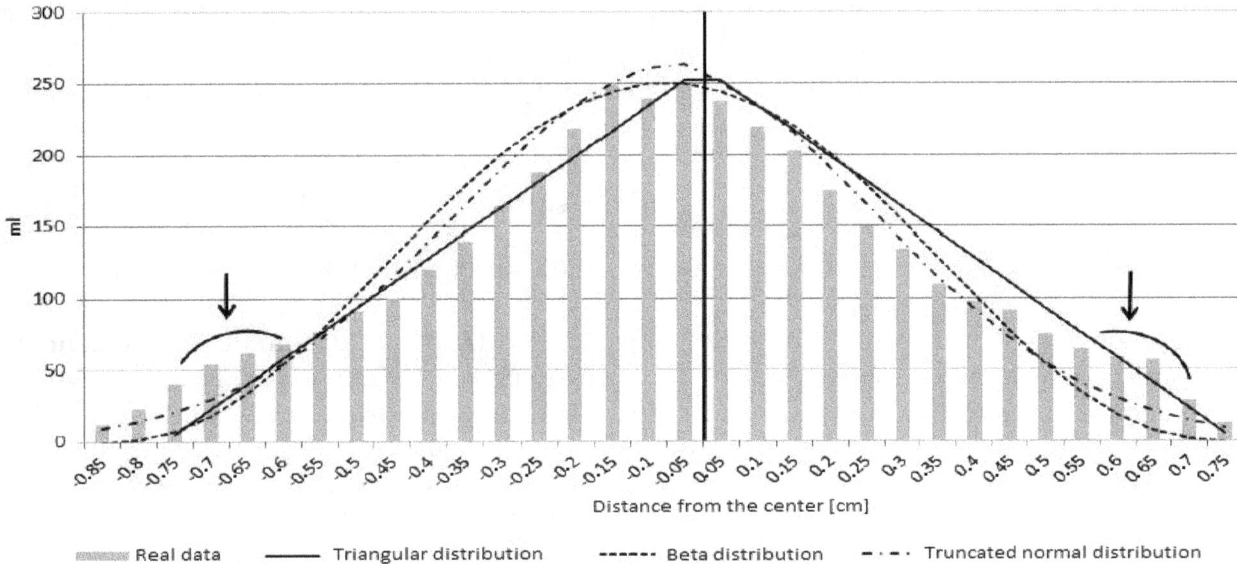

Figure 3. Example of the fit of results in selected models to real spray data [21].

An evident conclusion prompted by the analysis of this figure is the fact that none of the characterized models generate results that match real data adequately.

2.3. Development of a New Model of Liquid amount Distribution

Based on the determinants and data presented in the previous subsection, an original density function of a new distribution model describing the accumulation of liquid amounts after the spraying process was defined for a single nozzle as follows:

$$f(x) = \begin{cases} a(x-0.5)^p, & x \in (0.5;0.9) \\ b(x-1.25)^q + c, & x \in< 0.9;1.6) \\ (-1)^p a(x-2)^p, & x \in < 1.6;2) \\ 0, & x \in (-\infty;0.5 > \cup < 2;\ +\infty) \end{cases} \quad (5)$$

where:

x is the real variable with values from the spray range (0.5; 2) [m];

$a > 0$ is a shape parameter;

$b < 0$ is a shape parameter;

$c > 0$ is a shift parameter;

$p, q \geq 2$ are shape parameters (p = natural number, q = even number)

By increasing the values of parameters a, p, or q, the shape of the density curve becomes leptokurtic (thin). Reversely, i.e., when the values of any of the three parameters decrease, the shape becomes more flattened.

The opposite is noted in the case of parameter b, i.e., the higher its value, the greater the flattening of the curve in an appropriate range.

The method for the construction of the density function presented above ensures symmetry of distribution. The new model is symmetrical and is a probability density function.

Using the optimization method, i.e., the Microsoft Excel Solvers tool, the shape of parameters p, q, and b was selected in a way providing a properly set value of the coefficient of variation for the liquid amount after the spraying process.

After the determination of equations required for the correct determination of the probability distribution for a single nozzle, an original model was developed for the determination of the liquid distribution after simultaneous spraying with 25 nozzles in order to reflect the real work of the sprayer.

Therefore, based on the developed form of the density function, a function reflecting the amount of the liquid at any point in the spray area was defined:

$$g(x) = \sum_{n=0}^{24}\left[a(x-0.5-0.5n)^p + (-1)^p a(x-2-0.5n)^p + b(x-1.25-0.5n)^q + c\right] \quad (6)$$

The arguments of the function are points across the entire width of the sprayed area located at a distance of 0.05 m from each other.

The total width of the area sprayed by the boom is identical to that of all liquid-collecting containers aligned and assigned numbers from 1 to 290. According to ISO 16122-2: 2015 (E), the data range taken into account in the calculation of the coefficient of variation of the sprayed liquid distribution includes the amounts of liquid collected in containers located in the center of the spray area width for the second nozzle mounted on the boom to the center of the spray area width for the penultimate nozzle on the boom. Therefore, the sum range in Equation (6) corresponds to containers 36 to 256.

Based on the method of calculation of the coefficient of variation, the function dependent on all parameters a, b, c, p, and q is defined with the following formula:

$$V = \frac{S}{\overline{g}} \cdot 100\% \quad (7)$$

where the following equations are valid for i representing natural numbers and denoting the consecutive numbers of the liquid-accumulating containers after the spraying process:

$$\overline{g} = \frac{1}{221}\sum_{i=36}^{256} g(0.05i) \quad (8)$$

$$s^2 = \frac{1}{221}\sum_{i=36}^{256} \left(g(0.05i) - \overline{g}\right)^2 \quad (9)$$

and $\overline{g}$ and s^2 are the mean and standard deviation, respectively, for the values of function $g(x)$.

Next, all parameters of the probability distribution were selected for function V to reach the minimum, i.e., to simulate the work of a field boom equipped with the new (model) nozzles, thus providing the most uniform spray of the treated surface, assuming values of 5%, 7%, and 10%, and simulating the work of a field boom with nozzles characterized with an adequate wear degree.

The flow rate in all nozzles was fixed and the extension of the spray swath from virtual boom sprayer could be unlimited, because any number of nozzles may be added.

If we consider the case that the spacing of the grooves on the measurement table is 5 cm, then in the model we have, $a = 15.352$; $b = -2.024$; $c = 1.23$ for $p = 3$, $q = 2$, and $V = 10\%$.

On the over hand, if we consider that the spacing of the grooves on the measurement table is 10 cm, then in the model we have, $a = 14.87$; $b = -2.55$; $c = 1.259$ for $p = 3$, $q = 2$, and $V = 10\%$.

3. Results

To check the goodness of the fit of the data obtained with the new model to real results, the RS-MM 110 03 nozzles were used in such a way that each of them multiplied 25 times formed a single virtual field boom. The STATISTICA 13 program, supported by the Statsoft company from Poland in 2013, was used to perform the linear regression analysis of data generated by the model in comparison with the measurements provided by the virtual field booms. Figure 4 shows an example of a correlation diagram of the amount of liquid after the spraying process from the boom equipped with the new nozzles (model generating $V = 2.16\%$) and from a boom equipped with worn nozzles (generating $V = 9.37\%$ wear degree) with a fitted regression line.

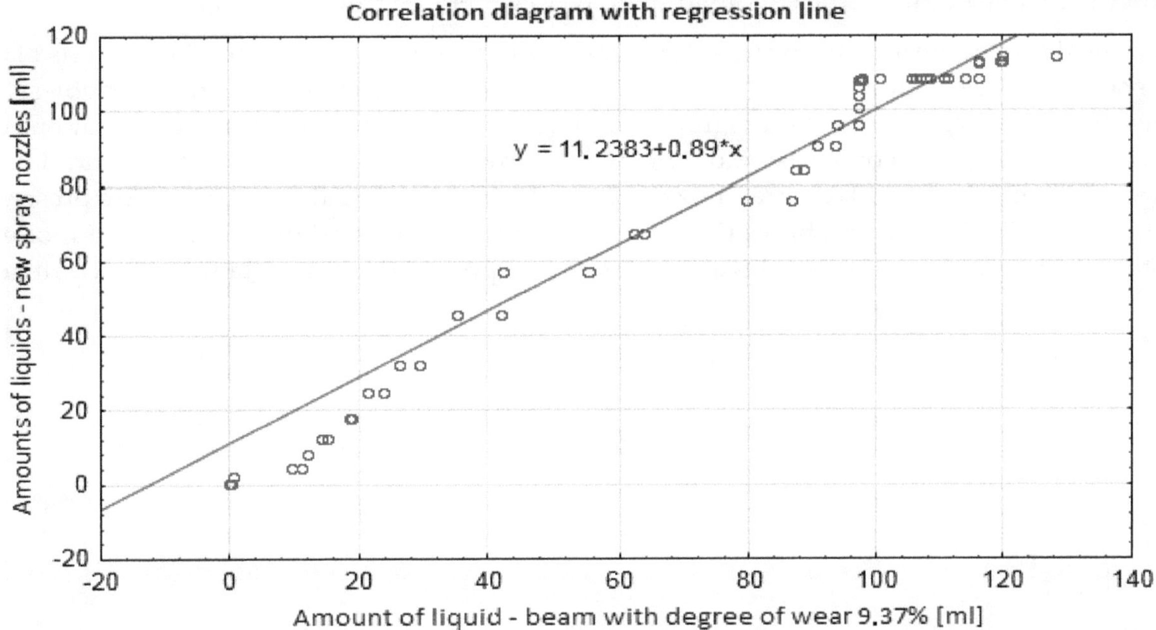

Figure 4. Correlation diagram of the amounts of liquid after the spray from a boom characterized by a 9.37% wear degree (axis OX) and from a boom equipped with the new spray nozzles (axis OY).

We analyzed the values of coefficients of correlation and coefficients of determination from each regression analysis of the data generated by the model and those obtained with the use of virtual field booms equipped with worn nozzles.

The coefficients of correlation have a value of approximately 0.95 and higher, which indicates a strong correlation (dependence) between the model and experimental data.

The coefficients of determination, i.e., the measure of the fit of the model to the experimental data, are higher than 0.9, which indicates that the model efficiently explains the distribution of the experimental data.

Therefore, the developed model is characterized by a very high goodness of fit to the experimental data.

The new model can be used for the description of the distribution of the liquid amounts after the spraying process, even with a very low value of the coefficient of variation equal to 2.16%.

This result should be regarded by manufacturers of sprayers as a guarantee of high uniformity of surface coverage.

Additionally, to test the compatibility of the data from the new distribution with the experimental data, the chi-square test for observed and expected values was applied, besides regression analysis.

A null hypothesis of the compatibility of the examined distributions with an alternative hypothesis of the absence of compliance was adopted. The critical value for the chi-square distribution with 28 degrees of freedom was 41.333 at the significance level $\alpha = 0.05$. The values of the test statistics for the results of each nozzle RS-MM 110 03 duplicated on the boom in relation to the data obtained with the new model. Experimental values from each RS-MM 110 03 nozzle correspond to an individual deposition in each groove of the distribution test bench. The higher the value of the χ^2 test statistic is, the more the nozzle, which generates a virtual boom sprayer, is worn out and, in consequence, the higher the value of the coefficient of variation (CV) generated from the boom.

A critical area at the significance level $\alpha = 0.05$ is the range <41.337; $+\infty$). The analysis of all the results shown in Table 4 indicates that there is no ground for the rejection of the null hypothesis about the consistency of distributions at the significance level of 0.05. Hence, the tested distributions can be considered identical. For comparison, data generated by the model (for $V = 2.16\%$) and data from measurements of RS-MM 110 03 spray nozzles constituting the virtual field booms are presented in a graph (Figure 5). The Figure shows the difference between the level of uniformity of model data simulating the work of the new nozzles and of the model provided by the data from the worn nozzles.

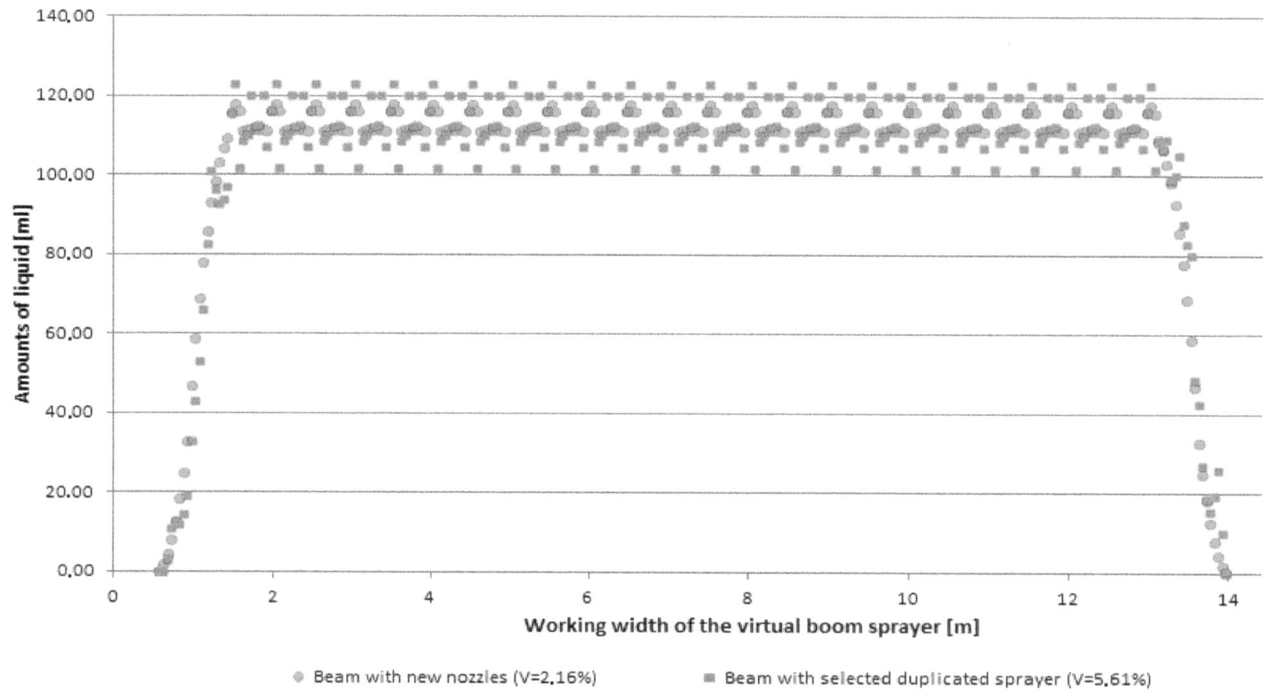

Figure 5. Liquid amounts from the boom equipped with the new nozzles and from the selected virtual boom.

Additionally, we considered the fact that the total amounts of liquid accumulated along the entire booms are the same in both distributions.

These calculations conducted for the measurement table with a 10 cm spacing of grooves, reflecting conditions of nozzle tests and analyses performed by producers, were analogous, although with one exception. Every second value of variable x, denoting the amount of liquid accumulated on the 55th centimeter, 65th centimeter, up to 13.95 meters of the length of the virtual field boom, was deleted from all values of the measurement table with grooves located 5 cm apart from each other.

Data provided by the proposed model generating the values of the coefficient of variation of $V = 2.16\%$ and $V = 10\%$, which simulated spraying performed with both a new nozzle and an adequately worn nozzle, were scaled to obtain the same total amounts of sprayed liquid. Figures 6 and 7 present the amounts of liquid obtained at different values of model parameters for two nozzles with normalized flow rates (the new nozzle and another one with a moderate or permissible degree of wear).

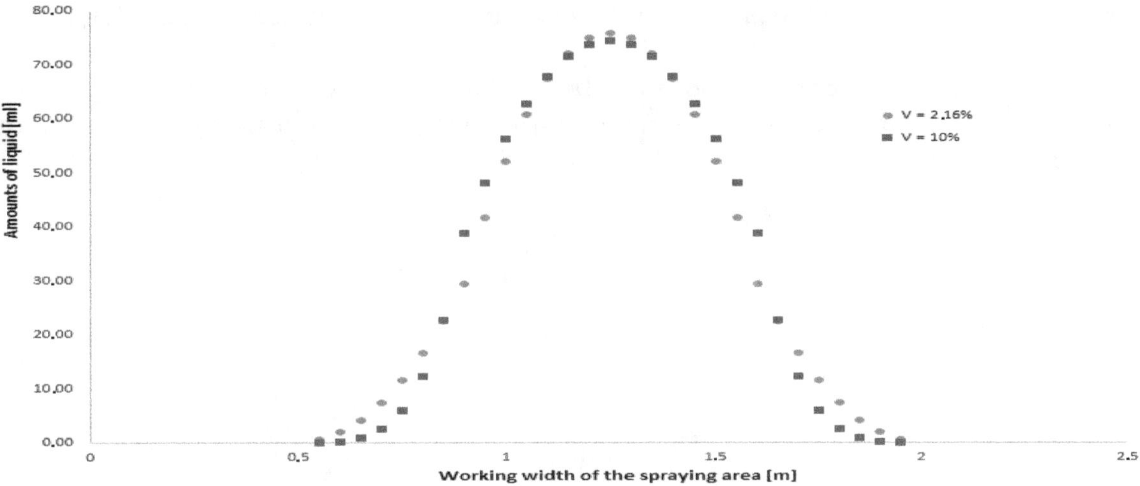

Figure 6. Amounts of liquid sprayed with one nozzle generating $V = 2.16\%$ (new) and $V = 10\%$ ($p = 4$, $q = 2$).

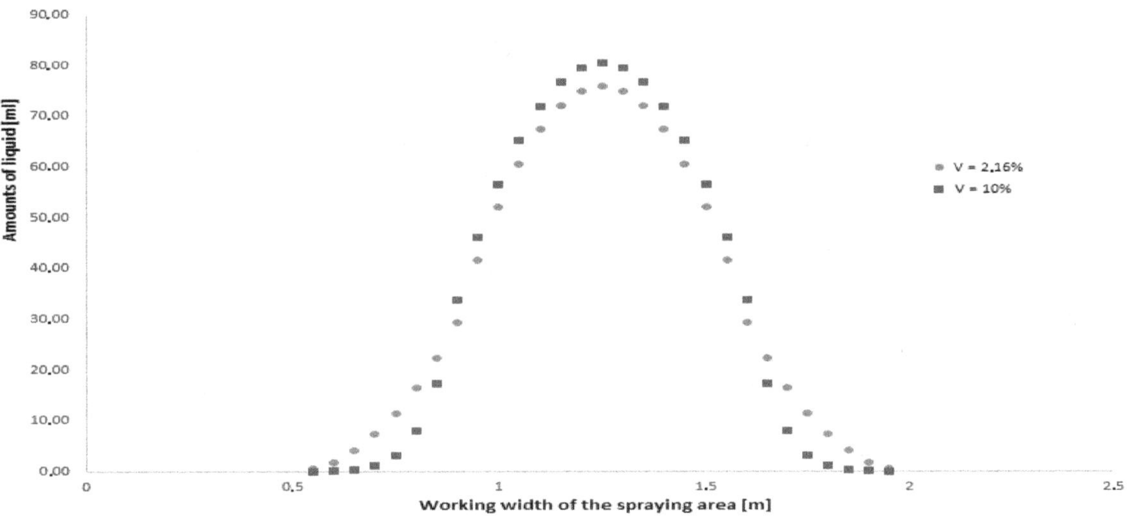

Figure 7. Amounts of liquid sprayed with one nozzle generating $V = 2.16\%$ (new) and $V = 10\%$ ($p = 3$, $q = 2$).

The analysis of the data presented in the graphs revealed the mean ranges of deviations of the amounts of sprayed liquid (Tables 1 and 2), which reflect the moderate or permissible wear of the new (model) nozzle.

Table 1. Changes in the amounts of liquid sprayed by the new nozzle yielding the relevant values of the coefficient of variation ($p = 4$, $q = 2$).

Range [m]	No. of Containers	Mean Change [mL]	Coefficient of Variation
<0.6; 0.85>	12–17	−4	
<0.9; 1.6>	18–32	+3.4	10%
<1.65; 1.9>	33–38	−4	

Table 2. Changes in the amounts of liquid sprayed by the new nozzle yielding the relevant values of the coefficient of variation ($p = 3$, $q = 2$).

Range [m]	No. of Containers	Mean Change [mL]	Coefficient of Variation
<0.6; 0.85>	12–17	−5.6	
<0.9; 1.6>	18–32	+4.7	10%
<1.65; 1.9>	33–38	−5.6	

In other words, the data in the tables facilitate observation and inference on the identification of the adequate nozzle wear.

The tables and graphs presented above can be used for the development of a diagnostic model showing ranges of deviations in the amounts of liquid sprayed by the new nozzle that causes permissible wear, i.e., at the coefficient of variation of 10% (Figures 8 and 9).

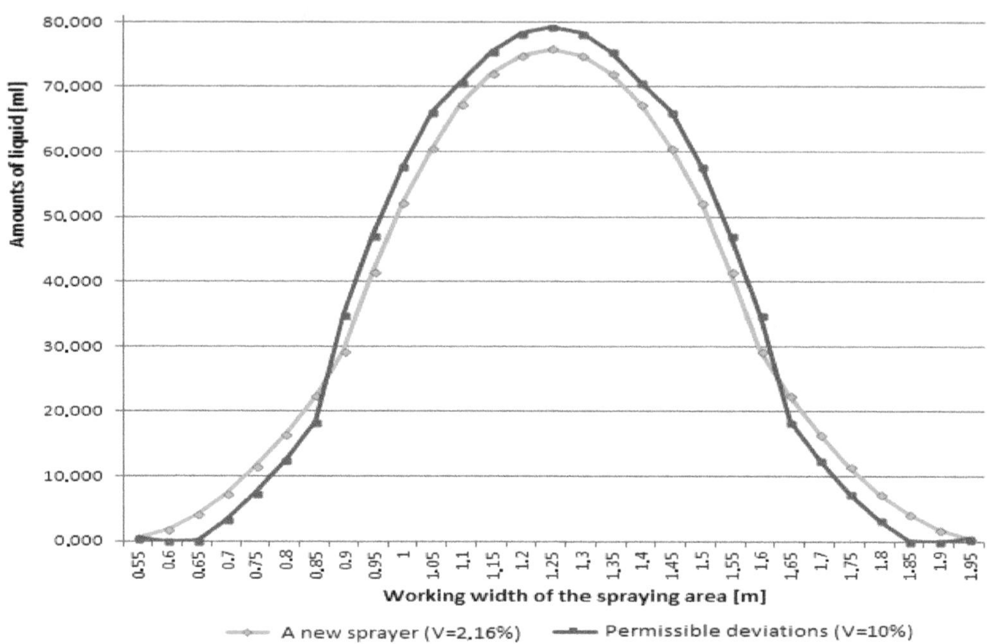

Figure 8. Permissible average ranges of deviations of the amounts of liquid sprayed by the new nozzle ($p = 4$, $q = 2$).

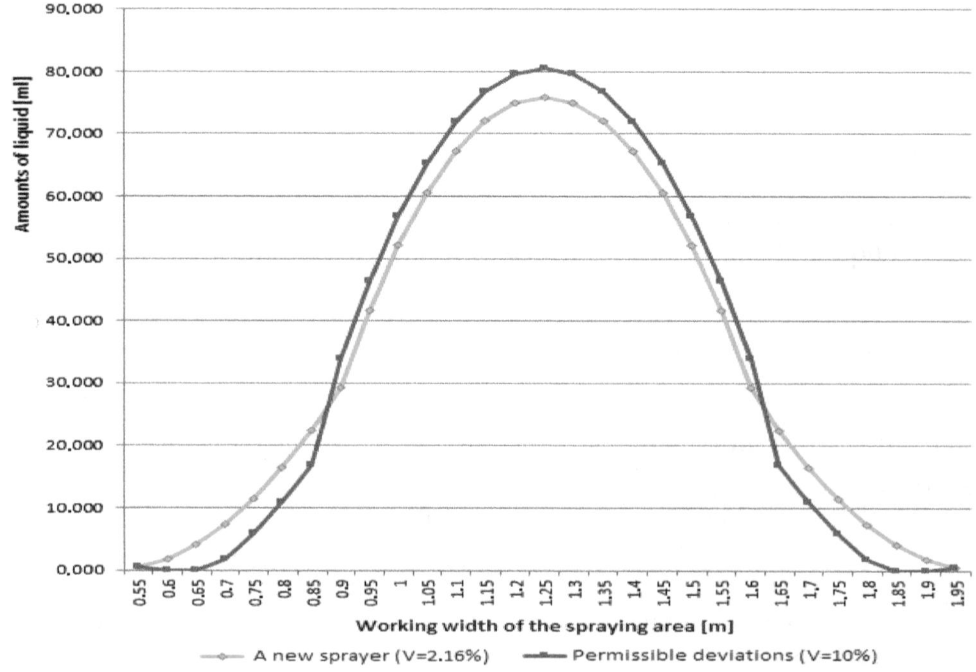

Figure 9. Permissible average ranges of deviations of the amounts of liquid sprayed by the new nozzle ($p = 3$, $q = 2$).

To present the conditions under which producers test new nozzles, a similar analysis of data was carried out for a model simulating measurements from the table with 10 cm groove spacing.

The simulation amounts of liquids from the new nozzle were compared to those from the nozzle with 10% wear (Figure 10).

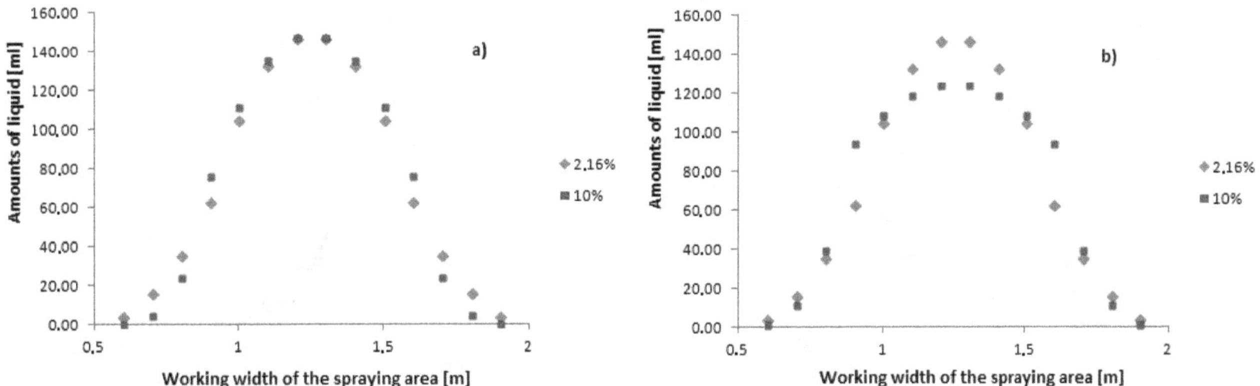

Figure 10. Amounts of liquid sprayed with one nozzle generating a value of the coefficient of variation of $V = 2.16\%$ and $V = 10\%$ on the boom: (**a**): $p = 4, q = 2$; (**b**): $p = 3, q = 2$.

Next, following the results presented above, as in the case of the measurement table with 5 cm groove spacing, Tables 3 and 4 were compiled to show the simulations of the moderate or permissible nozzle wear. Examples (Figure 11) were prepared to present the mean ranges of deviations indicating the wear of the analyzed nozzle in accordance with the adopted model.

Table 3. Changes in the amounts of liquid sprayed by the new nozzle yielding the relevant values of the coefficient of variation ($p = 4, q = 2$).

Range [m]	No. of Containers	Mean Change [mL]	Coefficient of Variation
<0.6; 0.9>	6–9	−9.8	
<1; 1.5>	10–15	+3.8	10%
<1.6; 1.9>	16–19	−9.8	

Table 4. Changes in the amounts of liquid sprayed by the new nozzle yielding the relevant values of the coefficient of variation ($p = 3, q = 2$).

Range [m]	No. of Containers	Mean Change [mL]	Coefficient of Variation
<0.6; 0.7>	6–8	−3.1	
<0.8; 1>	9–10	+13.9	
<1.1; 1.4>	11–14	−17.7	10%
<1.5; 1.7>	15–16	+13.9	
<1.8; 1.9>	17–19	−3.1	

The analysis of the results shown in Tables 1–4 and in Figures 5–11 allows the conclusion that the greater amounts of liquid in the central spray area and the lower amounts on its left and right side indicate the higher degree of nozzle wear. This facilitates the assessment of the degree of wear relative to the quality of the produced model.

Based directly on the data obtained without the application of the new model, the amounts of liquid from two adjacent containers were added, which yielded results that could be obtained at the 10 cm groove spacing. Hence, the spraying treatment exhibited greater uniformity, as the minimum value of the coefficient of variation was 1.85%, which was lower than in the initial case by ca. 0.3%.

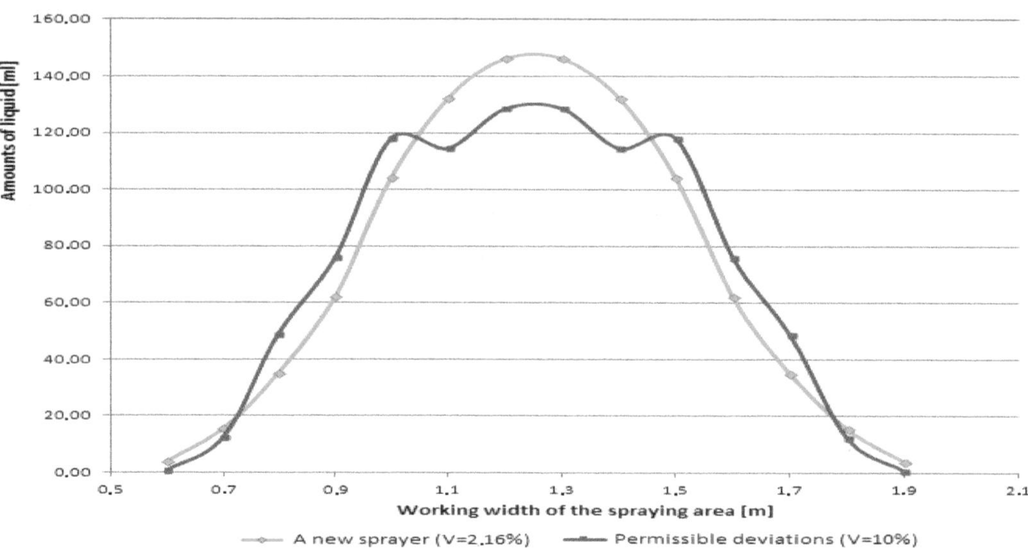

Figure 11. Permissible mean ranges of deviations for the amounts of liquid sprayed by the new nozzle ($p = 3$, $q = 2$).

Additionally, similarly lower values were obtained for each nozzle wear degree, which are summarized in Table 5.

Table 5. Comparison of the coefficients of variation for different nozzle wear degrees at the 5 cm and 10 cm groove spacing.

		Mean Wear		Permissible Wear	
p	2	3	4	3	4
q	2	2	2	2	2
V_{10cm}	1.85%	4.74%	4.61%	9.43%	9.52%
V_{5cm}	2.16%	5%	5%	10%	10%

These findings confirm the thesis that a greater distance between the grooves reduces the value of the coefficient of variability of data [22].

4. Discussion

The spraying process is applied in various areas, for example in fuel spraying [23], in the precipitation profiles of a fixed spray-plate sprinkler [24], in aerial spray application based on thermal imaging technology [25], and in evaluating irrigation system performance, which mainly depends on the uniformity of water application [26] or by using an unmanned aerial vehicle [27]. Another improvement introduced to examine the uniformity of spraying coverage was adding adjuvant to a liquid [28].

There are many factors that affect the uniformity of spray liquid coverage from spray nozzles, such as height of boom and nozzle pressure. The range of the coefficient of variation due to the height of the nozzle ranges from 8% to 17.6%, due to nozzle pressure from 7.6% to 20.3%, and for fixed height and nozzle pressure from 7.2% to 21.3% [29].

Coefficients of variation, in the case of repetitions performed with the same technique, were sometimes quite large, ranging from 7.5% to 24.0% [30], but we achieved with the same approach a value of this coefficient equal to 1.85%, as in Table 5.

When the technique of pulse width modulation is used, the coefficient of variation is around 10%. Increasing the signal means increasing the value of this factor [31]. In the same technique when for various kinds of nozzles the fixed pressure is equal to 207, 276, or 476 kPa, the values of the coefficient of variation range from 5.3% to 20.1% [32].

The best results of the uniformity were obtained in a wind tunnel and had values of CV from 0.5% to 7.6%, where 75% of all nozzle types tested had CV values below 4%. However, the use of water alone, like in our research, caused the largest differences in CV values [33]. In relation to the working width of the entire boom sprayer, the generated CV values were generally around 10% [34].

Considering the literature on the subject of the uniformity of field spraying and looking at the results achieved there, our model is the first one thanks to which we are able to forecast the permissible wear of the atomizer nozzle and then determine its usefulness.

5. Conclusions

The tools were designed and developed for the description and characterization of the distribution of the liquid amounts sprayed on an area with a specified size with the use of flat fan nozzles.

An original probability density function was proposed, which efficiently reflects the accumulation of the amounts of liquid sprayed from a single nozzle.

A new model of the distribution of the amounts of liquid after the spraying process was developed, yielding a minimum value of the coefficient of variation for the amount of liquid that ensures a very high uniformity of surface coverage.

The generation of the model of the distribution of sprayed liquid amounts simulating the operation of the new nozzle should help to maintain the accepted level of spraying uniformity as long as possible, which will allow the optimization of the process.

By the determination of the distribution of the spray ensured by the new nozzle with the use of the model, the level of nozzle wear can be assessed.

The application of the solution presented in this study will facilitate nozzle quality control consisting of the effective monitoring of wear degree and will contribute to the extension of nozzle service life, in compliance with accepted standards.

The results obtained from the model were standardized to the 10 cm spacing of the grooves on the measurement table, i.e., the distance for which manufacturers carry out tests and analyses of nozzles. This mode of presentation of the results will facilitate quality control aimed at the optimization of the field spraying process.

The solution of the research problem yielded a new diagnostic model as a tool for atomization and spray quality control and the assessment of nozzle wear. The model will have a positive effect on the quality, duration, and optimization of the field spraying process as well as the condition of the natural environment through the possibility of the application of an appropriate amount of sprayed agents.

Author Contributions: P.A.K. proposed the mathematical model and verified obtained data. I.K.-B. generally corrected the manuscript. S.P. provided data and formulated technical conclusions.

References

1. The European Parliament and the Council of the European Union. Directive 2009/128EC establishing a framework for Community action to achieve the sustainable use of pesticide. *Off. J. EU* **2009**, *309*, 71–86.
2. Herbst, E.; Herbst, K. Ernst Herbst Pruftechnik e.K.-Plant protection equipment, test engineering, agricultural technology. *Julius-Kühn-Archiv* **2010**, *426*, 127.
3. Lodwik, D.; Pietrzyk, J. Automated Test Station for Transverse Spray Non-Uniformity. *J. Res. Appl. Agric. Eng.* **2013**, *58*, 103–106.
4. International Standarization Organization (ISO). ISO 16119. In *Agriculture and Forest Machinery–Environmental Requirements for Sprayers—Part 2*; ISO: Geneva, Switzerland, 2013.
5. Mawer, C.J.; Miller, P.C.H. Effect of roll angle and nozzle spray pattern on the uniformity of spray volume distribution below a boom. *Crop Prot.* **1989**, *8*, 217–222. [CrossRef]
6. Nowakowski, T. Interaction between selected spraying parameters on the variability coefficient of liquid transverse distribution. *Agric. Eng.* **2007**, *3*, 135–141.
7. Nowakowski, T.; Chlebowski, J. The impact of liquid pressure and design of fan atomizers on spraying angle. *Agric. Eng.* **2008**, *1*, 319–323.

8. Nuyttens, D.; Baetens, K.; De Schampheleire, M.; Sonck, B. Effect of nozzle type, size and pressure on spray droplet characteristics. *Biosyst. Eng.* **2007**, *97*, 333–345. [CrossRef]
9. Parafiniuk, S.; Sawa, J.; Huyghebaert, B. The evaluation of the technical condition of the field toolbar of the spraying machine with the use of the method of survey of selected sprayers. *Agric. Eng.* **2011**, *5*, 207–215.
10. Parafiniuk, S.; Tarasińska, J. Work simulation of the sprayer field boom with the use of R program. *J. Cent. Eur. Agric.* **2013**, *14*, 166–175. [CrossRef]
11. Rojek, G. Analysis Of The Spray Distribution And Coverage In Variable Working Conditions of Selected Nozzles. Ph.D. Thesis, University of Life Sciences in Wroclaw, Wroclaw, Poland, 2013.
12. Szewczyk, A.; Wilczok, G. Theoretical and actual liquid distribution for selected atomizer setting parameters. *Agric. Eng.* **2007**, *8*, 265–271.
13. Zhu, H.; Rowland, D.L.; Dorner, J.W.; Derksen, R.C.; Sorensen, R.B. Influence of Plant Structure, Orifice Size, and Nozzle Inclination on Spray Penetration into Peanut Canopy. *Trans. ASAE* **2002**, *45*, 1295–1301.
14. Szewczyk, A.; Wilczok, G. Theoretical description of sprayed liquid distribution in conditions of frontal air stream operation. *Agric. Eng.* **2008**, *5*, 292–299.
15. Wilczok, G. Analysis of Spray Liquid Distribution during Spraying in Variable Working Conditions of Sprayers. Ph.D. Thesis, University of Life Sciences in Wroclaw, Wroclaw, Poland, 2008.
16. Świechowski, W.; Hołownicki, R.; Doruchowski, G.; Godyń, A. Comparison of the methods evaluating flat spraying nozzles. *Probl. Agric. Eng.* **2006**, *4*, 5–12.
17. Nation, H.J. Spray nozzle performance and effects of boom height on distribution. In *Departmental Note n° DN/S/777/1925*; National Institute of Agricultural Engineering Silsoe: Wrest Park, Silsoe, Bedford, UK, 1976; unpublished.
18. Mawer, C.J. *The Effect of Nozzle Characteristics and Boom Attitude on the Volume Distribution below a Boom*; Div. Note DN 1462; AFRC Institute of Engineering Research: Silsoe, Bedford, UK, 1988.
19. Leunda, P.; Debouche, C.; Caussin, R. Predicting the transverse volume distribution under an agricultural spray boom. *Crop Prot.* **1990**, *9*, 111–114.
20. Sinfort, N.; Bellon, V.; Sevila, F. Image analysis for in-flow measurement of particle size. *Food Control* **1992**, *3*, 84–90. [CrossRef]
21. Huyghebaert, B. Verification of Measurement Methods of Flat Fan Nozzles Working Parameters Used in Agriculture. Ph.D. Thesis, University of Life Sciences in Lublin, Lublin, Poland, 2015.
22. Sawa, J.; Kubacki, K.; Huyghebaert, B. Equivalence of the criteria of assessing results of tests in legalizing crop sprayers. *Agric. Eng.* **2001**, *4*, 1.
23. Li, T.; Nishida, K.; Hiroyasu, H. Droplet size distribution and evaporation characteristics of fuel spray by a swirl type atomizer. *Fuel* **2011**, *90*, 2367–2376. [CrossRef]
24. Sayyadi, H.; Nazemi, A.H.; Sadraddini, A.A. Characterising droplets and precipitation profiles of a fixed spray-plate sprinkler. *Biosyst. Eng.* **2014**, *119*, 13–24. [CrossRef]
25. Jiao, L.; Dong, D.; Feng, H.; Zhao, X.; Chen, L. Monitoring spray drift in aerial spray application based on infrared thermal imaging technology. *Comput. Electron. Agric.* **2016**, *121*, 135–140. [CrossRef]
26. Irmak, S.; Odhiambo, L.O.; Kranz, W.L.; Eisenhauer, D.E. Irrigation Efficiency and Uniformity, and Crop Water Use Efficiency. *Biol. Syst. Eng. Pap. Publ.* **2011**, *451*, 1–8.
27. Xue, X.; Lan, Y.; Sun, Z.; Chang, C.; Hoffmann, W.C. Develop an unmanned aerial vehicle based automatic aerial spraying system. *Comput. Electron. Agric.* **2016**, *128*, 58–66. [CrossRef]
28. Griesang, F.; Decaro, R.A.; dos Santos, C.A.M.; Souza Santos, E.; de Lima Roque, N.H.; da Costa Ferreira, M. How Much Do Adjuvant and Nozzles Models Reduce the Spraying Drift? Drift in Agricultural Spraying. *Am. J. Plant Sci.* **2017**, *8*, 2785–2794. [CrossRef]
29. Sehsah, E.M.E.; Kleisinger, S. Study of some parameters affecting spray distribution uniformity pattern. *MJ Agric. Eng.* **2009**, *26*, 69–93.
30. Nuyttens, D.; Zwertvaegher, I.K.A.; Dekeyser, D. Spray drift assessment of different application techniques using a drift test bench and comparison with other assessment methods. *Biosyst. Eng.* **2017**, *154*, 14–24. [CrossRef]
31. Mangus, D.L.; Sharda, A.; Engelhardt, A.; Flippo, D.; Strasser, R.; Luck, J.D.; Griffin, T. Analyzing the nozzle spray fan pattern of an agriculture sprayer using puls width modulation technology to generate an on-ground coverage map. *Trans. ASABE* **2017**, *60*, 315–325. [CrossRef]

32. Butts, T.R.; Luck, J.D.; Fritz, B.K.; Hofmann, W.C.; Kruger, G.R. Evaluation of spray pattern uniformity using three unique analyses as impacted by nozzle, pressure, and pulse-width modulation duty cycle. *Pest. Manag. Sci.* **2019**, *75*, 1875–1886. [CrossRef] [PubMed]
33. Ferguson, J.C.; O'Donnel, C.C.; Chauhan, B.S.; Adkins, S.W.; Kruger, G.R.; Wang, R.; Ferreira, P.H.U.; Hewitt, A.J. Determining the uniformity and consistency of droplet size across spray drift reducing nozzles in a wind tunnel. *Crop Prot.* **2015**, *76*, 1–6. [CrossRef]
34. Balsari, P.; Gil, E.; Marucco, P.; van de Zande, J.C.; Nuyttens, D.; Herbst, A.; Gallart, M. Field-crop-sprayer potential drift measured using test bench: Effects of boom height and nozzle type. *Biosyst. Eng.* **2017**, *154*, 3–13. [CrossRef]

10

Research on the Work Process of a Station for Preparing Forage

Andrzej Marczuk [1], Wojciech Misztal [1,*], Sergey Bulatov [2], Vladimir Nechayev [3] and Petr Savinykh [4]

1 Department of Agricultural, Forestry and Transport Machines, Faculty of Production Engineering, University of Life Sciences in Lublin, 28 Gleboka Street, 20-612 Lublin, Poland; andrzej.marczuk@up.lublin.pl
2 Department of Technical Service, SBEI HE Nizhniy Novgorod State Engineering and Economic University, 22a Oktyabrskaya Street, Knyaginino 606340, Russia; bulatov_sergey_urevich@mail.ru
3 SBEI HE Nizhniy Novgorod State Engineering and Economic University, 22a Oktyabrskaya Street, Knyaginino 606340, Russia; nechaev-v@list.ru
4 Federal State Budget Scientific Institution. Federal Agricultural Research Centre of the North-East named after Rudnitskiy N.V. 166a Lenin Street, Kirov 610007, Russia; peter.savinyh@mail.ru
* Correspondence: wojciech.misztal@up.lublin.pl

Abstract: Forage from grain plays a special role in animal nutrition because it constitutes feed with a high content of readily available carbohydrates. Unfortunately, the equipment used to prepare forage is often manufactured without the necessary justification and confirmation of the declared sizes and indicators of the work process. This forms the basis for our theoretical and experimental studies. Research has been carried out to provide justification of the design and operating parameters of the patented station for producing forage from cereal crops. This article describes the technology for preparing forage from grain and provides a detailed description of the station used and the principle of its operation. During the experiments, we studied the influence of the angle α of setting the grid-work (plate) and the distance S from the nozzle to the grid-work on the quality of forage. Qualitative, quantitative, and energy indicators have been evaluated using up-to-date measuring instruments and equipment. The method is described, and the studied factors and evaluation criteria for the preparation of forage from grain are indicated. The forage quality results are presented, as determined by the content of whole grains in it via the residue on a sieve with a sieve size of 3 mm when preparing it with a different combination of the studied factors. The analysis of the energy consumption results of the process of preparing forage from grain under various operating conditions of the plant is shown. As a result, the optimal location parameters of the passive grinder have been found, allowing to obtain high-quality forage with minimal power consumption of the electric motor. A grid-work should be used as a grinder. Its installation angle should be 30°, and the distance between the grid-work and the nozzle should be 205 mm. With this combination of parameters, the specific energy consumption is minimal and amounts to 41.5 W·h/L.

Keywords: forage from grain; cereal grain; energy consumption

---

1. Introduction

Obtaining high performance indicators for farm animals is determined by a scientifically validated diet and feeding regime, which must include properly processed cereal grains [1–6]. Preparation of feed directly on the farm will give a positive result if a producer has his/her own grain raw materials. With this knowledge, agricultural producers are trying to find feed preparation equipment that is optimal in terms of quality and quantity, separately or as part of a line, depending on the volume of production [7]. These are mainly crushers or flatteners for producing dry concentrates, which are the most known and proven [8–13]. However, studies show that processing carried out using this type of equipment does not ensure full utilization of the energy value of the grain [11–17]. As previous studies [16,18–22] show, it is possible to obtain feed in the form of forage from grain with a high content of readily available carbohydrates (this kind of feed has great value, primarily for dairy cows). Unfortunately, corresponding equipment is manufactured for such needs without the necessary justification and confirmation of the declared sizes and indicators of the work process [21–24]. Recognizing the practical importance of addressing these issues, theoretical and experimental studies have been carried out to provide justification for the design and operating parameters of the plant for producing forage from cereal crops on the basis of an agreement between the leading manufacturer of animal feed equipment "Doza-Agro" LLC and the Nizhniy Novgorod State Engineering and Economic University (NNSEEU). The article presents the data reflecting the key research insights obtained in the framework of R&D.

2. Materials and Methods

To study the process of preparing forage from grain, a laboratory-scale plant has been developed consisting of a frame on wheels, a tank with a built-on passive grinder, a 1SM65-50-160/2 centrifugal pump and a NGD-1.1 disperser, material pipelines, and a control cabinet (Figure 1). The measuring station operates as follows. Water is poured into the tank based on the required volume of prepared forage at the outlet, which is 2:1 with respect to the grain. The amount of water poured was recorded by a water meter SVK 15-3-8 (Figure 2). Next, the centrifugal pump is turned on, and the heating process to 30 °C takes place. When this temperature is reached, the tank is uniformly filled with pre-processed and cleaned grain at the given water:grain ratio. Next, enzymes are added. Through special inspection ports in the built-in grinder (chopper), the uniformity and homogeneity of the working mass are visually monitored, and, upon reaching a temperature of 60 °C, the pump is switched off. The prepared mass is allowed to infuse for one hour for the more efficient operation of enzymes. The result is a high-carbohydrate forage from grain recommended for feeding farm animals as a single feed or as a part of feed mix.

Figure 1. The measuring station for preparing forage from grain.

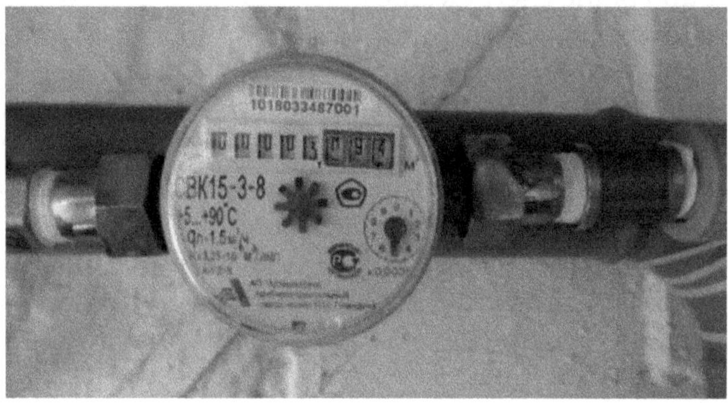

Figure 2. Water meter SVK 15-3-8 used for recording the amount of water poured into the tank of the measuring station.

While the plant pump is operating, the grain is crushed by its blades. To accelerate the process of grain grinding and to obtain a more even forage composition, it was proposed to install a passive grinder in the form of a grid-work (grill) or a plate in the upper part of the tank (Figure 3). During the experiments, the influence of the angle α of setting the grid-work (plate) and the distance S from the nozzle to the grid-work (Figure 4) on the quality of forage have been studied. It was modified by moving the grid-work (plate) between the holes of the fixture of the laboratory-scale plant (Figure 5). The distance S between the grid-work (plate) and the nozzle was adjusted by moving the grid-work (plate) along the slots in the cover (Figure 6).

a

b

Figure 3. Passive grain grinder made in the form of (**a**) grid-work and (**b**) plate.

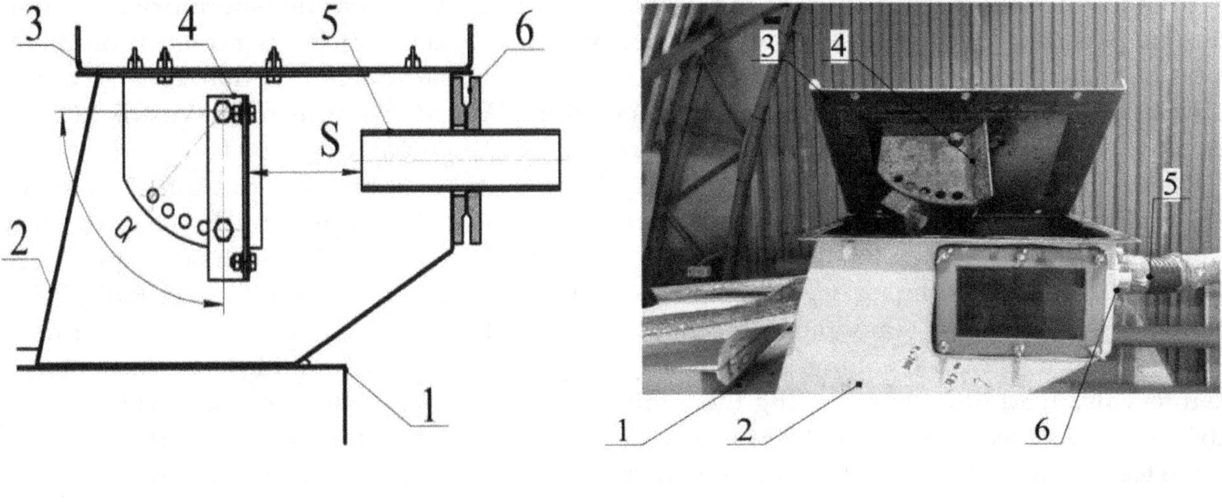

Figure 4. A device for adjusting the inclination angle of the grid-work (plate) and its distance to the nozzle. (**a**) Diagram; (**b**) General vie: (1) tank with a built-in passive grinder, (2) case of the passive grinder, (3) cover of the passive grinder, (4) grid-work (plate), (5) nozzle, (6) nozzle flange.

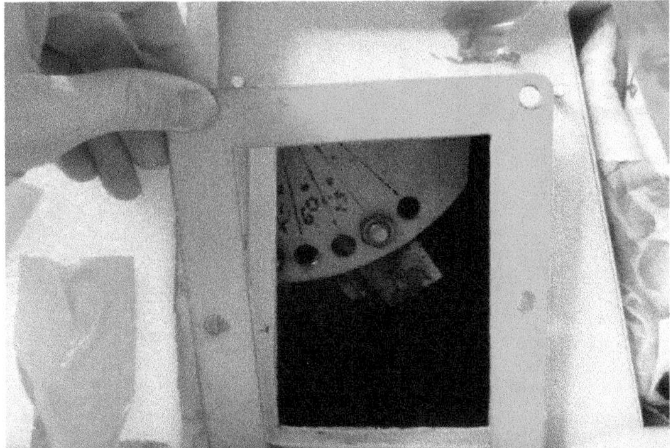

Figure 5. The system for adjusting the inclination angle of the grid-work (plate).

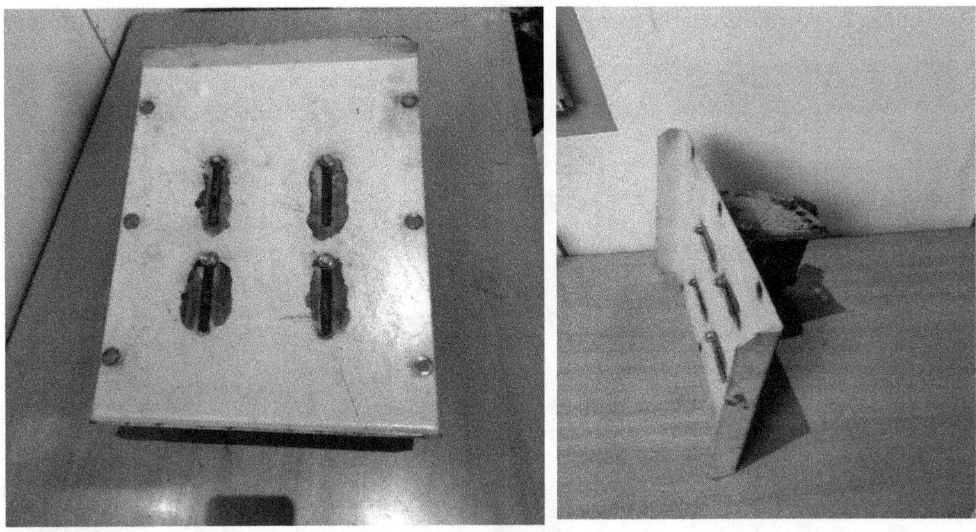

Figure 6. Cover with slots for adjusting the clearance S between the grid-work (plate) and the nozzle.

To determine the performance of the proposed plant under scientific laboratory conditions, a technique was developed to assess the influence of design and operating parameters on the work process of preparing forage from grain. Wheat was used as the source raw material. To carry out the fermentation process, the multi-enzyme composition MEC-AC-3 produced by Vostok LLC, Kirov Region (Russian Federation), was used. The volume of forage prepared was 50 L.

Routine of the experiment was as follows. First, 17 kg of grain was put into 33 L of water pre-heated to a temperature of 30 °C, and 80 g of enzyme was added, previously weighed with the accuracy of up to 0.01 g using the Biomer VK-300.01 laboratory balance (Novosibirsk) (Figure 7). From that moment, the process timing began using the HUAWEI P20 lite stopwatch (China). Every 7 min 30 s before the working mass temperature reached 60 °C, a test batch was taken from the tank: the tank cover was opened, and a 1 L volume bucket was placed under a stream and filled up. Within the same time intervals, in addition to recording the temperature values, the power consumed by the pump electric motor was recorded using the Mastech MS2203 clamp meter (Pittsburg U.S.) (Figure 8). After that, the feed sample was poured onto a sieve with a sieve size of 0.5 mm. Free water was removed by vigorous stirring with a brush. The resulting mass was poured into a plastic bag, which was assigned an information tag (Figure 9). Three to five packets with samples were obtained before the temperature reached 60 °C. After that, the pump was turned off, and one hour was timed until the forage from grain was completely prepared.

Figure 7. Weighing the multi-enzyme composition MEC-AC-3 using the VK-300.01 laboratory balance.

Figure 8. Measuring the power consumption of the electric motor using the Mastech MS2203 clamp meter.

Figure 9. Bags with samples of forage from grain.

Temperature of the forage from grain (and water) was recorded using the TST81 sensor, the data from which was transmitted to the TL-11-250 temperature controller (Figure 10). During the experiment, the dynamics of temperature changes was recorded from the digital display of the temperature controller (Figure 10).

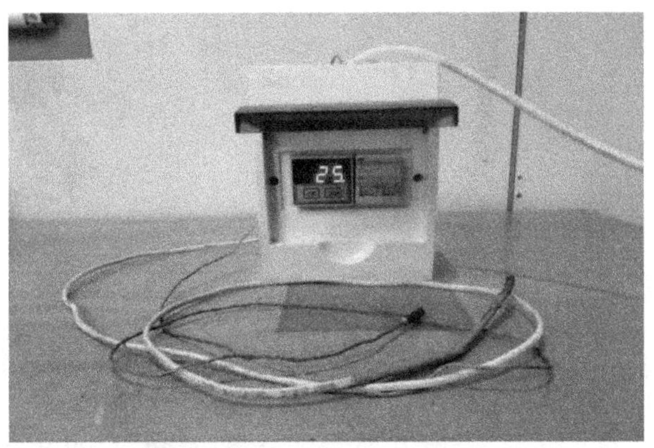

a

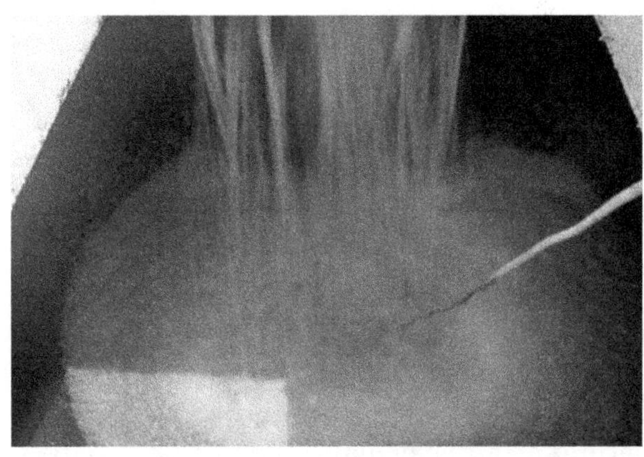

b

Figure 10. Temperature controller TL-11-250 with a temperature sensor TST81 (**a**) and recording the temperature of the forage from grain (**b**).

Next, a 100 g sample weight was selected from each bag and distributed on the sieve with a sieve size of 3 mm. After that, the sample was rinsed out in a bath with water. Next, the sieve residue was weighed, and the number of whole grains was counted.

An hour later, a control sample was taken from the premixed forage from grain. To determine the uniformity, screen sizing was collected using sieves of the following sizes: 2; 1.4, 1; 0.5 mm and the bottom. Then, 100 g of the sample weight was fed to the upper sieve, and then the plansifter (RL brand) was turned on for 5 min (Figure 11). First, the dry sieves were weighed, and their mass was determined. Then, they were weighed with the residue, and the difference was determined. For best data reliability, this analysis was performed in triplicate.

Figure 11. Laboratory plansifter RL brand with a set of sieves for determining the granulometric composition of the forage.

For energy estimation of the plant operation, specific energy consumption was used referred to a unit volume of the working mixture, w1:

$$w1 = \frac{W \cdot \tau}{V}, \quad (1)$$

where

W is the average power consumption, kW;
τ is the process time, s;
V is the volume of water and working mixture, l.

3. Results

After the experiments, tables and graphs were constructed that characterize the change in the quality of forage from grain depending on the installation parameters (α and S) of the grid-work and the time t of the fermentation process (Tables 1 and 2, Figures 12–15). According to the results of experiments and calculations according to the formula (1), histogram 16 was constructed. Histogram 16a shows what full electric motor power is spent on preparing the forage. Figure 16b presents the change in the specific costs of electricity calculated using the expression (1). The constructed histograms show how the power consumption of the electric motor changes when the angle α of setting the grid-work (plate) and the distance S between the grid-work (plate) and the nozzle are modified.

Regardless of the angle α of setting the grid-work and the distance S, the number of whole grains decreases to 0 after 1350 s (22.5 min) of the plant operation (Figure 12). After 900 s (15 min) of the plant operation, the minimum number of whole grains is observed when the grid-work is installed according to options 1 and 3 (Figure 12). That is, according to the indicator m1 "the number of whole grains in the forage," we can recommend the grid-work installation scheme using options 1 and 3.

The number of whole grains reaches the value of 0 in 1350 s (in 22.5 min) when the plate is installed at an angle $\alpha = 90°$ at S = 140 mm (Figure 13). In other cases, in order to avoid the presence of whole grains in the forage, the operation time of the plant should be at least 30 min. The graph also shows that with decreasing angle α, the number of whole grains in the forage increases. This is explained by the fact that, when the angle α is decreased, the impact force of the grain on the plate goes down. From the analysis of the graphs presented in Figure 13, we can conclude that the best results of preparing forage using a plate are obtained when it is installed according to options 3, 4, and 5 (Figure 13).

The best performance in terms of the indicator "residue on the sieve with a sieve size of 3 mm" is also achieved when the plate is installed according to options 3, 4, and 5 (Figure 14). In these cases, the m2 indicator is close to zero after 1800–1850 sec of the plant operation. When the plate is installed at an angle of 30 and 45° for a time t = 1800 s, m2 equals 0.1% and 0.4%, respectively, and a zero value is reached only after 2080 and 1989 s, respectively. That is, in terms of the indicator "residue on a sieve with a diameter of 3 mm," it can be recommended to prepare forage when the plate is installed according to options 3 and 5.

The average power consumed by the engine while preparing forage did not exceed 5 kW. The lowest value was recorded when the grid-work was installed at an angle of 30° and at a distance of 205 mm from the nozzle—it was 4 kW. In general, the average engine power during the preparation of forage using a grid-work is lower than when using a plate. When using the grid-work, it did not exceed 4.5 kW, and when using the plate, it varied from 4.51 to 5 kW (Figure 16a).

Change in specific energy consumption required to prepare 1 L of forage from grain is shown in Figure 16b. In general, it can be seen that when a grid-work is used, the specific energy consumption is lower than when a plate is used. Based on the findings presented, we can recommend the use of a grid-work installed at an angle of 30° at a distance of 140 and 205 mm from the nozzle, or a plate installed at an angle of 90° at a distance of 140 mm from the nozzle. With this combination of parameters, the minimum specific energy consumption is observed, which amounts to 41.5–48.4 W·h/L. In other cases, there is an increase in this indicator to 50–52 W·h/L (Figure 15c).

Table 1. The change in the quality of forage from grain depending on the installation parameters (α and S) of the grid-work (where M is average, SD is standard deviation, and SKE is skewness factor).

Czas [s]	Number of whole grains in molasses when prepared using a grid-work (pcs.)			The residue on the sieve with a sieve size of 3 mm when prepared using a grid-work (%)		
	M	SD	SKE	M	SD	SKE
1—distance from the grid-work to the nozzle is 140 mm, grid-work inclination angle is 30 degrees						
450	916	5.18	−0.68	35.95	0.95	0.18
900	1	0.63	0	0.8	0.13	0
1350	0	0	0	0.1	0.05	0
1800	0	0	0	0	0	0
1857	0	0	0	0	0	0
2—distance from the grid-work to the nozzle is 140 mm, grid-work inclination angle is 45 degrees						
450	780	2.9	0.73	28.81	2.21	−0.01
900	2	0.63	0	1.17	0.07	−0.06
1350	0	0	0	0.43	0.1	0
1800	0	0	0	0	0	0
1984	0	0	0	0	0	0
3—distance from the grid-work to the nozzle is 205 mm, grid-work inclination angle is 30 degrees						
450	255	2.28	−0.91	7.45	0.69	−0.6
900	1	0.63	0	0.75	0.05	−1.1
1350	0	0	0	0.17	0.01	0
1800	0	0	0	0.01	0	0
1998	0	0	0	0	0	0
4—distance from the plate to the nozzle is 205 mm, grid-work inclination angle is 45 degrees						
450	653	3.52	−0.61	24.26	2.21	0.19
900	4	1.67	−1.15	1.55	0.18	−0.6
1350	0	0	0	0.07	0.01	0
1800	0	0	0	0.01	0	0
1878	0	0	0	0	0	0

Table 2. The change in the quality of forage from grain depending on the installation parameters (α and S) of the plate (where: M—average, SD—standard deviation, SKE—skewness factor).

Czas [s]	Number of whole grains in molasses when prepared using a plate [pcs.]			The residue on the sieve with a sieve size of 3 mm when prepared using a plate [%]		
	M	S	SKE	M	S	SKE
1—distance from the plate to the nozzle is 140 mm, plate inclination angle is 30 degrees						
450	615	68.77	0.56	51.21	4.86	−0.81
900	56	10.53	0.68	2.7	0.21	0.47
1350	20	3.16	0	0.59	0.04	0
1800	0	0	0	0.4	0.08	0
2080	0	0	0	0	0	0
2—distance from the plate to the nozzle is 140 mm, plate inclination angle is 45 degrees						
450	41	4	0.98	28.29	1.16	−0.44
900	9	2.28	1.21	4.9	0.3	−0.19
1350	2	0.63	0	0.67	0.06	0
1800	0	0	0	0.09	0.02	0
1989	0	0	0	0	0	0
3—distance from the plate to the nozzle is 140 mm, plate inclination angle is 90 degrees						
450	680	41.42	−2.13	44.53	3.55	−0.85
900	3	1.1	1.36	1.02	0.13	−0.1
1350	0	0	0	0.08	0.02	0
1800	0	0	0	0.04	0	0
1851	0	0	0	0.01	0	0
4—distance from the plate to the nozzle is 205 mm, plate inclination angle is 90 degrees						
450	400	4.69	−0.26	400	15.61	0.72
900	5	2.76	−0.09	5	1.67	−0.38
1350	1	5.18	0	1	0.46	0
1800	0	2.83	0	0	0	0
1852	0	6.42	0	0	0	0
5—distance from the plate to the nozzle is 305 mm, plate inclination angle is 90 degrees						
450	400	1.41	2	400	17.52	1.93
900	12	4.05	3	12	2.53	2.1
1350	1	4.69	4	1	0.38	−0.25
1800	0	3.85	5	0	0	0
1852	0	7.21	6	0	0	0

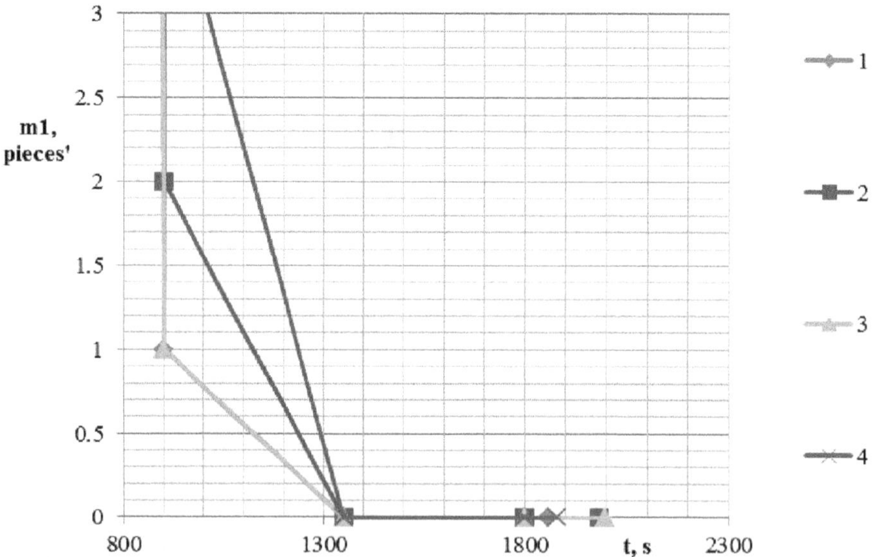

Figure 12. Number of whole grains in molasses when prepared using a grid-work: 1—the distance from the grate to the nozzle is 140 m the angle of inclination of the grate is 30 degrees; 2—the distance from the grate to the nozzle is 140 mm, the angle of inclination of the grate is 45 degrees; 3—distance between the grate and nozzle 205 mm, angle of inclination of the grate 45 degrees; 4—distance between the grate and nozzle 205 mm, angle of inclination 30 degrees.

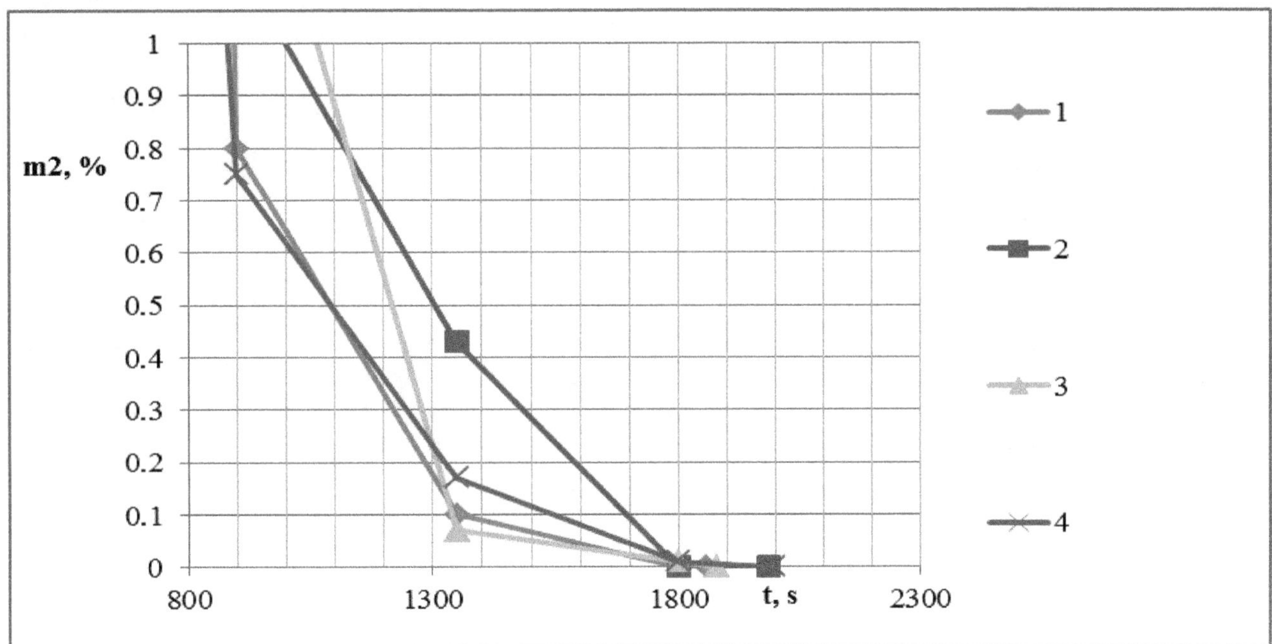

Figure 13. The residue on the sieve with a sieve size of 3 mm when preparing forage using a grid-work: 1—distance from the grid-work to the nozzle is 140 mm, grid-work inclination angle is 30 degrees; 2—distance from the grid-work to the nozzle is 140 mm, grid-work inclination angle is 45 degrees; 3—distance from the plate to the nozzle is 205 mm, grid-work inclination angle is 45 degrees; 4—distance from the grid-work to the nozzle is 205 mm, grid-work inclination angle is 30 degrees.

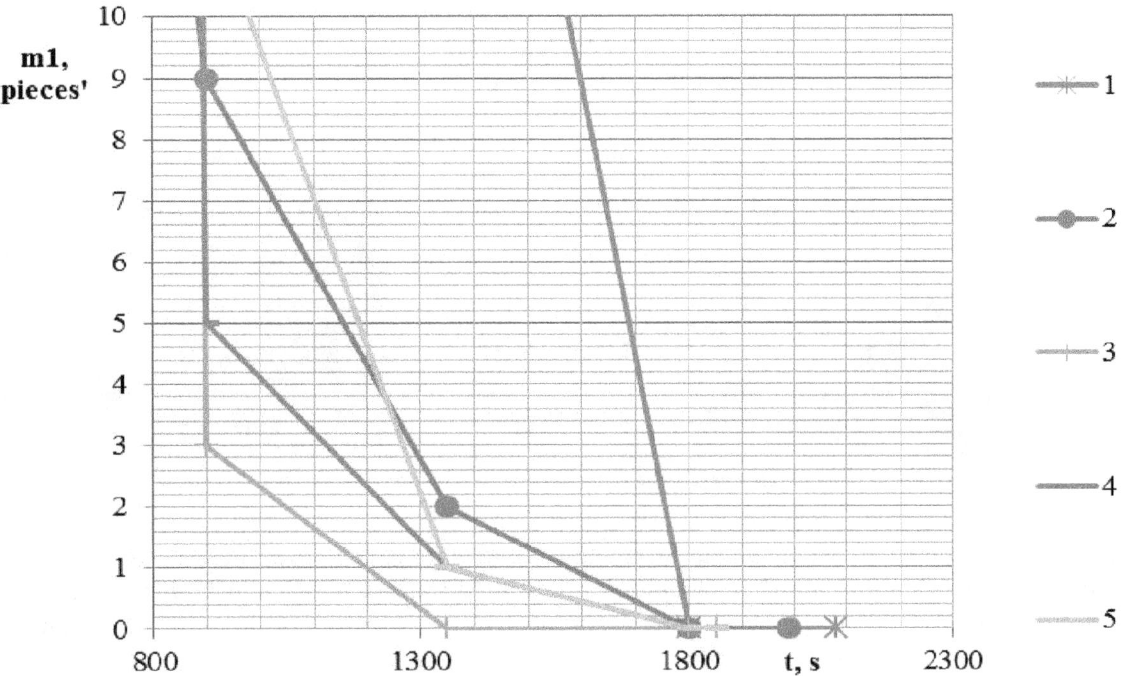

Figure 14. The number of whole grains in forage when it is prepared using a plate: 1—distance from the plate to the nozzle is 140 mm, plate inclination angle is 30 degrees; 2—distance from the plate to the nozzle is 140 mm, plate inclination angle is 45 degrees; 3—distance from the plate to the nozzle is 140 mm, plate inclination angle is 90 degrees; 4—distance from the plate to the nozzle is 205 mm, plate inclination angle is 90 degrees; 5—distance from the plate to the nozzle is 305 mm, plate inclination angle is 90 degrees.

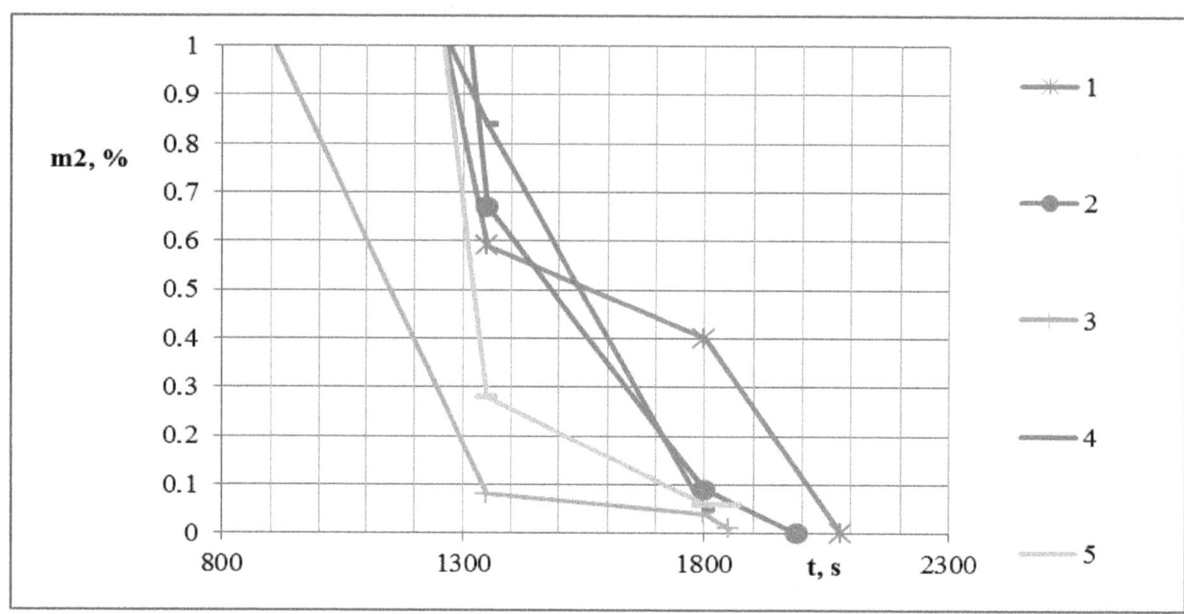

Figure 15. Residue on the sieve with a sieve size of 3 mm when preparing forage using a plate: 1—distance from the plate to the nozzle is 140 mm, plate inclination angle is 30 degrees; 2—distance from the plate to the nozzle is 140 mm, plate inclination angle is 45 degrees; 3—distance from the plate to the nozzle is 140 mm, plate inclination angle is 90 degrees; 4—distance from the plate to the nozzle is 205 mm, plate inclination angle is 90 degrees; 5—distance from the plate to the nozzle is 305 mm, plate inclination angle is 90 degrees.

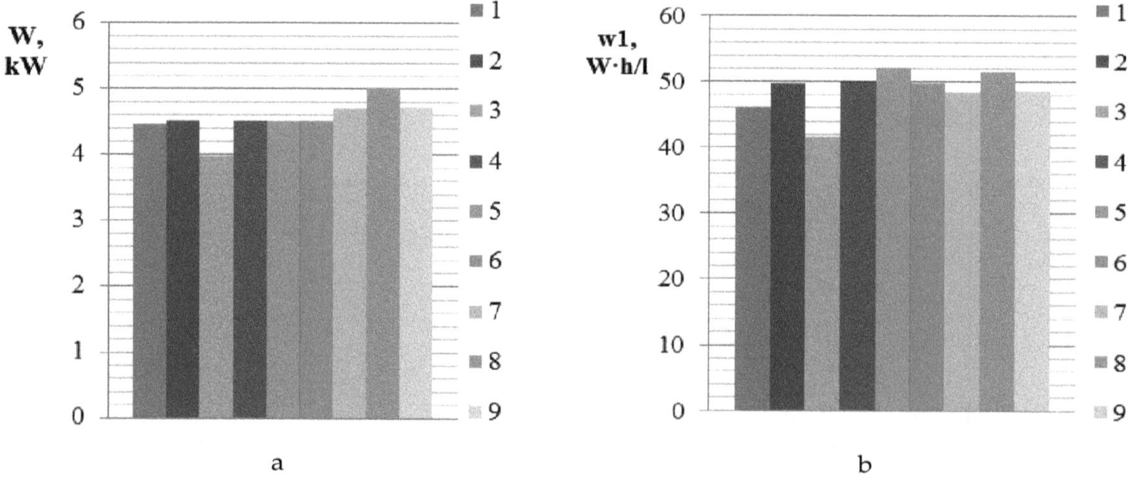

Figure 16. The influence of the studied parameters on: a—power consumption of the electric motor; b—specific energy consumption; 1—distance from the grid-work to the nozzle is 140 mm, grid-work inclination angle is 30 degrees; 2—distance from the grid-work to the nozzle is 140 mm, grid-work inclination angle is 45 degrees; 3—distance from the plate to the nozzle is 205 mm, grid-work inclination angle is 45 degrees; 4—distance from the grid-work to the nozzle is 205 mm, grid-work inclination angle is 30 degrees; 5—distance from the plate to the nozzle is 140 mm, plate inclination angle is 30 degrees; 6—distance from the plate to the nozzle is 140 mm, plate inclination angle is 45 degrees; 7—distance from the plate to the nozzle is 140 mm, plate inclination angle is 90 degrees; 8—distance from the plate to the nozzle is 205 mm, plate inclination angle is 90 degrees; 9—distance from the plate to the nozzle is 305 mm, plate inclination angle is 90 degrees.

4. Conclusions

As a result of studies of the working process of the developed patented plant for preparing forage from grain, and taking into account the aggregate estimate of the quality indicators of forage and the cost of electricity for its preparation, the optimal location parameters of the passive grinder have been found, allowing us to obtain high-quality forage with minimal power consumption of the electric motor. A grid-work should be used as a grinder. Its installation angle should be 30°, and the distance between the grid-work and the nozzle should be 205 mm. With this combination of parameters, the specific energy consumption is minimal and amounts to 41.5 W·h/L.

Author Contributions: Conceptualization, A.M. and S.B.; methodology, S.B and P.S.; software, V.N.; validation, A.M., S.B.; formal analysis, P.S.; writing—original draft preparation, S.B., P.S., V.N.; writing—review and editing, A.M., W.M. All authors have read and agreed to the published version of the manuscript.

Acknowledgments: We are thankful to the Editor and the reviewers for their valuable comments and detailed suggestions to improve the paper.

References

1. Djuragic, O.; Levic, J.; Serdanovic, S.; Lević, L. Evaluation of homogeneity in feed by method of microtracers. *Arch. Zootech.* **2009**, *12*, 85–91.
2. Flizikowski, J.; Tomporowski, A. Movement characteristics of multi-disc cereal crusher. *Przemysł Chem.* **2013**, *92*, 498–503.
3. Flizikowski, J.; Sadkiewicz, J.; Tomporowski, A. Use characteristics of six-roll milling of grained raw materials for the chemical and food industry. *Przemysł Chem.* **2015**, *94*, 69–75.
4. Tomporowski, B.; Flizikowski, J.; Kruszelnicka, W. A new concept for a cylindrical plate mill. *Przemysł Chem.* **2017**, *96*, 1750–1755.
5. Svihus, B.; Kløvstad, K.H.; Perez, V.; Zimonja, O.; Sahlström, S.; Schüller, R.B.; Prestløkken, E. Physical and nutritional effects of pelleting of broiler chicken diets made from wheat ground to different coarsenesses by the use of roller mill and hammer mill. *Anim. Feed Sci. Technol.* **2004**, *117*, 281–293. [CrossRef]
6. Sysuyev, V.A.; Aleshkin, A.V.; Savinykh, P.A. Feed-processing machines. In *Theory, Development, Experiment*; North-Eastern Zonal Agricultural Research & Development Institute: Kirov, Russia, 2008; Volume 1.
7. Ali, E.; Morad, M.M.; Fouda, T.; Elmetwalli, A.; Derbala, A. Development of a local animal feed production line using programming control system. *MISR J. Agric. Eng.* **2016**, *33*, 181–194.
8. Bulgakov, V.; Pascuzzi, S.; Ivanovs, S.; Kaletnik, G.; Yanovich, V. Angular oscillation model to predict the performance of a vibratory ball mill for the fine grinding. *Biosyst. Eng.* **2018**, *171*, 155–164. [CrossRef]
9. Kusmierczak, S.; Majzner, T. Comprehensive approach to evaluation of degradation in chosen parts of energy equipment. *Eng. Rural Dev. Proc.* **2017**, *12*, 673–679.
10. Bulgakov, V.; Holovach, I.; Bandura, V.; Ivanovs, S. A theoretical research of the grain milling technological process for roller mills with two degrees of freedom. *INMATEH Agric. Eng.* **2017**, *52*, 99–106.
11. Savinykh, P.; Kazakov, V.; Czerniatiev, N.; Gerasimova, S.; Romanyuk, V.; Borek, K. Technology and equipment for obtaining starch syrup with ground and whole cereal grain. *Agric. Eng.* **2018**, *22*, 57–67. (In Russian)
12. Vaculik, P.; Maloun, J.; Chladek, L.; Poikryl, M. Disintegration process in disc crushers. *Res. Agric. Eng.* **2013**, *59*, 98–104. [CrossRef]
13. Yancey, N.; Wright, C.T.; Westover, T.L. Optimizing hammer mill performance through screen selection and hammer design. *Biofuels* **2013**, *4*, 85–94. [CrossRef]
14. Ghorbani, Z.; Masoumi, A.; Hemmat, A. Specific energy consumption for reducing the size of alfalfa chops using a hammer mill. *Biosyst. Eng.* **2010**, *105*, 34–40. [CrossRef]
15. Linke, B.S.; Dornfeld, D.A. Application of axiomatic design principles to identify more sustainable strategies for grinding. *J. Manuf. Syst.* **2012**, *31*, 412–419. [CrossRef]
16. Sysuev, V.; Savinyh, P.; Saitov, V.; Gałuszko, K.; Caban, J. Optimization of structural and technological parameters of a fermentor for feed heating. *Agric. Eng.* **2015**, *12*, 99. (In Russian)
17. Motovilov, K.; Motovilov, O.; Aksyonov, V. *Nanobiotechnology in the Production of Forage from Grain for Animal*

Husbandry: A Monograph; Publishing House of the Novosibirsk State Agrarian University: Novosibirsk, Russia, 2015. (In Russian)

18. Donkova, N.; Donkov, S. Biotechnology for the production of sugars from grain raw materials. *Bull. Krasn. State Agrar. Univ.* **2014**, *6*, 211–213. (In Russian)
19. Volkov, V. Efficiency of up-to-date equipment for the production of forage from grain. *World Sci. Cult. Educ.* **2013**, *1*, 351–354. (In Russian)
20. Aksyonov, V. Up-dating the lines for obtaining feed forage from grain raw materials using cost-effectiveness analysis. *Bull. Krasn. State Agrar. Univ.* **2011**, *12*, 220–224. (In Russian)
21. Novosibirsk Prototype Manufacturer of Non-Standard Equipment-Selmash LLC. Available online: http://noezno.ru/equipment/korm/kip-06/kip-06 (accessed on 12 April 2019).
22. Agrotechnology. Available online: http://www.stav-agro.ru/index.php/katalog/urva-250.html (accessed on 15 May 2018).
23. Agrobase. Available online: https://www.agrobase.ru (accessed on 12 April 2019).
24. Short Feeding of Cattle. Available online: http://agrokorm.info/ru/kormoagregat-mriya-05/1/ (accessed on 12 April 2019).

11

Biochars Originating from Different Biomass and Pyrolysis Process Reveal to have Different Microbial Characterization: Implications for Practice

Wioletta Żukiewicz-Sobczak [1], Agnieszka Latawiec [2,3,4], Paweł Sobczak [5,*], Bernardo Strassburg [2,3,4], Dorota Plewik [1] and Małgorzata Tokarska-Rodak [1]

[1] Pope John Paul II State School of Higher Education in Biala Podlaska, Sidorska 95/97, 21-500 Biala Podlaska, Poland; wiola.zukiewiczsobczak@gmail.com (W.Z.-S.); dorotaplewik@gmail.com (D.P.); rodak.malgorzata@gmail.com (M.T.-R.)

[2] International Institute for Sustainability, Estrada Dona Castorina 124, Horto, Rio de Janeiro 22460-320, Brazil; alatawiec@gmail.com (A.L.); bernardobns@gmail.com (B.S.)

[3] Department of Geography and the Environment, Rio Conservation and Sustainability Science Centre, Pontifical Catholic University of Rio de Janeiro, Rio de Janeiro 22451-900, Brazil

[4] Department of Production Engineering, Logistics and Applied Computer Science, Faculty of Production and Power Engineering, University of Agriculture in Kraków, Balicka 116B, 30-149 Kraków, Poland

[5] Department of Food Engineering and Machines, University of Life Sciences in Lublin, Akademicka 13, 20-950 Lublin, Poland

* Correspondence: pawel.sobczak@up.lublin.pl

Abstract: Sustainable technologies are increasingly promoted in various production areas. Protection of natural resources, as well as rational waste management, may lead to better optimization of technologies. Biochar, a product of pyrolysis of organic residues has found wide applications in waste management, agriculture, energy and construction industry. In the present study biochar samples produced in Poland and in Brazil were analysed for microbial content using three substrates: Plate Count Agar, Malt Agar, and Potato Agar. Both qualitative and quantitative measurements were done. Microscopic analysis of the biochar structure was also performed. We found that microbial cultures in both biochars represented a wide range of biodiversity of microorganisms genera and species. We demonstrate that the biochar samples differ depending on the botanical origin as well as on the production technology. Structure of the tested samples also varied depending on the botanical origin. Sample 1-PL (pine) was characterised by a compact and regular structure, while sample 2-PL (oak) showed porous and irregular structure. Sample from Brazil (1-BR) showed a more delicate structure than Polish biochars. Obtained properties may suggest a range of implications for practice.

Keywords: biochar; microbiological analysis; structure; implications for practice

---

1. Introduction

Biochar, carbonous substrate originating from the process of pyrolysis [1] has been demonstrated to have various potential benefits such as increasing agricultural yields [2–4], soil and sediment remediation [5] or improving animals health in livestock farming [6]. There has been also a recent interest in using biochar in construction [7] with potential benefits for diminishing humidity inside buildings, thermal isolation and fungus development control [8–12]. Biochar or microporous carbon is also used for humidity control, a property that has been extensively researched by Nakano et al. (1996) and Abe et al. (1995). Humidity control by microporous carbon can reduce asthma and dermatitis occurrences by controlling the growth of mould and ticks [10–13]. However, there is little research that looks into a range of biochars, identifying and quantifying microbial colonies. Although the number and the type of microbial colonies in biochar have various implications, including crop interactions and safety, the microbial composition remains an under-investigated aspect of biochar research. Of particular interest, are the bacteria from family Bacillus. They are common throughout a range of environments and tolerate acid and alkaline conditions. On the account of the possibility to create endospores they can survive in extreme environments and are commonly isolated from unfresh and spoiling food products. Majority of the Bacillus species are safe for people with the exception of B. anthracis and B. cereus. Due to their widespread persistence and resistance, these bacteria are used in commercial production of enzymes, antibiotics, insecticides, vitamins and metabolites such as hyaluronic acid (used for example in cosmetics) [14,15]. Enzymes such as α-amylases, able to hydroliseα-1,4-glycoside bonds in starch, originating from Bacillus sp. are commonly used in food, chemical and textile industry as well as for the production of pharmaceutics [16–18].

Another important characterization of biochar is their fungus content, such as Alternaria, Aspergillus, Candida, Cladosporium, Penicillium and Rhizopus. Although favouring humid environments, Aspergillus and Penicillium, are able to survive in dry conditions. A range of fungus metabolites are antibiotics, such as the metabolites of Penicillium puberulum. An important feature of fungus is their mycotoxin production (such as alphatoxins), on the account of their cancerogenic and teratogenic properties, even in small concentrations [19,20]. They are degraded in an alkaline environment and under UV radiation [21–25]. The implications of microbial and fungus content in biochar can be far-reaching. Even though some of these organisms may remain inactive due to its limitations in accessibility, the presence of microbes and fungus may have both positive and negative implications, and their use will vastly depend on destination. Biochars rich in desired bacteria may be used as a substrate in in-vitro production whereas those with limited undesired organisms may be used for construction or in the food industry. However, the microbial characterization is not yet commonplace and the properties of different biochars from different biomass and different pyrolysis processes have been rarely reported while the microbial community composition associated with biochar is poorly understood [26,27].

Here we present the qualitative and quantitative comparison of two dissimilar biochars, focusing on the microbial content, and discuss the implications for the use of these biochars in practice. The biochars were produced in different pyrolysis processes: biochar from Brazil was produced in home-made stove adopted usually by the farmers in their on-site production for their agricultural use or for charcoal sale. Biochar from Poland was industrially produced from a large-scale energy plant. Our results have important implications for the use of biochar in practice and are the first that discuss in comprehensive manner biochars from dramatically different production processes and different biomass with different potential applications.

The aim of the study was to compare Brazilian and Polish biochar samples in terms of the microbial content and the structure of the surface. In addition, an attempt was made to assess the properties of test samples in terms of their potential use in practice.

2. Materials and Methods

The research material consisted of two types of biochar coming from Brazil and Poland:

1. biochar from Brazil obtained from Gliricidia plant (1-BR),
2. biochar from Poland obtained from Pine woodchips (1-PL),
3. biochar from Poland obtained from Oak woodchips (2-PL).

Biochar samples provided for the research were placed in sterile conditions. Similarly, throughout the study, extensive precautions were put in place to limit microbial contamination.

1. Brazilian biochar characterization

The Brazilian biochar was produced in home-made drum stove from Gliricidia sepium at the temperature of approximately 350 °C. Gliricidia is a commonly used plant in organic farming for nitrogen-fixing. It is also used as 'living fence' at agricultural farms and pasturelands. Despite its benefits, it grows fast and may provide excessive shadow, limiting plant growth. In certain circumstances, there is, therefore, a need for cutting off the upperground branches that do not have an alternative use. Moreover, at the site where the biochar was produced (Brazilian Agricultural Research Station, Embrapa Agrobiologia, Seropedica km 47), Gliricidia become an invasive species that disseminated rapidly entering native forest fragments. The source of biomass for biochar production in Brazil, therefore, did not compete with alternative biomass uses, as at times criticised in literature [28]. Branches of Gliricidia were dried for two weeks before pyrolization. Cut and dry biomass was put into the stove, fired up and closed with drum cup and isolated with a layer of sand. The pyrolization time was 24 h. After that time biochar was cooling for another 24 h (with open stove) and sieved at 4 mm (Figure 1).

biochar	pH H2O	C	H	N	H/C
1. Brazil	8.5	60.4	2.47	1.18	0.041

Figure 1. Steps of production: Rio de Janeiro state where biochar was produced, Gliricidia tree, dry Gliricidia before being cut at put into the stove, drum stove, biochar (before being sieved), table with biochar properties [4].

2. Polish biochar characterization

Biochar was produced according to Fluid SA company technology. The process consisted in thermal refining of plant biomass and other post-production biomass residues through their autothermal roasting at the temperature 260 °C in reduction atmosphere and without the use of additional energy, catalysts, and chemical additives.

Carbonising products consisted mainly of biochar (from 65% to 70% of energy compared to the energy of the feed) and process gases (from 20% to 35% of energy compared to the energy of the feed). During the process, a significant increase of the carbon element (C) in relation to the biomass of the feed was recorded (1.5 to 2.0 times) as well as an increase of energy density, on average by 4.0 times, reduction of the amount of hydrogen (H), on average by 2.5 times, and reduction of the amount of oxygen (O_2), on average by 3.0 times. Polish samples of biochar were produced from pine woodchips (1-PL) and oak woodchips (2-PL).

The research was carried out at the Innovation Research Center (CBNI) and Regional Center of Agriculture, Environmental and Innovative Technology (EAT) Pope John Paul II State School of Higher Education in Biała Podlaska, Poland.

2.1. *Microbiological Analysis of Biochar Samples Consisted of Qualitative and Quantitative Determination of Bacteria and Fungi Species*

For the research three standardized substrates were used in the assessment process:

(a) The total number of microorganisms cultivated on PCA substrate (PLATE COUNT LAB-AGAR™) in temperature of 30 °C and time period 72 h.

(b) The total number of fungi colonies cultivated on PDA substrate (POTATO DEXTROSE LAB-AGAR™) in temperature of 30 °C and time period 72 h, after that time the temperature was decreased to 21 °C for the next 72 h.

(c) The total number of fungi colonies cultivated on MA substrate (MALT EXTRACT LAB-AGAR™) in a temperature of 24 °C and time period 144 h.

The substrates used had been purchased from BIOMAXIMA S.A. The dilution method PN-EN ISO 7218 [29] was used with the dilutions of 10^{-1}, 10^{-2}, 10^{-3}, 10^{-4}, and 10^{-5}.

The qualitative and quantitative assessment was done on the basis of microbial flora species composition using macroscopic and microscopic methods, as well as taxonomic keys and atlases. It was expressed in CFU/g units [30–35].

2.2. *Microscopic Examination of the Structure*

Examination of the biochar sample structure was done using a research microscope Nikon Eclipse E-200 with fluorescence attachment and SCA image analysis.

3. Results

In the sample, 1-BR cultured on PCA substrate dominated strains of Bacillus sp. (the average number of colonies from UNC to 204 colonies for dilutions ranging from 1 to 10^{-5}. The plates were overgrown on the entire surface with the colonies of Bacillus sp. On two plates confluent growth was observed, which made it impossible to perform a quantitative count (Table 1).

Table 1. Microbiological analysis of the culture from the biochar sample from Brazil (1-BR) cultivated on PCA (PLATE COUNT LAB-AGAR™) substrate (CFU/g).

		I	II	III	Average Number of Colonies
Dilution	1	Bsp. (UNC)	Bsp. (UNC)	Bsp. (UNC)	UNC
	10^{-1}	Bsp (UNC)	Bsp (UNC)	Bsp (UNC)	UNC
	10^{-2}	Bsp. (214)	Bsp. (194)	Bsp (UNC)	204.0
	10^{-3}	Bsp. (36)	Bsp. (62)	Bsp (UNC)	49.0
	10^{-4}	Bsp. (3)	Bsp. (6)	Bsp. (23)	10.7
	10^{-5}	Bsp. (2)	0	Bsp. (4)	2.0

I, II, III—repetitions; (UNC)—in brackets the uncountable colonies; (NC)—in brackets the number of colonies (NC); Bsp.—Bacillus sp.

On malt substrate (Table 2) strains of Bacillus sp. were identified for the dilution of 1 and 10^{-1}. Furthermore, in the second repetition for the dilution of 10^{-1}, the following strains were isolated: Aspergillus versicolor (UNC), Aspergillus flavus (UNC), Aspergillus fumigatus (UNC), Aspergillus niger (3), Aspergillus candidus (2), Aspergillus nidulus (7), Penicillium sp. (UNC), Rhizopus sp. (2), while during the third repetition: Aspergillus versicolor (UNC), Aspergillus fumigatus (8), Aspergillus niger (8), and Penicillium sp. (UNC). For the dilution of 10^{-2}, in the first repetition, besides Bacillus sp. the following were isolated: Aspergillus versicolor (UNC), Aspergillus nidulus (1), Aspergillus flavus

(1), Aspergillus niger (5), Rhizopus sp. (UNC), in the second repetition: Aspergillus versicolor (18), Aspergillus fumigatus (1), Penicillium sp. (8), while in the third repetition: Aspergillus versicolor (3), and an overgrowth of Bacillus sp. on the entire surface plate was recorded. In the case of dilution of 10^{-3}, on average 7.3 colonies of Aspergillus sp., Aspergillus versicolor, Aspergillus flavus, and Penicillium sp. were isolated, as well as Bacillus sp., in the case of which confluent growth was observed.

Table 2. Microbiological analysis of the culture from the biochar sample from Brazil (1-BR) cultivated on MALT substrate (CFU/g).

		I	II	III	Average Number of Colonies
	1	Bsp (*)	Bsp (*)	Bsp (*)	-
	10^{-1}	Bsp (*)	Aver (UNC) Afl (UNC) Afum (UNC) Anig (3) Acan (2) Anid (7) Psp. (UNC) R (2) Σ(UNC)	Aver (UNC) Afum (8) Anig (8) Psp (UNC) Σ(UNC)	UNC
Dilution	10^{-2}	Aver (UNC) Afl (1) Anig (5) Anid (1) R (UNC) Bsp (*) Σ(UNC)	Aver (18) Afum (1) Psp (8) Σ(27)	Aver (36) Asp. (1) Psp (6) R (2) Σ(45)	36.0
	10^{-3}	Aver (6) Asp. (1) Bsp (*) Σ(7)	Aver (5) Afl (1) Asp. (1) Psp (1) Σ(8)	Aver (5) Psp (1) Σ(7)	7.3
	10^{-4}	Bsp (*)	Aver (2) Σ(2)	Aver (1) Σ(1)	1.0
	10^{-5}	0	0	0	0

I, II, III—repetitions; (UNC)—in brackets the uncountable colonies; (NC)—in brackets the number of colonies (NC); Σ (NC)—total numbers of colonies; Aver—Aspergillus versicolor; Afl—Aspergillus flavus; Afum—Aspergillus fumigatus; Anig—Aspergillus niger; Acan—Aspergillus candidus; Anid—Aspergillus nidulus; Asp.—Aspergillus sp; Psp.—Penicillium sp; R—Rhizopus sp; Bsp (*)—plate overgrown by Bacillus sp.

For the dilution of 10^{-4}, in the first repetition, Bacillus sp. was isolated, characterised by a confluent growth, while in the second and third repetition Aspergillus versicolor species were recorded with the average number of colonies equal to 1.0. After averaging the readings the average number of fungi colonies ranged from UNC to 45. On POTATO substrate (Table 3) for the dilution of 1, no colonies of microorganisms were recorded.

In the case of dilutions from 10^{-1} to 10^{-5}, in all repetitions, strains of Bacillus sp. overgrown on the entire surface of the plate were observed, which made a quantitative count impossible to perform. Additionally, the following genera or species were isolated: Aspergillus sp., Aspergillus niger, Aspergillus flavus, Aspergillus versicolor, Aspergillus candidus, Aspergillus fumigatus, as well as Penicillium sp. (the average number of colonies ranging from 0.3 for the dilution of 10^{-4} to 3.7 for the dilution of 10^{-3}). However, for the dilution of 10^{-5}, no colonies of microorganisms were isolated. In the case of sample no. 1-PL cultured on PCA substrate (Table 4) dominated strains of Bacillus sp.

Table 3. Microbiological analysis of the culture from the biochar sample from Brazil (1-BR) cultivated on POTATO substrate [CFU/g].

		I	II	III	Average Number of Colonies
Dilution	1	0	0	0	0
	10^{-1}	Bsp (*)	Anig (2) Afl (2) Asp. (2) Bsp (*) Σ(6)	Aver (UNC) Acan(2) Anig (1) Afl (UNC) Σ(UNC)	3.0
	10^{-2}	Bsp (*)	Aver (2) Acan(1) Afl (1) Σ(4)	Aver (3) Bsp (*) Σ(3)	2.3
	10^{-3}	Bsp (*)	Bsp (*)	Aver (6) Acan(1) Anig (1) Afum (1) Afl (1) Psp (1) Bsp (*) Σ(11)	3.7
	10^{-4}	Bsp (*)	Bsp (*)	Psp (1) Bsp (*) Σ(1)	0.3
	10^{-5}	0	0	0	0

I, II, III—repetitions; (UNC)—in brackets the uncountable colonies; (NC)—in brackets the number of colonies (NC); Σ (NC)—total numbers of colonies; Aver—Aspergillus versicolor; Afl—Aspergillus flavus; Afum—Aspergillusfumigatus; Anig—Aspergillus niger; Acan—Aspergillus candidus; Anid—Aspergillus nidulus; Asp.—Aspergillus sp; Psp.—Penicillium sp; R—Rhizopus sp; Bsp (*)—plate overgrown by Bacillus sp.

Table 4. Microbiological analysis of the culture from the biochar sample from Poland (1-PL, 2-PL) cultivated on PCA substrate (CFU/g).

		I		II		III		Average Number of Colonies	
	Biochar	**Pine 1-PL**	**Oak 2-PL**	**Pine 1-PL**	**Oak 2-PL**	**Pine 1-PL**	**Oak 2-PL**	**Pine 1-PL**	**Oak 2-PL**
Dilution	1	Bsp. (UNC)	0	Bsp. (UNC)	Afl (1)	Bsp. (UNC)	0	UNC	0
	10^{-1}	Bsp (421)	0	Bsp (443) Micro (1) Σ(444)	0	Bsp (771)	0	545.3	0
	10^{-2}	Bsp. (61)	0	Bsp. (54)	0	Bsp (124)	0	79.7	0
	10^{-3}	Bsp. (6)	0	Bsp. (3)	0	Bsp (10)	0	6.3	0
	10^{-4}	0	0	0	0	Bsp. (1)	0	0.3	0
	10^{-5}	0	0	0	0	0	0	0	0

I, II, III—repetitions; (UNC)—in brackets the uncountable colonies; (NC)—in brackets the number of colonies (NC); Σ (NC)—total numbers of colonies; Bsp.—*Bacillus* sp; Micro—*Micrococcus* sp; Afl—*Aspergillus flavus*.

The average number of colonies ranged between 545.5 for the dilution of 10^{-1} and for the dilution of 1 also uncountable growth of Micrococcus sp. For the dilution of 10^{-5}, no colonies of microorganisms were isolated. On MALT substrate (Table 5), for the dilution of 1, strains of Aspergillus flavus were isolated, and the average number of colonies count was 1.7.

Table 5. Microbiological analysis of the culture from the biochar sample from Poland (1-PL, 2-PL) cultivated on MALT substrate (CFU/g).

		I		II		III		Average Number of Colonies	
	Biochar	**Pine 1-PL**	**Oak 2-PL**	**Pine 1-PL**	**Oak 2-PL**	**Pine 1-PL**	**Oak 2-PL**	**Pine 1-PL**	**Oak 2-PL**
Dilution	1	Afl (2)	0	0	0	Afl (1) # (2) Σ(3)	0	1.7	0
	10^{-1}	0	0	0	0	0	Psp. (1)	0	0.3
	10^{-2}	0	0	0	0	0	0	0	0
	10^{-3}	0	0	0	0	0	0	0	0
	10^{-4}	0	0	0	0	0	0	0	0
	10^{-5}	0	0	0	0	0	0	0	0

I, II, III—repetitions; (UNC)—in brackets the uncountable colonies; (NC)—in brackets the number of colonies (NC); Σ (NC)—total numbers of colonies; Afl—Aspergillus flavus; Psp.—Penicillium sp; #—unidentified fungi colony.

Within the range of dilutions between 10^{-1} and 10^{-5}, no colonies of microorganisms were isolated. On POTATO substrate (Table 6) strains of Bacillus sp. overgrown on the entire plate were observed (average number of colonies from UNC for the dilution of 1 to 1 colony for the dilution of 10^{-4}). For the dilution of 10^{-5}, no colonies of microorganisms were isolated. In the case of sample no. 2-PL on PCA substrate for the dilution of 1, only in the second of the three repetitions, 1 colony of Aspergillus flavus was isolated. On the plates, with the dilutions ranging from 10^{-1} to 10^{-5}, no colonies of microorganisms were isolated. On the MALT substrate for the dilution of 10^{-1} 1 colony of Penicillium sp. was isolated. Meanwhile, for the dilutions ranging from 10^{-2} to 10^{-5} no colonies of microorganisms were isolated. On POTATO substrate, for the dilution of 10^{-1}, 1 colony Aspergillus versicolor was isolated, while for the dilution of 10^{-2} 1 colony of unidentified fungi was recorded. On the plate, with the dilution of 10^{-3}, no microorganisms were recorded. In contrast, on the plate with the dilution of 10^{-4} one yeast colony was isolated. On the plate, with the dilution of 10^{-5}, no microorganisms were recorded. After the microbiological analysis of biochar samples, it was noted that sample 2-PL was the least microbiologically polluted. However, it should be stated that in Polish samples biologically different microorganisms were present. In the sample, 1-PL dominated strains of Bacillus sp. While in sample 2-PL dominated, though in small amount, strains of fungi species: Aspergillus flavus, Aspergillus versicolor, and Penicillium sp. In the case of biochar sample from Brazil, the same microorganisms were present as the ones isolated and identified in Polish samples.

Table 6. Microbiological analysis of the culture from the biochar sample from Poland (1-PL, 2-PL) cultivated on POTATO substrate (CFU/g).

		I		II		III		Average Number of Colonies	
	Biochar	Pine 1-PL	Oak 2-PL	Pine 1-PL	Oak 2-PL	Pine 1-PL	Oak 2-PL	Pine 1-PL	Oak 2-PL
Dilution	1	Bsp (*)	0	Bsp (*)	0	Bsp (*)	0	0	0
	10^{-1}	Bsp (*>100)	0	Bsp (*>100)	Aver (1)	Afl (1) Bsp (*>100)	0	0.3	0.3
	10^{-2}	Bsp (*32)	# (1)	Bsp (*43)	0	Bsp (*54)	0	0.3	0
	10^{-3}	Bsp (*1)	0	Bsp (*5)	0	Bsp (*9)	0	0	0
	10^{-4}	0	0	Bsp (*1)	0	0	## (1)	0.3	0
	10^{-5}	0	0	0	0	0	0	0	0

I, II, III—repetitions; (UNC)—in brackets the uncountable colonies; (NC)—in brackets the number of colonies (NC); Σ (NC)—total numbers of colonies; Aver-Aspergillus versicolor; Afl—Aspergillus flavus; Bsp (*)—plate overgrown by Bacillus sp; #—unidentified fungi colony; ##—unidentified yeast colony.

The microscopic analysis, carried out using research microscope Nikon Eclipse E-200 with fluorescence attachment and SCA image analysis, shows the structure of biochar 1-BR (Figure 2), biochar 1-PL (Figure 3) and biochar 2-PL (Figure 4). Biochar 1-BR presents irregular structure, but this is not as dense as in the case of biochar 2-PL (oak). The structure of biochar sample 1-BR (Brazil) is undulating and porous. However, this sample seems to have the most irregular structure when compared with samples 1-PL (pine) and 2-PL (oak).

Figure 2. Microscopic images of biochar from Brazil (1-BR). (**a**) magnification 300×, (**b**) magnification 500×, (**c**) magnification 1000×, (**d**) magnification 2500×.

Figure 3. Macroscopic images of biochar from Poland (1-PL). (**a**) magnification 300×, (**b**) magnification 500×, (**c**) magnification 1000×, (**d**) magnification 2500×.

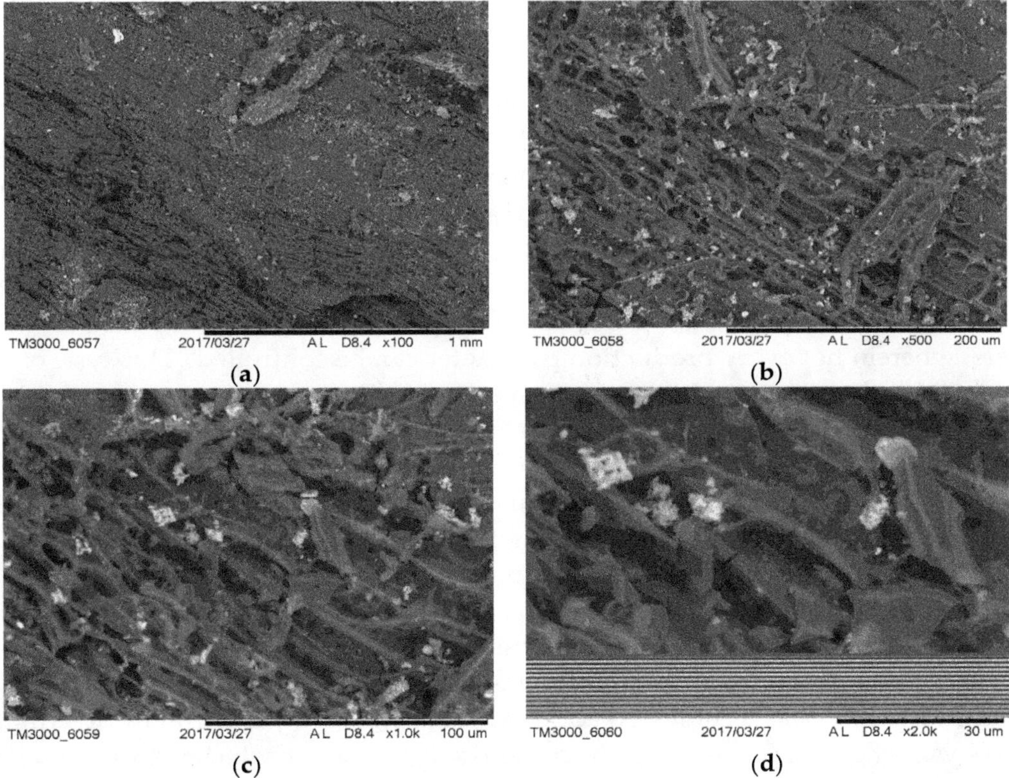

Figure 4. Microscopic images of biochar from Poland (2-PL). (**a**) magnification 100×, (**b**) magnification 500×, (**c**) magnification 1000×, (**d**) magnification 2000×.

Oak (sample 2-PL), being hardwood, in the microscopic images shows a hard and porous structure, while pine (sample 1-PL) has regular, slightly porous, and delicate structure.

4. Discussion

The microbiological analysis of the three biochar samples demonstrated that samples 1-PL and 2-PL were less microbiologically contaminated than sample 1-BR. In the case of sample 1-BR, similar microorganisms were isolated as from samples 1-PL and 2-PL. In sample 1-BR, among others, the following were present: Bacillus sp. Aspergillus sp. Aspergillus niger, Aspergillus flavus, Aspergillus versicolor, Aspergillus candidus, Aspergillus fumigatus, as well as Penicillium sp. and Rhizopus sp. Such a broad range of microorganisms was isolated neither from sample 1-PL nor 2-PL. At the same time, there is a certain similarity between the biochar from Poland and Brazil, namely the strains isolated from Brazilian biochar were present in two types of biochar from Poland, originating from two botanically different sources. When analysing the two samples from Poland opposite observations should be made, namely that biologically different microorganisms were present in these two samples. In sample 1-PL (pine) strains Bacillus sp. dominated. In sample 2-PL (oak) dominated, though in small amounts, strains of fungi: Aspergillus flavus, Aspergillus versicolor, as well as Penicillium sp. In the sample from Brazil strains Bacillus sp., Aspergillus sp., and Rhizopus sp. were the most numerous ones. After the microbiological and microscopic analysis, it can be concluded that the two samples have very interesting characteristics that can be used depending on the purpose. It was observed that sample 2-PL was less microbiologically contaminated. However, it should be stated that in the tested samples biologically different microorganisms were identified. In the sample 1-PL strains Bacillus sp. dominated. In the sample 2-PL dominated, thought in small amount, strains of fungi: Aspergillus flavus, Aspergillus versicolor, as well as Penicillium sp. Additionally, differences were observed in the microscopic images, where sample 1-PL (pine) was characterised by a compact and regular structure, while sample 2-PL (oak) had irregular and porous structure. At the same time, the significantly differing structure was observed in the case of biochar 1-BR (Brazil). It was more delicate than the other test samples. The porous structure and specific surface of biochar are its most important physical properties and they are responsible for the course of various processes in the soil. Biochar has high internal porosity, which affects the water absorption, sorption capacity, and retention of nutrients in the soil. Therefore, it improves physicochemical properties of the soil, facilitates the use of nutrients by plants, and prevents nutrients leaching. Biochar may improve soil structure (water-air properties) [36,37]. Presence of certain genera and species in the test samples may be associated with the generic traits of the raw materials from which they were derived. In addition, the manner, in which test samples were produced, might also be important. The biochar from Poland came from controlled, patented, and repeatable production. On the other hand, the biochar from Brazil was produced in a simple way reflecting the possibility to be repeated by the farmers, wherein not every production parameter can be controlled. This type of dependency is directly proportional to the microbiology of test samples. Summing up, all the samples have very interesting properties that may be used depending on the intended purpose. Implications for practice are now widely used in many research centres around the world. Herrman et al. (2019) concluded that the addition of biochar from rubber tree caused a raise of the soil pH value as well as its nutrient content. In the research cited higher sensitivity to the activity of applied biochar was recorded in the case of fungi occurring in the soil as compared to bacteria present there [38]. Similar studies were performed in Bulgaria [39] using biochar in the cultivation of wheat and maize in crop rotation. The stimulating effect of biochar on soil microflora was observed, in particular on the number of bacteria. The research was considered to be a promising method for preserving soil fertility. Azis et al. (2019) concluded that in the case of soil on which wheat is cultivated addition of biochar, in experimentally determined amount, improves the physicochemical properties of the soil and soil microflora, which in turn has a positive effect on the productivity of the soil and increased yields [40]. In other studies, Hardy et al. investigated the long-term application of biochar in order to improve the quality of soil microorganisms. For this purpose, specialized models imitating soil conditions were designed. Biochar was shown to modify the quantity and quality of soil microorganisms, however, in order to properly design a successful modification of the soil using biochar its properties should be considered in conjunction with the soil conditions [41]. Unfortunately, in the available literature of the subject, there is no similar research with which one could compare the results obtained in this work. Based on the definition of sustainable development, strategies of environmental protection should consist of continuous, integrated and anticipatory actions undertaken in production processes. Such actions lead to an increase in production efficiency and reduce risks to humans and the natural environment. Sustainable production aims at the manufacture of goods which uses processes limiting environmental pollution. In this context, biochar production based on waste-free use of organic biomass has the traits of sustainable development.

5. Conclusions

1. In the biochar sample from Brazil (1-BR), both bacteria: Bacillus sp., and fungi: Penicillium sp., Aspergillus versicolor, Aspergillus flavus, Aspergillus fumigatus, Aspergillus niger, Aspergillus candidus, Aspergillus nidulus, and Rhizopus sp. were observed and identified.

2. In biochar sample from pine woodchips (1-PL) colonies of Bacillus sp. and Micrococcus sp. bacteria were observed and identified.

3. In biochar sample from oak woodchips (2-PL) fungi: Aspergillus flavus, Aspergillus versicolor, and Penicillium Sp. were observed and identified.

4. On the basis of microscopic images, diverse structures of the samples were recorded. Sample 1-PL (pine) is characterised by a compact and regular structure, while sample 2-PL (oak) shows porous and irregular structure. Sample from Brazil (1-BR) is characterised by a structure more delicate than these of the remaining test samples.

5. Samples coming from controlled production (1-PL, 2-PL) were less microbiologically contaminated than biochar sample from Brazil (1-BR).

6. The results obtained can be used as relevant utilitarian data for implications for practice.

Author Contributions: Conceptualization, W.Ż.-S. and P.S. and A.L.; methodology, W.Ż.-S. and D.P.; prepared the materials, D.P.; W.Ż.-S.; analysed the data, W.Ż.-S.; M.T.-R.; wrote the paper, W.Ż.-S.; P.S., A.L.; B.S. visualization, P.S. All authors have read and agree to the published version of the manuscript.

Acknowledgments: The authors would like to cordially thank the President of Fluid SA, eng. Jan Gładki and International Institute for Sustainability in Rio de Janeiro, Brazil for help in this work.

References

1. Lehmann, J.; Joseph, S. *Biochar for Environmental Management: Science and Technology*; Routledge: Abingdon, UK, 2012.
2. Filiberto, D.M.; Gaunt, J.L. Practicality of biochar additions to enhance soil and crop productivity. *Agriculture* **2013**, *3*, 715–725. [CrossRef]
3. Latawiec, A.E.; Strassburg, B.B.N.; Junqueira, A.B.; Araujo, E.; de Moraes, L.F.D. Biochar amendment improves degraded pasturelands in Brazil: Environmental and cost-benefit analysis. *Sci. Rep.* **2018**, *9*, 1–12. [CrossRef]
4. Castro, A.; da Silva Batista, N.; Latawiec, A.; Rodrigues, A.; Strassburg, B.; Silva, D.; Araujo, D.; de Moraes, L.F.D.; Guerra, J.G.; Galvao, G.; et al. The effects of *Gliricidia*-derived biochar on sequential maize and bean farming. *Sustainability* **2018**, *10*, 578. [CrossRef]
5. Hale, S.E.; Cornelissen, G.; Werner, D. Sorption and remediation of organic compounds in s oils and sediments by (activated) biochar. In *Biochar for Environmental Management Science, Technology and Implementation*; Ohannes Lehmann, J., Stephen, J., Eds.; Earthscan: London, UK, 2015.
6. Schmidt, H.; Hagemann, N.; Draper, K.; Kammann, C. The use of biochar in animal feeding. *PeerJ* **2019**, *7*, e7373. [CrossRef]

7. Schmidt, H.P. Novel uses of biochar. In Proceedings of the USBI North American Biochar Symposium, Center for Agriculture, University of Massachusetts, Amherst, MA, USA, 13–16 October 2013.
8. Abe, I.; Hitomi, M.; Ikuta, N.; Kawafune, I.; Noda, K.; Kera, Y. Humidity-control capacity of microporous carbon. *J. Urban Living Health Assoc.* **1995**, *39*, 333–336.
9. Nakano, T.; Haisbi, T.; Mizuno, T.; Takeda, T.; Tokumoto, M. Improvements of the under floor humidity in woody building and water content of wood material. *Mokuzai Kogyo* **1996**, *51*, 198–202.
10. Morita, H. The effects of humid controlling charcoal on the environ-mental antigenic allergy. In *Proceeding of the 35th Annual Meeting of Japanese Society for Dermatoallergology, Yokohama, Japan, 16–17 July2005*; Japanese Society for Dermatoallergology: Wakayama, Japan, 2005; p. 115.
11. Taketani, T. Evaluation of the effect of humid controlling charcoal on the infantile bronchial asthma. *Allergy* **2006**, *55*, 467.
12. Riva, L.; Nielsen, H.K.; Skreiberg, Ø.; Wang, L.; Bartocci, P.; Barbanera, M.; Bidini, G.; Fantozzi, F. Analysis of optimal temperature, pressure and binder quantity for the production of biocarbon pellet to be used as a substitute for coke. *Appl. Energy* **2019**, *256*, 113933. [CrossRef]
13. Souradeep, G.; Harn, W.K. Factors determining the potential of biochar as a carbon capturing and sequestering construction material: Critical review. *J. Mater. Civ. Eng.* **2017**, *29*, 04017086. [CrossRef]
14. Schallmey, M.; Singh, A.; Ward, O.P. Developments in the use of *Bacillus species* for industrial production. *Can. J. Microbiol.* **2004**, *50*, 1–17. [CrossRef]
15. Harwood, C.P.; Cranenburgh, R. Bacillus protein secretion: An unfolding story. *Trends Microbiol.* **2008**, *16*, 73–79. [CrossRef]
16. Aiyer, P.; Modi, H. Isolation and screening of alkaline amylase producing bacteria *Bacillus Licheniformis*. *Asian J. Microbiol. Biotechnol. Environ. Sci.* **2005**, *7*, 895–897.
17. De Souza, P.M.; de Oliveira e Magalhães, P. Application of microbial A -amylase in industry—A review. *Braz. J. Microbiol.* **2010**, *41*, 850–861. [CrossRef]
18. Sailas, B.; Smitha, R.B.; Jisha, V.N.; Pradeep, S.; Sajith, S.; Sreedevi, S.; Priji, P.; Unni, K.N.; Sarath Josh, M.K. A monograph on amylase from *Bacillus* spp. *Adv. Biosci. Biotechnol.* **2013**, *4*, 227–241.
19. Hussein, H.S.; Brasel, J.M. Toxicity, metabolism and impact of mycotoxins on humans and animals. *Toxicology* **2001**, *167*, 101–134. [CrossRef]
20. Kołaczyńska-Janicka, M. Mikotoksyny—Realne zagrożenie. *Kukurydza* **2006**, *1*, 59–62.
21. Bittencourt, A.B.F.; Oliveira, C.A.F.; Dilkin, P.; Correa, B. Mycotoxin occurrence in corn meal and flour traded in São Paulo, Brazil. *Food Control* **2005**, *16*, 117–120. [CrossRef]
22. Cegielska-Radziejewska, R.; Lesnierowski, G.; Kijowski, J. Antibacterial activity of hen egg white lysozyme modified by thermochemical technique. *Eur. Food Res. Technol.* **2009**, *228*, 841–845. [CrossRef]
23. Cegielska-Radziejewska, R.; Leśnierowski, G.; Kijowski, J.; Szablewski, T.; Zabielski, J. Effects of treatment with lysozyme and its polymers on the microflora and sensory properties of chilled chicken breast muscles. *Bull. Vet. Inst. Pulawy* **2009**, *53*, 455–461.
24. Ghali, R.; Hmaissia-Khlifa, K.; Ghorbel, H.; Maaroufi, K.; Hedili, A. Incidence of aflatoxins, ochratoxin A and zearalenone in tunisian foods. *Food Control* **2008**, *19*, 921–924. [CrossRef]
25. Łozowicka, B. Zanieczyszczenia chemiczne w żywności pochodzenia roślinnego. *Prog. Plant Prot.* **2009**, *49*, 2071–2080.
26. Dai, Z.; Barberán, A.; Li, Y.; Brookes, P.C.; Xu, J. Bacterial community composition associated with pyrogenic organic matter (biochar) varies with pyrolysis temperature and colonization environment. *mSphere* **2017**, *2*, e00085-17. [CrossRef]
27. Russo, G.; Verdiani, G.; Anifantis, A.S. Re-use of agricultural biomass for nurseries using proximity composting. *Contemp. Eng. Sci.* **2016**, *9*, 1151–1182. [CrossRef]
28. Rodriguez, A.; Latawiec, A.E. Rethinking organic residues: The potential of biomass in Brazil. *Mod. Concep. Dev. Agron.* **2018**, *1*, 1–5. [CrossRef]
29. *PN-EN ISO 7218Microbiology of Food and Animal Feeding Stuffs—General Rules for Microbiological Examination*; Polish Committee for Standardization: Warsaw, Poland, 2013.
30. Ramirez, C. *Manual and Atlas of the Penicillia*; Elsevier Biomedical Press: Amsterdam, The Netherlands, 1982.
31. Baran, E. *Zarys Mikologii Lekarskiej*; Volumed: Wrocław, Poland, 1998.
32. Larone, D.H. *Medically Important Fungi. A Guide to Identification*, 5th ed.; ASM Press: Washington, DC, USA, 2011.

33. Kwaśna, H.; Chełkowski, J.; Zajkowski, P. *Grzyby*; Instytut Botaniki PAN: Kraków, Poland, 1991; Volume XXII.
34. Krzyściak, P.; Skóra, M.; Macura, A.B. *Atlas Grzybów Chorobotwórczych Człowieka*; MedPharm: Wrocław, Poland, 2011.
35. Samson, R.A.; Hoekstra, E.S.; Frisvad, J.C.; Filtenborg, O. *Introduction to Food- and Airborne Fungi*, 6th ed.; Centraalbureau voor Schimmelcultures: Utrecht, The Netherlands, 2002.
36. Bis, Z. Czy węgiel będzie dzieckiem Europy? *Nowa Energ.* **2013**, 2–3. Available online: http://kie.is.pcz.pl/images/wegiel_dzieckiem_europy.pdf (accessed on 15 September 2019).
37. Bis, Z. Biowęgiel-powrót do przeszłości, szansa dla przyszłości. *Czysta Energ.* **2012**, *6*, 28–31.
38. Herrmann, L.; Lesueur, D.; Robin, A.; Robain, H.; Wiriyakitnateekul, W.; Bräu, L. Impact of biochar application dose on soil microbial communities associated with rubber trees in North East Thailand. *Sci. Total Environ.* **2019**, *689*, 970–979. [CrossRef]
39. Petkova, G.; Nedyalkova, K.; Mikova, A.; Atanassova, I. Microbiological characteristics of biochar amended alluvial meadow soil. *Bulg. J. Agric. Sci.* **2018**, *24* (Suppl. S2), 81–84.
40. Aziz, S.; Laiba, Y.; Asif, J.; Uzma, F.; Zahid, Q.; Isfahan, T.; Syed, K.H.; Ali, M.I. Fabrication of biochar from organic wastes and its effect on wheat growth and soil microflora. *Pol. J. Environ. Stud.* **2020**, *29*, 1069–1076. [CrossRef]
41. Hardy, B.; Sleutel, S.; Dufey, J.E.; Cornelis, J.-T. The long-term effect of biochar on soil microbial abundance, activity and community structure is overwritten by land management. *Front. Environ. Sci.* **2019**, *7*, 110. [CrossRef]

12

Assessment of the Potential use of Young Barley Shoots and Leaves for the Production of Green Juices

Agata Blicharz-Kania [1], Dariusz Andrejko [1], Franciszek Kluza [1], Leszek Rydzak [1,*] and Zbigniew Kobus [2]

[1] Department of Biological Bases of Food and Feed Technologies, University of Life Sciences in Lublin, 20-612 Lublin, Poland

[2] Department of Technology Fundamentals, University of Life Sciences in Lublin, 20-612 Lublin, Poland

* Correspondence: leszek.rydzak@up.lublin.pl

Abstract: It is possible to use the aboveground parts of barley, which are cultivated as a forecrop. They are often simply composted or dried for bedding. It is worth trying other more effective methods of processing aboveground biomass. The aim of this study was to preliminary investigate the possibility of using young barley leaves and shoots for the production of green juice with potential health properties. The material was collected at days 7, 14, 21, and 28 after plant emergence. The length and strength of the shoots were measured and the pressing yield was calculated. The pH value and the content of protein, chlorides, and reducing sugars were also determined. The juice was additionally subjected to pasteurisation and freezing, and changes in pH and chlorophyll content occurring during storage were determined. The pressing yield of young barley leaves and shoots was estimated to be between 69% and 73%. The product was characterised by a high content of total protein (34.45%–51.81%$_{d.w.}$) and chlorophylls (6.62 mg·g^{-1}). The chlorophyll content declined during barley juice storage. Pasteurisation of the juice from young barley leaves does not induce statistically significant changes in the pH of the juice, but reduces the chlorophyll content. Our results revealed that the most effective way to preserve the green juice is by freezing. This process does not induce changes in juice acidity and only slightly reduces the chlorophyll content during storage of the product.

Keywords: juice; barley; pressing; protein; chlorophylls; green food

---

1. Introduction

In order to improve the soil quality and protect it from weeds and, consequently, obtain a better crop of the cultivated plant, producers often use forecrop cultivation. Depending on what kind of plant will be grown, an appropriate forecrop will be selected. For cereals, the best options are root crops (mainly potatoes), legumes, or rapeseed. However, when growing vegetables, such as tomatoes, carrots, or white cabbage, it is better to use a cereal forecrop, and rye, wheat, or barley can be used [1]. These cereals can be used via two techniques. The first method involves ploughing whole plants and using them as a fertilizer. Plants that are a forecrop can also be cut. Then, only roots and postharvest residues are ploughed. In the case of the latter method, it is possible to continue using the green parts of the plants, which are often simply composted or dried for bedding.

Recently, products from young barley leaves and shoots have gained popularity. They are available in a variety of forms, i.e., juice, tablets, or powder. Young parts of plants are a source of phenolic acids and many vitamins, e.g., C, E, and B group vitamins. Additionally, barley grass contains substantial amounts of carotenoids, folic acid, calcium, iron, magnesium, potassium, zinc, and copper. Importantly, products obtained from cereal leaves and shoots can be used as supplements in a high-protein diet. Barley grass contains approximately 30% of protein in dry matter [2–4]. The chemical index of nutritional value (Chemical Score, CS) is 41.44% (Methionine) [5]. Barley shoots and leaves are also a source of chlorides. Naturally occurring chlorides exert a beneficial effect on organism function. Chloride ions participate in the regulation of water, as well as electrolyte metabolism and maintaining acid–base balance [6].

One of the most important active compounds in "green food" is chlorophyll. Seed plants contain type A and B chlorophylls, which differ according to the type of substituent on the second pyrrole ring. However, these pigments are unstable. Many factors, e.g., UV radiation or pH and temperature changes, cause chlorophyll degradation in food products. Chlorophylls are a valuable source of magnesium; they can also improve metabolism and eliminate unpleasant mouth odours. Additionally, they have antibacterial and anti-inflammatory properties, remove toxins from the liver and blood [7–11], and even act as a haemoglobin substitute [12].

The consumption of green juices has therapeutic properties: it exerts an antidiabetic effect, regulates blood pressure, strengthens immunity, protects the liver, and has anti-acne and antidepressant activities. It also improves the function of the digestive tract and prevents hypoxia, cardiovascular diseases, fatigue, and constipation. Additionally, it alleviates atopic dermatitis and has anti-inflammatory, antioxidant, and anticancer effects [4]. Kubatka et al. [13] have demonstrated positive changes in tumour cells in rats treated with juice from young barley leaves. A significantly more pronounced effect of the therapeutic treatment was observed in a study group receiving a diet supplemented with the juice. It has also been confirmed that supplementation of the diet with barley leaf powder can relieve the clinical symptoms of diabetes [14]. Additionally, barley grass contains substantial amounts of dietary fibre (mainly an insoluble fraction), which has a positive effect on metabolism through regulating the appetite and, thus preventing the development of overweight. Son et al. [3] recommend enrichment of the diets of young children using valuable nutrients from young barley leaves.

Products from young barley leaves have been used in East Asia for a long time. Currently, they are available in supermarkets, as well as online, in the United States and many European countries. The increase in consumer awareness has contributed to an appreciation of the health benefits of "green food" [8]. A number of products based on young cereal leaves are recommended as dietary supplements and, hence, are currently being manufactured to be purchased in pharmacies.

There is a paucity of scientific publications confirming the health-promoting properties of the juice from young barley leaves. There are also no preliminary investigations describing the impact of production and processing on the quality of green juices. Furthermore, the relationship between the date of raw material harvesting and the pressing process, as well as the content of nutrients—including proteins—in the juice, has not yet been analysed. Another unexplored area is the changes occurring in the chlorophyll content of barley juice, depending on the thermal treatment. Knowledge of these relationships has great practical significance as it provides information on methods for the acquisition of a product with a high nutritional value and, at the same time, ensures the longest possible growth period for the plant (and, thus, the greatest mass of roots to be used as forecrop).

The aim of this work was to preliminary examine the efficiency of the process of pressing green juice from young barley leaves and shoots and to determine the chemical composition of the product obtained. An additional goal of the research was to compare the effect of preservation methods like pasteurization and freezing on the chlorophyll content and pH of barley juice.

2. Materials and Methods

2.1. Research Material

Barley grain cv. Kangoo was used for the investigations. The seeds, weighing 5 kg, were sown under laboratory conditions (Figure 1). No fertilisation was applied during the cultivation. The material was collected at days 7, 14, 21, and 28 after plant emergence. The crop area was divided into four parts, and these four parts were then divided again into 12 smaller ones. Three parts from the whole field were chosen at random for each series of investigation (different harvest time: 7, 14, 21, and 28 days). Each sample of shoots and leaves of young barley for juice production weighed 200 g.

Figure 1. Cultivation of barley.

2.2. *Measurement of Physical Properties*

Samples without mechanical damage were selected for determination of the length and strength of the barley. The length of five barley shoots was measured with the use of a calliper to the nearest 0.01 mm. The shoot strength was determined by applying a uniaxial tensile test to the leaves. The test was carried out using a Zwick/Roell Z0.5 materials testing machine (Zwick.Roell AG, BT1-FR0.5TN.D14, Ulm, Germany) equipped with a measuring head at a maximum force of 50 N (travel speed = 50 mm·min^{-1}). Tensile strength was applied to the material until rupture. TestXpert II software (Zwick/Roell AG, Ulm, Germany) was used to assess the force required for destruction of the barley grass.

2.3. *Pressing*

The juice was squeezed from barley grass (with a weight of 200 g for each repetition) using a press designed for pressing green plant leaves, i.e., Manual Juicer BL-30 (BioChef, Byron Bay, NS, Australia).

The pressing efficiency was calculated using the following equation:

$$W_j(\%) = \frac{M_j}{M_i} \cdot 100 \quad (1)$$

where:

W_j—pressing yield, %;
M_j—mass of juice after pressing, kg;
M_i—mass of input material, kg.

2.4. *Preserving Juice*

The juice obtained in the first harvest term (at seven days after plant emergence) was additionally subjected to pasteurisation for 10 min at 75 °C and freezing. Immediately after the thermal treatment, the product was cooled (in a blast chiller–freezer) to 3 °C. The other batch of juice was blast-frozen to a temperature of −18 °C. The material was stored under appropriate conditions, i.e., 4 °C in the case of the unprocessed (UJ) and pasteurised (PJ) juice and −18 °C in the case of the frozen juice (FJ). The study material was analysed after three (UJ and PJ) and seven (UJ, PJ, and FJ) storage days.

2.5. Determination of pH

Juice samples were placed in 50 mL beakers and the pH was recorded with a pH meter (model 780, Metrohm AG, Herisau Switzerland).

2.6. Measurement of Total Protein Content

The determinations were carried out using the Kjeldahl method and a Foss Kjeltec 8400 automatic distiller (Foss Anatytical AB, Höganäs, Sweden). The total protein content was calculated using a 6.25 conversion factor.

2.7. Determination of the Chloride Content

The chloride content was determined with the Mohr method using a TitraLab AT1000 Series automatic titrator (HACH Company, Willstätterstraße, Germany). The solution was titrated with a 0.1 N silver nitrate solution. The chloride content was given as g in 100 $g_{f.j.}$ (of fresh juice).

2.8. Determination of the Content of Reducing Sugars

The Lane–Eynon method was used to determine the content of reducing sugars. The material was extracted and deproteinised. The content of reducing sugars was determined in the obtained liquid by direct hot titration of a specific copper salt with the analysed sugar solution (against methylene blue as an indicator of the end of the reaction) [15,16].

2.9. Determination of Dry Matter Content

The moisture content of the research material was measured by drying 3 g of juice at 105 °C for 3 h. The measurements were carried out in triplicate.

2.10. Measurement of Chlorophyll Content

The juice was analysed for the content of chlorophylls A and B. The pigments were extracted with methyl alcohol. The chlorophyll content was measured using a UV–vis Helios Omega 3 spectrophotometer (Thermo Scientific, England). The measurement consisted of the determination of the absorbance (A) at different wavelengths (λ): 650 and 665 nm [17]. Next, the content of chlorophylls A and B and the total chlorophyll content were calculated with Equations (2)–(4):

Chlorophyll A content ($C_{chl(a)}$):

$$C_{chl(a)} = 16.5 \cdot A_{(665)} - 8.3 \cdot A_{(650)} \tag{2}$$

Chlorophyll B content ($C_{chl(b)}$):

$$C_{chl(b)} = 33.8 \cdot A_{(650)} - 12.5 \cdot A_{(665)} \tag{3}$$

Total chlorophyll content (C):

$$C = 4.0 \cdot A_{(665)} + 25.5 \cdot A_{(650)} \tag{4}$$

where:

$A_{(650)}$ = absorbance at a 650 nm wavelength;
$A_{(665)}$ = absorbance at a 665 nm wavelength.
The chlorophyll content was calculated in $mg \cdot g^{-1}$, taking into account the sample weight.

2.11. Statistical Analysis

The data were analysed statistically. A significance level of $\alpha = 0.05$ was assumed for inference. The analysis was carried out using ANOVA (StatSoft Polska, Poland) with post hoc tests for homogeneous groups based on Tukey's test. These groups comprised means between which no statistically significant difference was found at the assumed significance level, α.

The determinations were carried out in triplicate, except for leaf length and strength tests, which were repeated five times.

3. Results and Discussion

3.1. Characterisation of the Physical Traits of the Raw Material

Changes in the length of the barley shoots relative to the harvest date are shown in Table 1.

Table 1. Properties of the raw material relative to the harvest time of barley leaf harvesting.

Harvest Time (Day)	7	14	21	28	*p*-Value
Length of leaves (mm)	9.08 ± 1.14 [a]	15.48 ± 1.26 [b]	17.58 ± 1.44 [bc]	19.53 ± 1.35 [c]	<0.0001
Strength of leaves (N)	3.75 ± 0.25 [a]	3.27 ± 0.45 [ab]	3.20 ± 0.39 [ab]	3.01 ± 0.69 [b]	0.0144

[a,b,c,d] Means in the same line denoted by different letters were significantly different. The results are expressed as mean ± SD ($n = 5$).

The largest gain in the length of barley shoots was noted within seven days after emergence. In the following days, the rate of shoot growth declined. There were no statistically significant differences in the length of shoots between the material harvested at day 21 and day 28 after emergence. The length of shoots collected on day 21 and day 28 was 17.58 and 19.53 cm, respectively. The height of the plants was characteristic of unfertilised crops [18–20].

The strength of the barley shoots decreased over time. However, there were only significant changes in the tensile strength of the shoots in the material collected on day 28 of growth (in comparison to the material collected on day 7). Changes in the strength of cereal shoots are associated with the chemical composition, which is modified during plant growth [18].

3.2. The Pressing Efficiency

The effect of the harvest date on the pressing yield is shown in Figure 2.

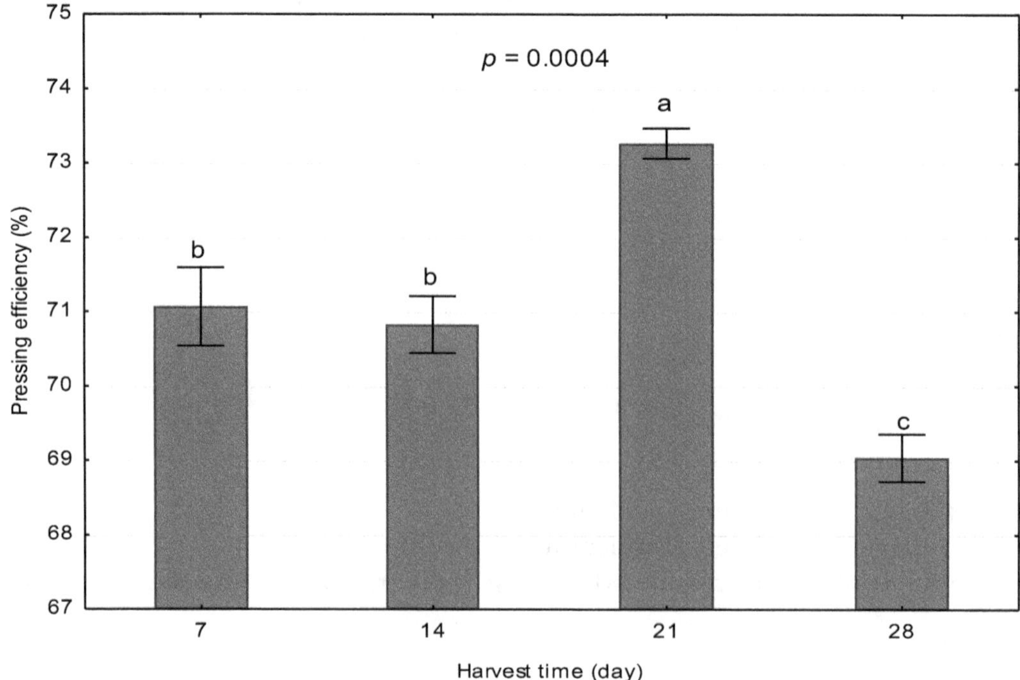

Figure 2. Pressing yield of the barley shoot and leaf juice relative to the harvest date.

The pressing yield ranged from 69.04% to 73.26%. It was obtained from 137 to 146 mL of juice (depending on the time of harvest). The highest value was noted during the processing of material harvested on cultivation day 21. The longer period of cultivation was associated with a statistically significant drop in the pressing yield. The pressing yield of barley leaves collected on day 28 was estimated at 69.4%. These results are consistent with data obtained by other authors. Paulíčková et al. [2] reported a pressing yield of 68% in a study that involved the extraction of juice from barley shoots. The decline in the pressing yield accompanying the longer harvesting period is probably caused by changes in the chemical composition, which lead to an increase in the fibre content.

3.3. Juice Acidity

Irrespective of the harvest date, the juice from the leaves and shoots of young barley had an acidic reaction (changes in pH are shown in Figure 3), shown by the significant decrease in pH values over time. The pH of the products pressed from leaves and shoots collected on day 7–28 ranged from 5.71 to 5.95. These values are similar to the pH of vegetable juices such as carrot or beetroot juice [21–23].

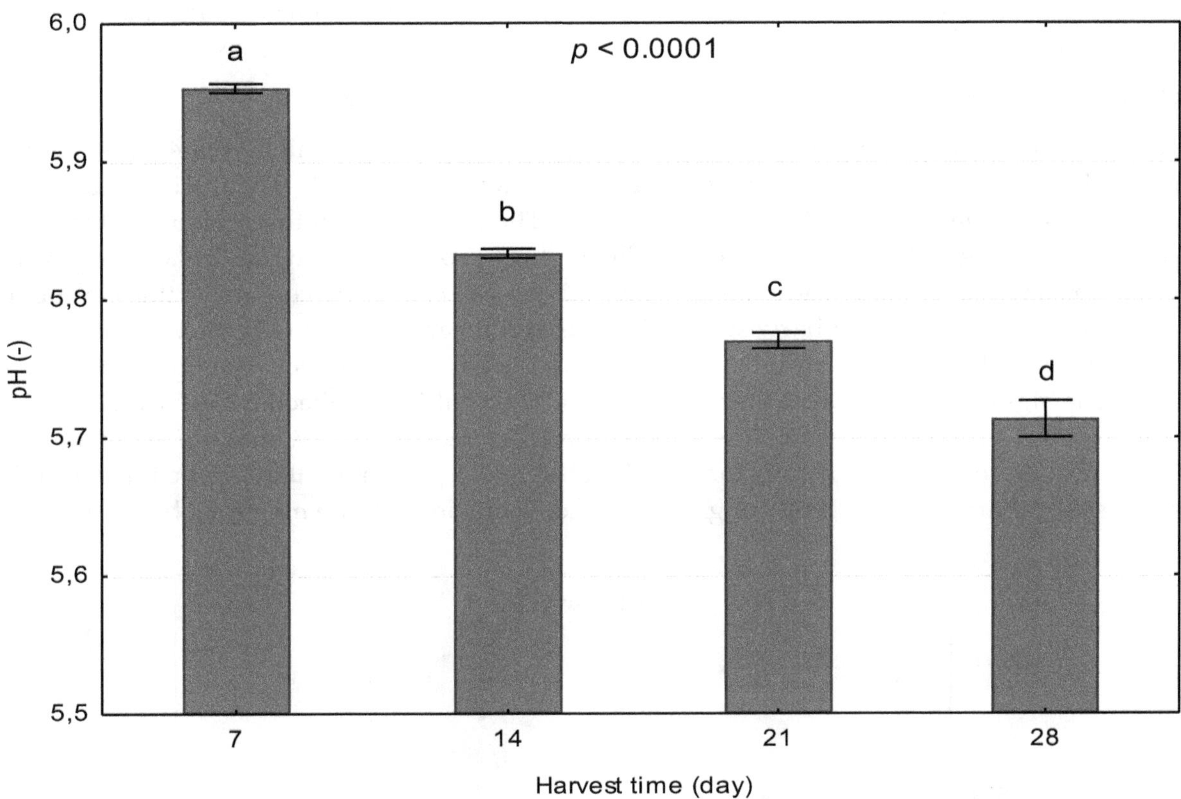

Figure 3. Impact of barley shoot harvesting time on juice pH.

The pH values of the juice from young barley leaves did not change significantly during storage (Figure 4). The statistical analysis only revealed significant differences in the pH value in the case of juice refrigerated for seven days. The pasteurisation and freezing processes did not change this parameter significantly. Juice acidity has an important effect on pigments and other ingredients (e.g., chlorophyll, carotenoids, anthocyanins, myoglobin, etc.) responsible for the colour of fruits, vegetables, and meat [21,24–26].

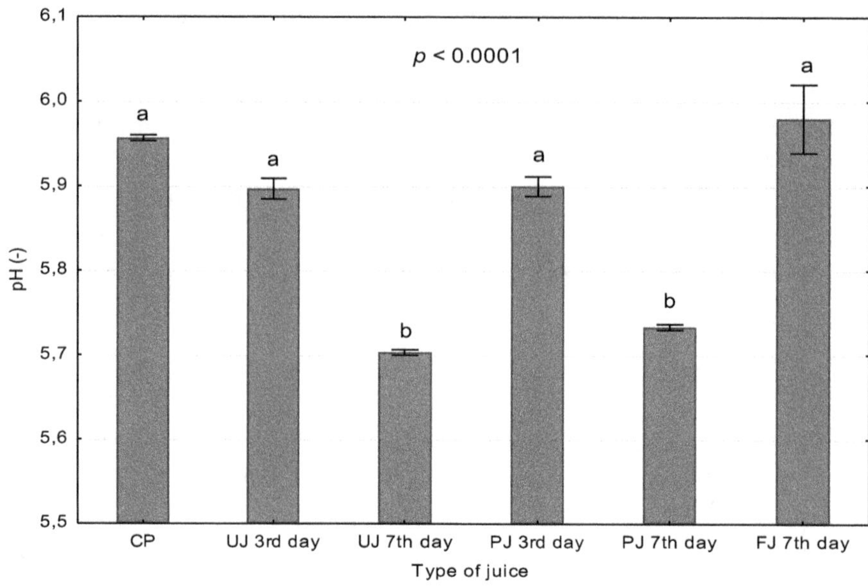

Figure 4. Impact of the conditions and length of storage on barley juice pH. (CP—control probe for fresh juice, UJ—unprocessed juice, PJ—pasteurised juice, FJ—frozen juice; 3rd day—after three storage days, 7th day—after seven storage days).

3.4. Total Protein Content

Protein content was expressed as a percentage of dry matter, which was 4.71% on average. Changes in the protein content of the analysed juices are shown in Figure 5. The total protein content in the juice increased along with the barley harvest date. The differences in the protein content between the harvest dates were statistically significant. The highest protein content, i.e., 51%, was determined in samples collected on day 21. The statistical analysis revealed that the content of this component in the juice produced after the next harvest (day 28) was significantly lower (42.68%). As demonstrated by Paulíčková et al. [2], the total amino acid content in juice from barley leaves and shoots decreased over time. The highest protein content recorded for the Malz cultivar (collected in phase I—DC 29) was 30.44%$_{d.m.}$ However, it should be noted that the authors collected the raw material at a later stage of barley growth. Therefore, these results may explain the lower protein content in the juice from barley leaves and shoots harvested on day 28 of growth (in comparison with the material obtained on day 21).

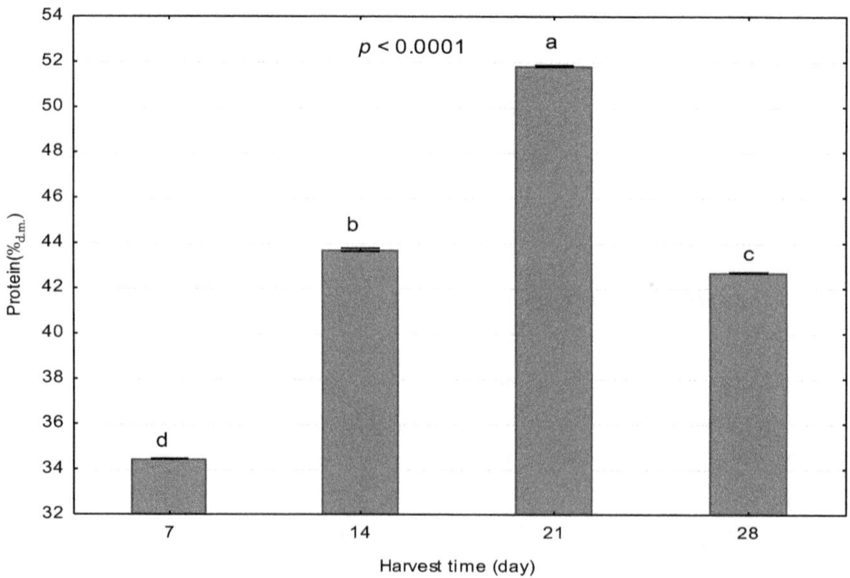

Figure 5. Changes in the protein content in barley shoot juice relative to the harvest date.

3.5. Chloride Content

The changes in chloride content in the analysed juices are shown in Figure 6. The chloride content in the juices was positively correlated with the length of barley growth. The statistical analysis demonstrated statistically significant differences in this parameter between the harvest dates. The chloride content ranged from 0.021 to 0.117 g·100 $g_{f.j.}^{-1}$ for juice pressed from barley leaves and shoots collected on days 7–28. The increase in the chloride content was clearly correlated with a decrease in the pH of the juice. Park et al. [27] also showed that the amount of chlorides in the aboveground parts of plants may depend on the type of fertilization used.

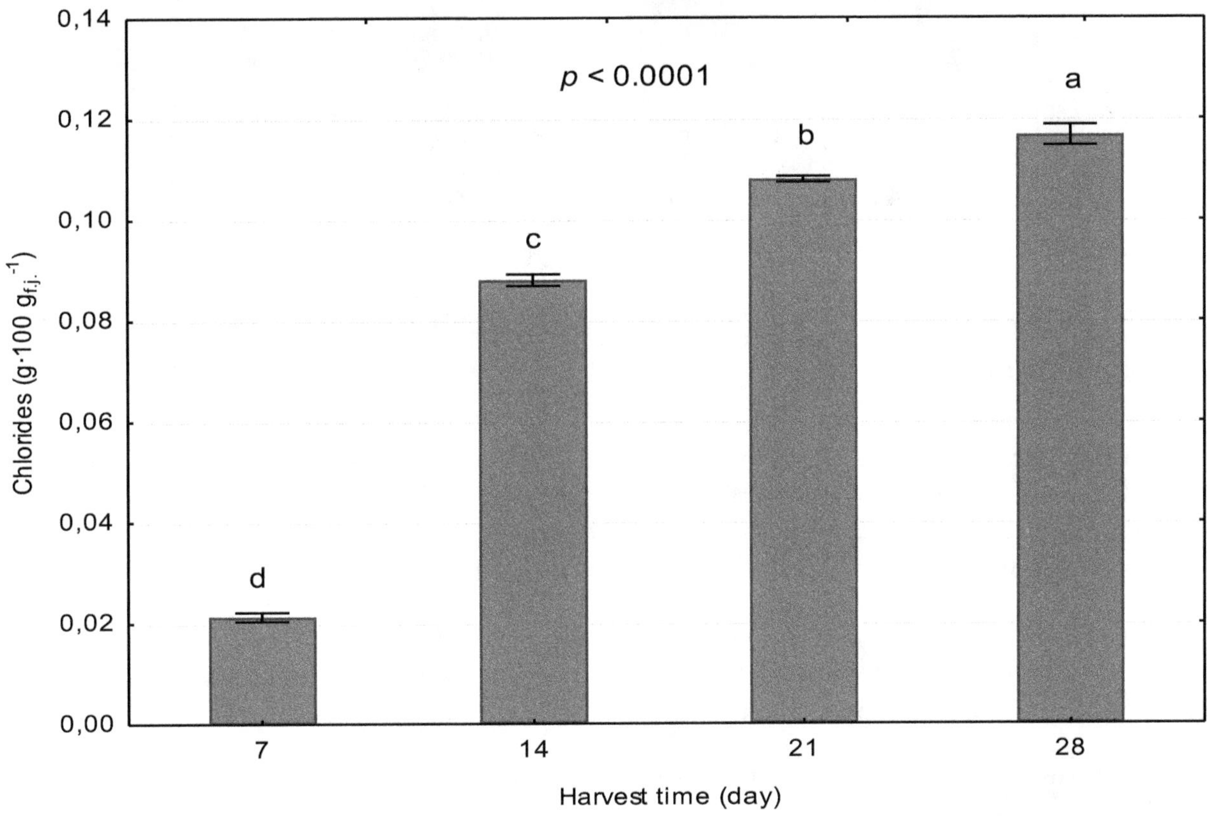

Figure 6. Changes in the chloride content in barley shoot juice relative to the harvest date.

3.6. Content of Reducing Sugars

Changes in the content of reducing sugars are shown in Figure 7. Their highest content was determined in the juice from barley leaves and shoots collected on day 21 of growth (8.20 g·100 $g_{d.m.}^{-1}$). The differences in the value of this parameter between products obtained from shoots collected on days 14 and 28 were not statistically significant. The statistical analysis confirmed the lower content of reducing sugars only for juice made from raw material collected on day 7. The study conducted by Paulíčková et al. [2] showed that the content of simple sugars varied, depending on the plant growth phase. Other factors that significantly determined the changes in the analysed parameter include the conditions of the soil and the variety of barley. Paulíčková et al. [2] demonstrated that the sugar content in most barley varieties steadily decreased throughout the growing season.

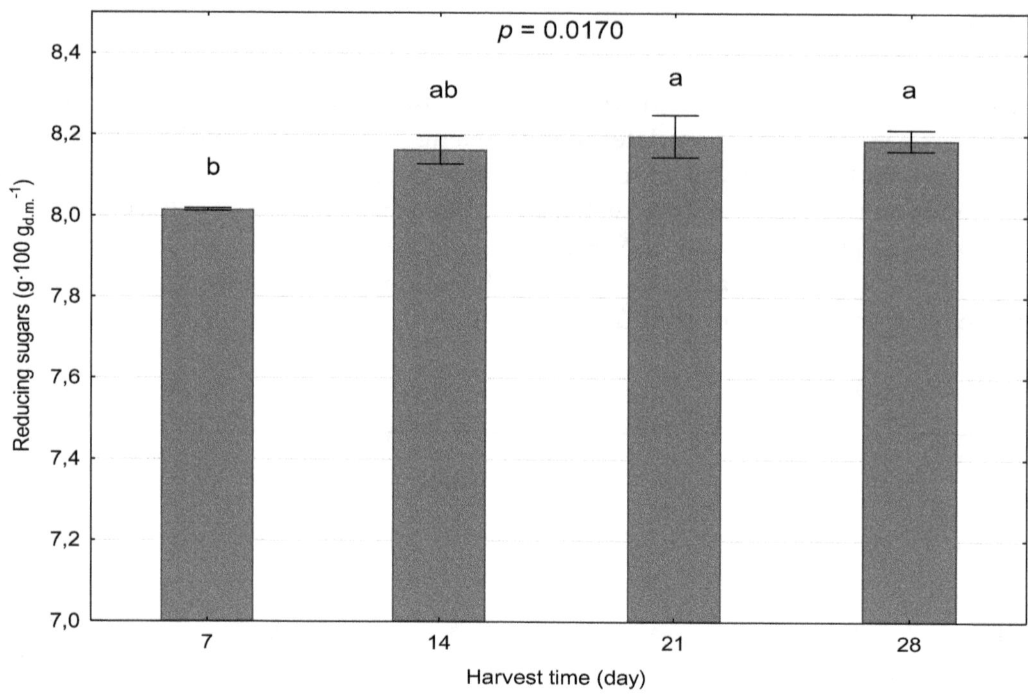

Figure 7. Changes in the content of reducing sugars in barley shoot juice, depending on the harvest date.

3.7. Chlorophyll Content

The chlorophyll content in the barley juice and the impact of the thermal treatment methods on changes in this parameter are shown in Table 2. The content of chlorophyll A and B and the total chlorophyll content changed significantly during storage. The highest determined content of total chlorophylls, i.e., 6.62 mg/g, was found in fresh juice. The chlorophyll contents in fresh raw material obtained by other authors ranged from 10.15 to 19.62 mg/g [28,29]. After seven days of storage, the total chlorophyll content was reduced by 37.9% (UJ), 42.12% (PJ), and 2.43% (FJ), in comparison with the untreated juice. It can thus be concluded from the present investigations that chlorophyll A is more sensitive to heat than chlorophyll B. Similar findings were reported by Weemaes et al. [30], who analysed the kinetics of chlorophyll degradation in thermally treated broccoli juice.

Table 2. Changes in the chlorophyll content in juice from young barley leaves during storage in various conditions ($p < 0.0001$).

Type of Heat Treatment/Ti-Me of Storage	Fresh (Control Probe)	Unprocessed (UJ)		Pasteurised (PJ)		Frozen (FJ)
		3rd Day	7th Day	3rd Day	7th Day	
Chlorophyll A (mg·g^{-1})	4.80 ± 0.01 [a]	4.15 ± 0.02 [c]	2.99 ± 0.01 [e]	3.80 ± 0.00 [d]	2.83 ± 0.015 [f]	4.66 ± 0.01 [b]
Chlorophyll B (mg·g^{-1})	1.53 ± 0.00 [a]	1.51 ± 0.00 [bc]	0.85 ± 0.01 [d]	1.50 ± 0.00 [c]	0.83 ± 0.00 [e]	1.51 ± 0.00 [b]
Total chlorophylls (mg·g^{-1})	6.62 ± 0.01 [a]	6.02 ± 0.04 [c]	4.11 ± 0.04 [e]	5.72 ± 0.02 [d]	3.83 ± 0.007 [f]	6.46 ± 0.04 [b]

[a,b,c,d,e,f] Means in the same line denoted by different letters were significantly different. The results are expressed as mean ± SD (n = 3).

The process of freezing had the lowest effect on changes in the content of chlorophyll A and total chlorophylls. By contrast, in the case of chlorophyll B, the least significant changes were noted for the unpasteurised product refrigerated for one day and in the case of the frozen juice stored for seven days. Paulíčková et al. [2] also confirmed the significant effect of thermal treatment of barley leaf juice (freezing, drying, and freeze-drying) on changes in the nutrient content. The analysis of their results also allows for the conclusion that freezing exerts the weakest effect on the quality of the product. Koca et al. [25] demonstrated a faster rate of degradation of chlorophylls at a lower pH value (from 5.5 to 7.5). Hence, the significantly lower chlorophyll content of juice stored for seven days may be caused by, e.g., changes in the pH of the product.

4. Conclusions

The present investigations confirm the potential benefits of consuming green juice from young barley shoots and leaves as part of a daily diet. The product obtained from the material harvested after 7, 14, 21, and 28 days of growth contains many valuable nutrients, e.g., a high level of total protein (with CS = 41.44 Meth). Additionally, high pressing yields of approximately 70% can be achieved.

The study has demonstrated that barley leaves and shoots harvested on day 21 of plant growth are the best raw material for the production of juice. The process of pressing material collected at this time exhibits the highest efficiency, and the juice contains the highest levels of protein and reducing sugars, as well as a high chloride content.

The most effective way to preserve the juice from young barley leaves is freezing. This process does not induce changes in juice acidity and only slightly reduces the content of chlorophylls A and B during storage of the product. Pasteurisation of the juice significantly reduces the chlorophyll content, but does not induce statistically significant changes in the pH of the juice.

The reported results are a preliminary study on the topic. However, it is necessary to continue research to determine the impact of other factors (e.g., barley varieties, growing conditions) on the quality of juices from young barley shoots and leaves. It is also possible scale-up the experiment using a higher number of samples, filed conditions, etc.

In addition, this way of using shoots and leaves of young barley will have economic importance. The costs of obtaining raw materials for juice production will be reduced. The use of shoots and leaves of young barley, which is grown as a forecrop, for the production of green juices, may have a beneficial effect on the development of sustainable crop production. This will allow more efficient use of the plants.

Author Contributions: Conceptualization, A.B.-K., D.A., F.K. and L.R.; methodology, A.B.-K. and D.A.; formal analysis, A.B.-K. and F.K.; investigation, A.B.-K. and L.R.; data curation, A.B.-K. and L.R.; writing—original draft preparation, A.B.-K. and Z.K.; writing—review and editing, D.A.; visualization, A.B.-K.; supervision, D.A.

References

1. Jabłońska-Ceglarek, R.; Rosa, R. Forecrop green manures and the size and quality of white cabbage yield. *Electron. J. Pol. Agric. Univ.* **2003**, *6*. Available online: http://www.ejpau.media.pl/volume6/issue1/horticulture/art-08.html (accessed on 21 July 2019).
2. PaulíčkoVá, I.; EhrENbErgEroVá, J.; FIEdlEroVá, V.; Gabrovska, D.; Havlova, P.; Holasova, M.; Vaculová, K. Evaluation of barley grass as a potential source of some nutritional substances. *Czech J. Food Sci.* **2007**, *25*, 65–72. [CrossRef]
3. Son, H.-K.; Lee, Y.-M.; Park, Y.-H.; Lee, J.-J. Effect of Young Barley Leaf Powder on Glucose Control in the Diabetic Rats. *Korean J. Community Living Sci.* **2016**, *27*, 19–29. [CrossRef]
4. Zeng, Y.; Pu, X.; Yang, J.; Du, J.; Yang, X.; Li, X.; Li, L.; Zhou, Y.; Yang, T. Preventive and Therapeutic Role of Functional Ingredients of Barley Grass for Chronic Diseases in Human Beings. *Oxidative Med. Cell. Longev.* **2018**, *2018*, 3232080. [CrossRef] [PubMed]
5. Barczak, B.; Nowak, K. Skład aminokwasowy białka biomasy jęczmienia ozimego (Hordeum vulgare L.) w zależności od stadium rozwoju rośliny i nawożenia azotem. *Acta. Sci. Pol. Agric.* **2008**, *7*, 3–15.
6. Jarosz, M.; Szponar, L.; Rychlik, E.; Wierzejska, R. Woda i elektrolity. In *Normy Żywienia Dla Popul. Pol. Nowelizacja*, 1st ed.; Jarosz, M., Ed.; Instytut Żywności i Żywienia: Warsaw, Poland, 2012; pp. 143–151.
7. García-Caparrós, P.; Almansa, E.M.; Chica, R.M.; Lao, M.T. Effects of Artificial Light Treatments on Growth, Mineral Composition, Physiology, and Pigment Concentration in Dieffenbachia maculata "Compacta" Plants. *Sustainability* **2019**, *11*, 2867. [CrossRef]
8. Kawka, K.; Lemieszek, M.K. Prozdrowotne właściwości młodego jęczmienia. *Med. Ogólna I Nauk. O Zdrowiu* **2017**, *23*, 7–12. [CrossRef]
9. Polak, R. Chlorofile jako naturalne źródło energii biomasy. *Przemysł Chem.* **2019**, *1*, 138–142. [CrossRef]

10. Pradhan, J.; Das, S.; Das, B.K. Antibacterial activity of freshwater microalgae: A review. *Afr. J. Pharm. Pharmacol.* **2014**, *8*, 809–818.
11. Kandhasamy, S.; Chin, N.L.; Yusof, Y.A.; Lai, L.L.; Mustapha, W.A.W. Effect of Blender and Blending Time on Color and Aroma Characteristics of Juice and Its Freeze-Dried Powder of Pandanus amaryllifolius Roxb. Leaves (Pandan). *Int. J. Food Eng.* **2016**, *12*, 75–81.
12. Qamar, A.; Saeed, F.; Nadeem, M.T.; Hussain, A.I.; Khan, M.A.; Niaz, B. Probing the storage stability and sensorial characteristics of wheat and barley grasses juice. *Food Sci. Nutr.* **2019**, *7*, 554–562. [CrossRef] [PubMed]
13. Kubatka, P.; Kello, M.; Kajo, K.; Kruzliak, P.; Výbohová, D.; Šmejkal, K.; Maršík, P.; Zulli, A.; Gönciová, G.; Mojžiš, J.; et al. Young barley indicates antitumor effects in experimental breast cancer in vivo and in vitro. *Nutr. Cancer* **2016**, *68*, 611–621. [CrossRef] [PubMed]
14. Son, H.-K.; Lee, Y.-M.; Lee, J.-J. Nutrient Composition and Antioxidative Effects of Young Barley Leaf. *Korean J. Community Living Sci.* **2016**, *27*, 851–862. [CrossRef]
15. Majumdar, T.; Wadikar, D.; Vasudish, C.; Premavalli, K.; Bawa, A. Effect of Storage on Physico-Chemical, Microbiological and Sensory Quality of Bottlegourd-Basil Leaves Juice. *Am. J. Food Technol.* **2011**, *6*, 226–234. [CrossRef]
16. Teixeira, E.M.B.; Carvalho, M.R.B.; Neves, V.A.; Silva, M.A.; Arantes-Pereira, L. Chemical characteristics and fractionation of proteins from Moringa oleifera Lam. leaves. *Food Chem.* **2014**, *147*, 51–54. [CrossRef]
17. Pielesz, A. Skład chemiczny algi brązowej Fucus vesiculosus L. *Postępy Fitoter.* **2011**, *1*, 9–17.
18. Briggs, D.E. *Barley*; Springer Science and Business Media: Berlin/Heidelberg, Germany, 2012; pp. 14–24.
19. Frank, A.B.; Bauer, A.; Black, A.L. Effects of Air Temperature and Fertilizer Nitrogen on Spike Development in Spring Barley. *Crop. Sci.* **1992**, *32*, 793–797. [CrossRef]
20. McMaster, G.S.; Wilhelm, W.W.; Frank, A.B. Developmental sequences for simulating crop phenology for water-limiting conditions. *Aust. J. Agric. Res.* **2005**, *56*, 1277–1288. [CrossRef]
21. Kırca, A.; Özkan, M.; Cemeroğlu, B.; Kirca, A.; Cemeroğlu, B. Effects of temperature, solid content and pH on the stability of black carrot anthocyanins. *Food Chem.* **2007**, *101*, 212–218. [CrossRef]
22. Nabrdalik, M.; Świsłowski, P. Microbiological Evaluation of Unpasterized Fruit and Vegetable Juices. In Proceedings of the ECOpole, Zakopane, Poland, 5–8 October 2016.
23. Yoon, K.Y.; Woodams, E.E.; Hang, Y.D. Fermentation of beet juice by beneficial lactic acid bacteria. *LWT Food Sci. Technol.* **2005**, *38*, 73–75. [CrossRef]
24. Czapski, J. Heat stability of betacyanins in red beet juice and in betanin solutions. *Eur. Food Res. Technol.* **1990**, *191*, 275–278. [CrossRef]
25. Koca, N.; Karadeniz, F.; Burdurlu, H.S. Effect of pH on chlorophyll degradation and colour loss in blanched green peas. *Food Chem.* **2007**, *100*, 609–615. [CrossRef]
26. Saguy, I. Thermostability of red beet pigments (betanine and vulgaxanthin–I): Influence of pH and temperature. *J. Food Sci.* **1979**, *44*, 1554–1555. [CrossRef]
27. Park, J.; Cho, K.H.; Ligaray, M.; Choi, M.-J. Organic Matter Composition of Manure and Its Potential Impact on Plant Growth. *Sustainability* **2019**, *11*, 2346. [CrossRef]
28. Cao, X.; Zhong, Q.; Wang, Z.; Zhang, M.; Mujumdar, A.S. Effect of microwave freeze drying on quality and energy supply in drying of barley grass. *J. Sci. Food Agric.* **2017**, *98*, 1599–1605. [CrossRef]
29. Cao, X.; Zhang, M.; Mujumdar, A.S.; Zhong, Q.; Wang, Z. Effect of nano-scale powder processing on physicochemical and nutritional properties of barley grass. *Powder Technol.* **2018**, *336*, 161–167. [CrossRef]
30. Weemaes, C.A.; Ooms, V.; Van Loey, A.M.; Hendrickx, M.E. Kinetics of Chlorophyll Degradation and Color Loss in Heated Broccoli Juice. *J. Agric. Food Chem.* **1999**, *47*, 2404–2409. [CrossRef]

13

Mechanical and Processing Properties of Rice Grains

Weronika Kruszelnicka [1,*], Andrzej Marczuk [2], Robert Kasner [1], Patrycja Bałdowska-Witos [1], Katarzyna Piotrowska [3], Józef Flizikowski [1] and Andrzej Tomporowski [1]

1 Department of Manufacturing Techniques, Faculty of Mechanical Engineering, University of Science and Technology in Bydgoszcz, 85-796 Bydgoszcz, Poland; robert.kasner@gmail.com (R.K.); patrycja.baldowska-witos@utp.edu.pl (P.B.-W.); fliz@utp.edu.pl (J.F.); a.tomporowski@utp.edu.pl (A.T.)
2 Faculty of Production Engineering, University of Life Sciences in Lublin, 20-612 Lublin, Poland; andrzej.marczuk@up.lublin.pl
3 Faculty of Mechanical Engineering, Lublin University of Technology; 20-618 Lublin, Poland; k.piotrowska@pollub.pl
* Correspondence: weronika.kruszelnicka@utp.edu.pl

Abstract: Strength properties of grains have a significant impact on the energy demand of grinding mills. This paper presents the results of tests of strength and energy needed the for destruction of rice grains. The research aim was to experimentally determine mechanical and processing properties of the rice grains. The research problem was formulated in the form of questions: (1) what force and energy are needed to induce a rupture of rice grain of the *Oryza sativa* L. of long-grain variety? (2) what is the relationship between grain size and strength parameters and the energy of grinding rice grain of the species *Oryza sativa* L. long-grain variety? In order to find the answer to the problems posed, a static compression test of rice grains was done. The results indicate that the average forces needed to crush rice grain are 174.99 kg m·s^{-2}, and the average energy is 28.03 mJ. There was no statistically significant relationship between the grain volume calculated based on the volumetric mass density V_ρ and the crushing energy, nor between the volume V_ρ and other strength properties of rice grains. In the case of Vs, a low negative correlation between strength σ_{min} and a low positive correlation between the power inducing the first crack were found for the grain size related volume. A low negative correlation between the grain thickness a_3, stresses σ_{min} and work W_{Fmax} was found as well as a low positive correlation between thickness a_3 and the force inducing the first crack $\boldsymbol{F_{min}}$.

Keywords: rice; grinding; compressive strength; rupture energy

---

1. Introduction

The strength properties of grains have a significant impact on the energy demand of grinders [1–4]. During grinding of grains in a five-disc mill, a complex state of stress occurs in the material, with shear and compressive stresses prevailing [5,6]. Identification of the forces causing grain cracking (rupture) can be considered as the first step to determine the energy demand in the grain grinding process [7,8]. Two cases can be distinguished: static squeezing of grains and shearing of grains [9]. In the case of static compression, in order to determine the forces and stress and consequently, the work (energy) needed to crush one grain and then more than a dozen grains, a static compression test can be carried out. The ranges of probable forces destroying the grain may be determined in an experimental manner, and subsequently, the energy ranges of destroying its structure [10].

The subject of research on the physical-mechanical properties of rice grains has already been addressed by researchers, because the specificity of rice grain processing and the energy demand of processing lines, e.g., grinding, drying, pelleting, etc., depends largely on these properties [11–17]. In their research, Zeng, et al. [18] focused on modeling cracking of rice grains depending on the grain moisture and load speed, using the Discrete Element Method (DEM). The impact of humidity on the strength properties of white rice was also addressed in the work of Sadeghi et al. [19], and brown rice in the work by Cao et al. [20] and Chattopadhyay et al. [21]. Buggenhout et al. [22] studied the influence of physicochemical properties, in particular, the impact of grain husks and moisture on cracking phenomena during rice processing. Esehaghbeygi et al. [23] in turn, analyzed the effect of drying temperature and kinetic energy during the drying process on grain susceptibility to breaking. Similar research was conducted by Sarker et al. [24], Tajaddodi et al. [25], Nasirahmadi et al. [26] and Bonazzi and Courtois [27]. The influence of grain orientation on its mechanical properties under load was analyzed, among others, by Li et al., Shu et al., and Zareiforoush et al. [28–30]. They showed that rice grains are more flexible in horizontal orientation based on the results of static compression and three-point bending tests. Zareiforoush et al. [31], based on the conducted research, found that increasing the speed of load during the compression test results in lowering crushing forces and energy.

From the point of view of energy demand, processing of grainy biomaterials, particularly grinding, the strength of grains plays the key role. [32]. Considering the process of grinding, e.g., by means of grinding machines or roller mills, these are the compressive loads which prevail in the grinded material, whereas permanent deformation (fragmentation) occurs after exceeding the load value corresponding to the compressive strength limit. Strength is closely related to the power necessary to cause the strain and the grinded material cross-section field (hence being dependent on its geometric features). Thus, material fragmentation occurs upon application of appropriate forces, which, in the system of grinding machines, roller mills, is performed by rotary motion of rollers. In such a case, the force is a direct effect of torques, which, in turn, are related to the power of the devices affecting the energy demand of the grain processing. In general, the higher force to be applied the higher power, that is, the machine energy, is needed. The aspects concerning calculation of energy demand for grain grinding systems are described in detail in [33].

The strength of grains depends on the type of material, especially on its internal structure (porosity), moisture, components of the grain, and biological properties [34]. In the case of biomass grains, a significant diversification in terms of dimensions, physical and strength properties, can be observed even within one grain species which, apart from biological characteristics, is conditioned by the weather conditions and the cultivation method [35–37] Earlier research has shown that the energy needed to grind hard materials with higher strength is larger [38–40]. It was also observed that along with a moisture increase, the energy consumption increases as well [18–20,36]. The internal structure of the grain endosperm and tegument has an impact on the strength properties and the energy needed for grinding. The endosperms which are characterized by higher glassiness are usually harder; thus, for permanent deformation, it is necessary to use higher forces which, in turn, results in an increased energy demand as compared to materials whose endosperm is less glassy [36,41,42]. The glassiness of the endosperm also has an influence on the material fragmentation efficiency and the size of particles after division—the higher glassiness, the easier to separate the endosperm from bran and the grain disintegrates into smaller parts. [38,42]. Tests of physical properties and grinding energy, carried out for wheat grains have also revealed that the grinding energy is proportional to the grinded material mass [35]. Dziki and Laskowski [37] indicate, using the example of wheat, that the values of work and force to be applied for crushing the grain increase along with the grain thickness growth.

This study contains an analysis of strength properties of rice grains, mainly forces causing the first violation of the grain structure (first crack) and grain stiffness.

The aim of the research is to experimentally determine the mechanical and processing properties (strength and energy properties) of granular biomass (rice) accepted for research in the project

"Intelligent monitoring of the grinding characteristics of grainy biomass". Determining the forces needed to break grains is of key importance when developing energy and environmental efficiency indicators for the grinding process and modeling grinding and crushing processes using the discrete element method DEM.

The research problem was formulated in the form of questions: (1) what strength and energy is needed to induce a rupture of rice grain of the species *Oryza sativa* L. long-grain variety? (2) what is the relationship between grain size and strength parameters and the energy of grinding rice grain of the species *Oryza sativa* L. long-grain variety?

2. Materials and Methods

2.1. Rice Grains Preparation

To determine the force needed to break the grain, a static compression test was carried out for 100 grains of rice of the species *Oryza sativa* L., a long-grained variety with a stabilized humidity equal to 13% ± 0.1%. *Oryza sativa* L., a long-grained variety of rice was accepted to be the research object due to its popularity, among others, in food industry [32]. Knowing the processing properties of this species, in particular, crushing energy, can significantly affect power demand of the processing devices, e.g., grinders. Samples of 100 individual rice grains were prepared and described by numbers (Figure 1). Then, three dimensions were measured with the vernier caliper: length, width and height of the grain.

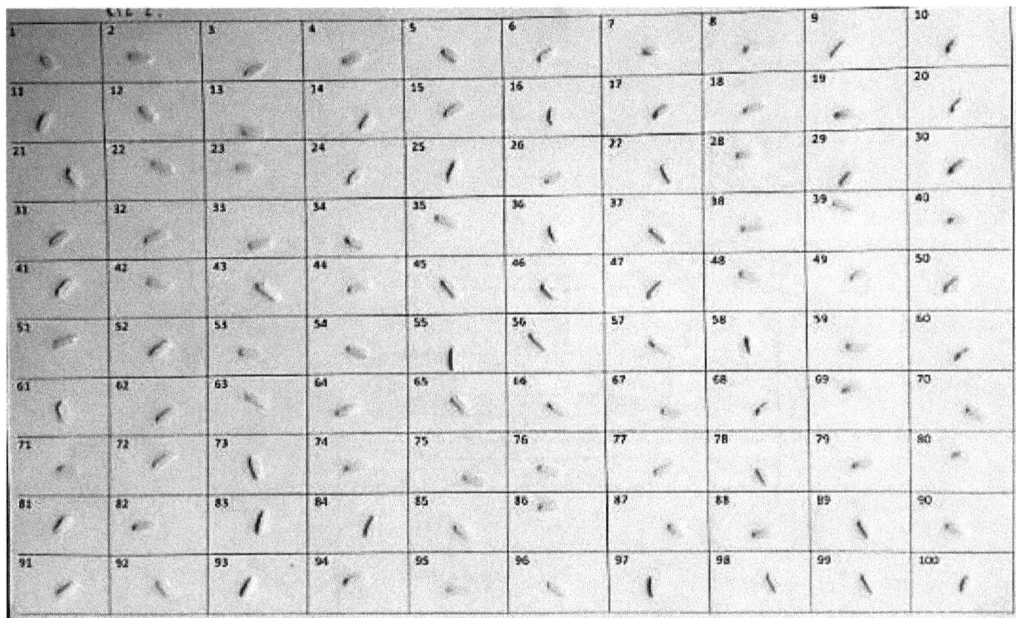

Figure 1. Rice grains prepared for compression test.

2.2. Test Stand

A static compression test was carried out on an Instron 5966 testing machine (Figure 2). The samples were placed in the machine in a horizontal orientation, in such a way that the dimensions a_1 and a_2 were the large axis and small axis of cross-sectional areas of the grain subjected to the load (Figure 3). The load speed was 2 mm·min^{-1}.

Figure 2. The Instron 5966 testing machine.

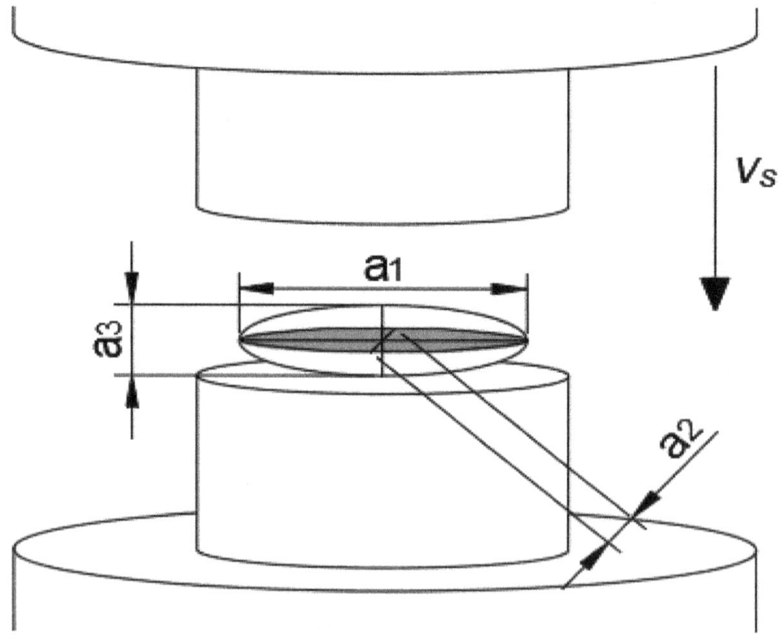

Figure 3. Position of the rice grain during the static compression test.

2.3. Research Methods

The analysis of the grain strength properties during a static compression test of a single grain was performed according to the plan shown in Figure 4. Three dimensions were measured with the vernier caliper: length, width and height of the grain.

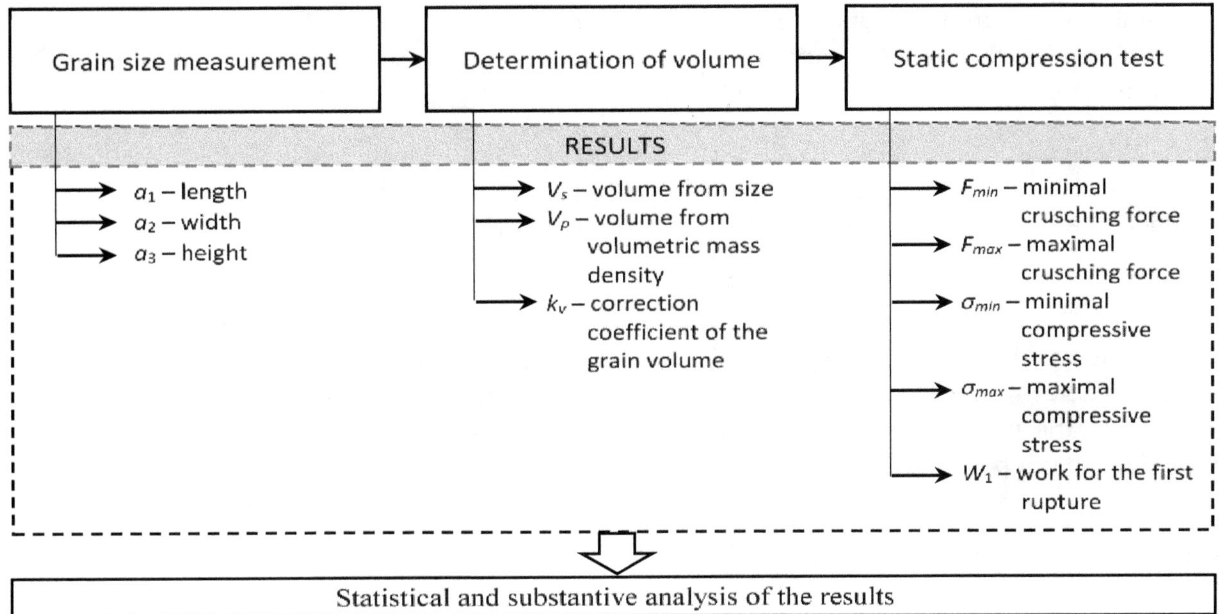

Figure 4. Plan and research program.

The volume of the studied seeds was determined in two ways: on the basis of the three measured dimensions a_1, a_2 and a_3 (V_s) and on the basis of the relationship of the grain mass and density (V_ρ).

The first method of determining the volume takes into consideration three basic grain sizes, i.e., the height and width length, and allows a simplified grain volume estimation with a certain error by aligning the grain shape to the cuboid with dimensions a_1, a_2, a_3 (Figure 5).

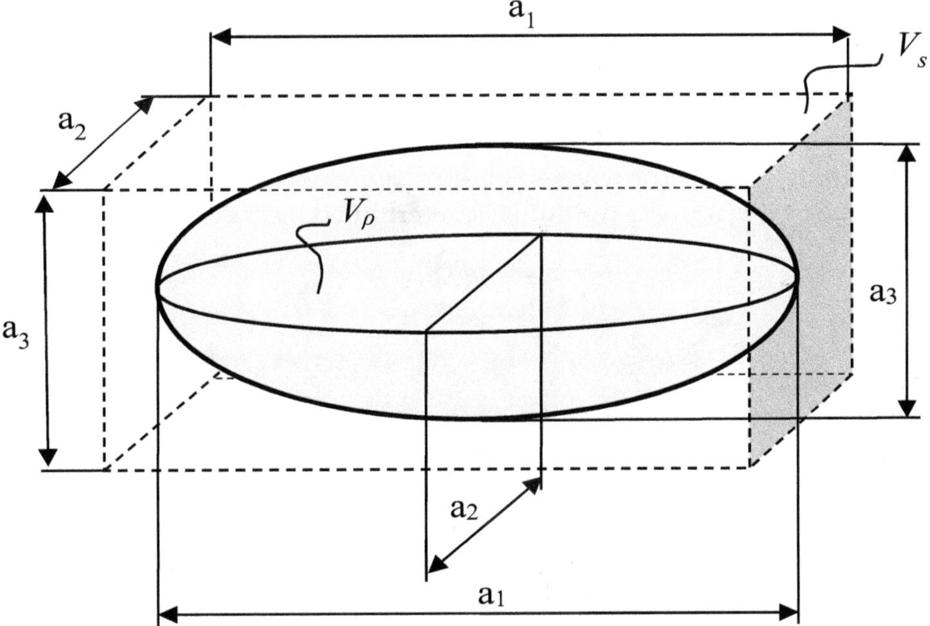

Figure 5. Graphical representation of grain volume determination based on the knowledge of three dimensions, a_1—length of the grain, mm, a_2—width of the grain, mm, a_3—height of the grain, mm, vs.—grain volume calculated based on three dimensions a_1, a_2, a_3 (volume of a cuboid with dimensions a_1, a_2, a_3), mm^3, V_ρ—grain volume calculated based on the volumetric mass density, mm^3.

For this interpretation, the formula for the estimated volume of grain will be:

$$V_s = a_1 \cdot a_2 \cdot a_3, \tag{1}$$

and equal to the volume of a cuboid circumscribed on the grain.

The second method of determining the volume is based on the knowledge of density ρ of the grain and its mass m. The grain density is determined from the dependence [43]:

$$\rho = m/V_{\rho}, \tag{2}$$

Hence, the volume determined on the basis of density V_{ρ}, will be [43]:

$$V_{\rho} = m/\rho, \tag{3}$$

The relationship between computational volume (V_s) and determined based on the grain density (V_{ρ}) can be determined by calculating the correction factor for the grain volume k_v taking into consideration the grain spherical and uneven shape. It can be determined experimentally and expressed by the dependence:

$$k_v = V_{\rho}/V_s, \tag{4}$$

hence:

$$V_{\rho} = k_v \cdot V_s, \tag{5}$$

The static compression test is mainly used for brittle materials, i.e., not showing significant plastic deformation. Rice can be considered as a fragile material with some (small) plastic deformability due to the internal structure that differentiates it from the cross-linked metal structure. A feature that characterizes fragile materials is compressive strength (R_c) [24]:

$$R_c = F_c/A_0, \tag{6}$$

where
F_c—the largest value of the compressive load at which the sample is crushed,
A_0—the initial cross-section of the sample.

If the compression diagram $l = f(F)$ has a part where shortening (Δl) is directly proportional to compressive force (F), then, on this basis, we determine Young's modulus (E) for this material. If this relationship is not directly proportional, then based on the first few results (where it is possible to assume that the material behaves linearly and elastically), we determine the mean value of Young's modulus (E). The value of the Young's modulus is determined by Hooke's law [24]:

$$E = F \cdot l/(\Delta l \cdot A_0), \tag{7}$$

where:
F—compressive force, kg·m·s^{-2},
Δl—sample shortening corresponding to force (F), m,
l—the initial length of the sample, m,
A_0—area of the initial sample cross-section, m^2.

During the compression, some force F affects the grain and causes it to break (displacement s), so the elementary work done over the grain by force F causing the crack can be determined [24,44]:

$$dW = F \cdot ds. \tag{8}$$

During compression, we deal with a variable force F and displacement $s \rightarrow 0$, then, this characterized work can be determined by integrating both sides of the equation [24]:

$$W = \int_{s_1}^{s_2} F ds \tag{9}$$

Then, the work during compression of one grain is the area under graph $F = f(s)$ (Figure 6).

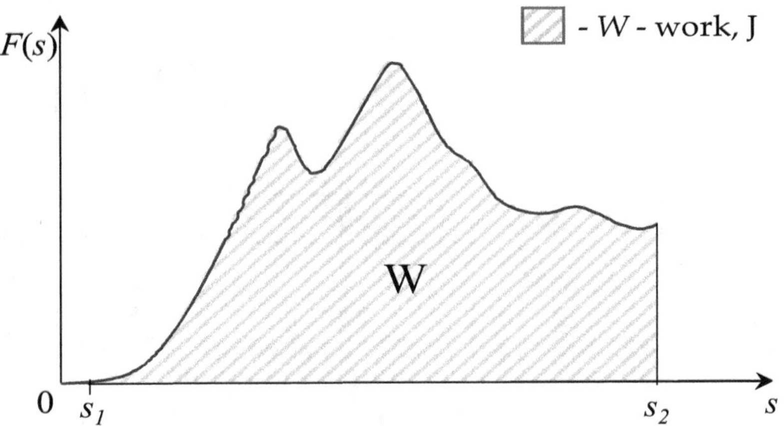

Figure 6. Graphic presentation of work during compression of grains, $F(s)$—compressive force, kg·m·s^{-2}, s—displacement, m, W—work, J, s_1—starting point of displacement, m, s_2—final displacement, m.

2.4. Analytical Methods

The statistical analysis tools available in MS Excel and Statistica were used in processing the results. Basic descriptive statistics of the examined physical and mechanical properties and rupture energy of rice grains were determined. The relationship between the grain volume and physical-mechanical properties and rupture energy was examined using Spearman's correlation analysis. A significance level $p < 0.05$ was adopted.

3. Results and Discussion

3.1. Results of Research on the Physical Properties of Rice Grains and Their Discussion

Firstly, the physical properties of rice grains were determined, such as grain length a_1, grain width a_2, grain height a_3, and grain volume determined based on the knowledge of grain mass and density V_ρ, volume determined on the basis of dimensions V_s, correction factor for volume k_v. The results of the tests after a basic statistical analysis are presented in Figures 7 and 8 in the form of a box plot.

Based on the analysis, it was found that the average grain length of rice was 6.38 mm, the average width was 1.91 mm, the average height was 1.51 mm. The average grain volume determined on the basis of the grain weight and density was equal to 14.82 mm^3, while the volume determined on the basis of grain size 18.44 mm^3. The correction factor for given volumes V_ρ and vs. assumed an average value of 0.82. The obtained results allowed to formulate the dependence of the real volume of the grain determined on the basis of density as a function of the volume determined on the basis of dimensions:

$$V_\rho = k_v \cdot V_s = 0.82 \cdot V_s \tag{10}$$

The results obtained for rice dimensions a_1, a_2, and a_3, are similar, although slightly smaller than those reported in the literature by other researchers, e.g., Sadeghi et al., Zareiforoush et al., and Zeng et al. [18,19,30]. Differences in dimensions may be caused primarily by the difference in the varieties and types of rice grains studied, the country of origin (the study of rice originated from Burma), grain humidity and growing conditions of grains (e.g., extensive, intensive cultivation, weather conditions that affect the grain size). The presented grain size results are an indispensable element of building models in computer simulations based on the DEM discrete element method [18], e.g., in the RockyDEM environment.

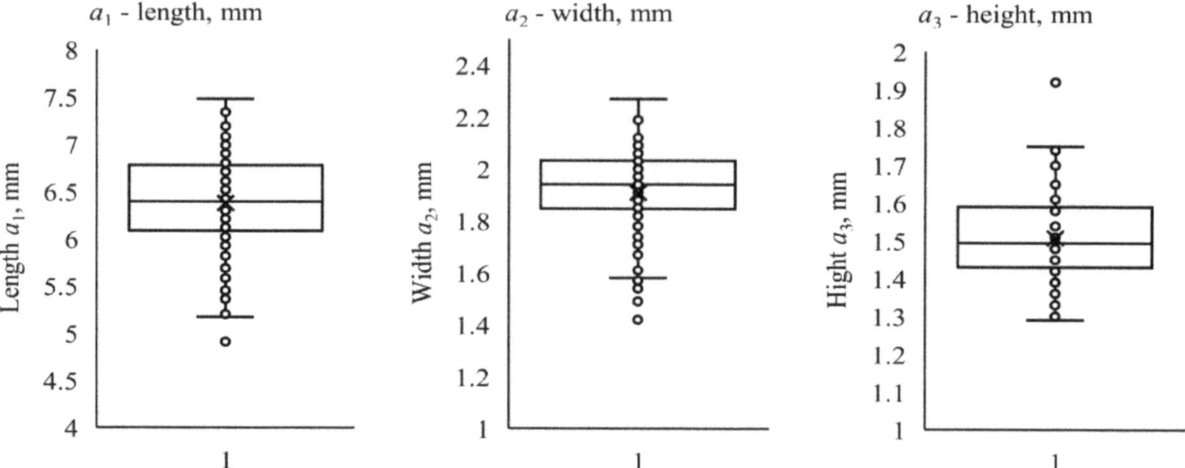

Figure 7. Results of the rice grains size determination.

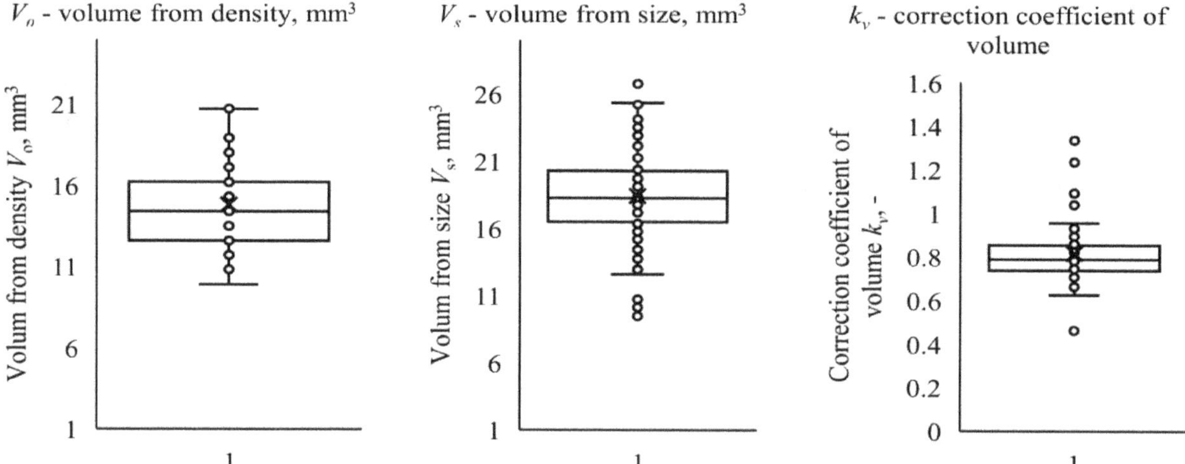

Figure 8. Results of rice grains volume determination.

3.2. The Results of Testing the Strength Properties of Rice Grains and Their Discussion

Figure 9 presents examples of graphs showing the characteristic of rice grain cracks during a compression test. Based on the presented curves, it can be stated that for each grain, the crack proceeded differently. This is caused by differences in internal structure between each grain, which is characteristic for biomaterials. However, noticeable are the characteristic points in the force-displacement graph marked as F_{min} and F_{max}. The point marked as F_{min} symbolizes the first crack of the grain, while the F_{max} point corresponds to the forces causing the breakdown of the grain into smaller fragments (Figure 10). Similar conclusions are presented in the work of Sadeghi et al. [19].

For rice grains, the forces inducing fracture F_{min} of the grain structure were within the range (35.86–198.71) kg m s^{-2}. The maximum forces F_{max} during the crack growth for rice were within the range (70.05–535.74) k g·m s^{-2}. Stresses σ_{min} for rice were in the range (2.21–17.38) MPa. Stresses σ_{max} for rice were in the range (5.00–29.61) MPa. Work W_{Fmin} for rice was in the range (1.88–56.55) mJ and W_{Famx} in the range (2.53–98.93) mJ (Figure 11).

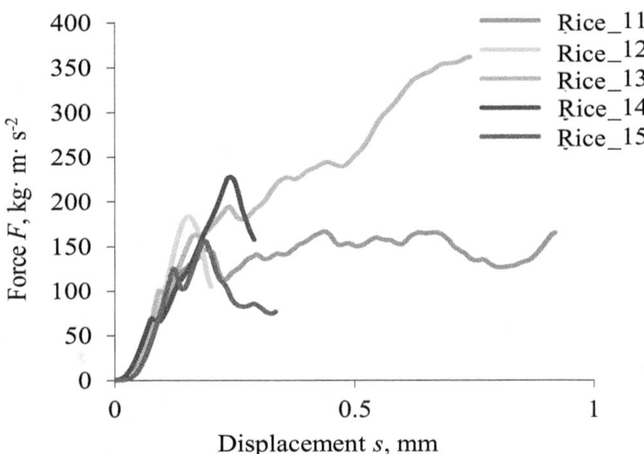

Figure 9. Exemplary curves for five rice grains from 100 tested illustrating the course of the rice grain compression process in the force-displacement coordinate.

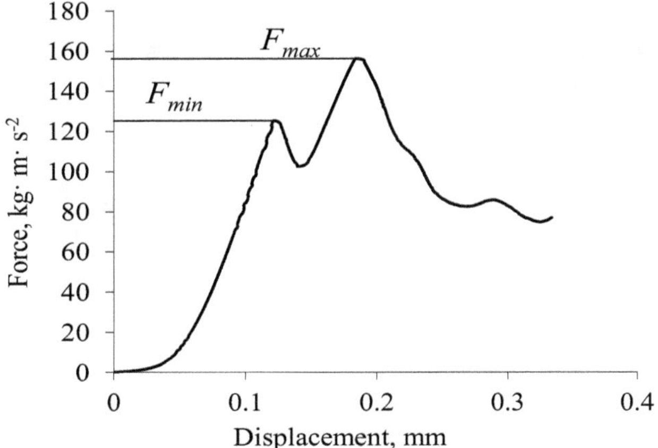

Figure 10. The cracking process for rice grains in function $F(s)$.

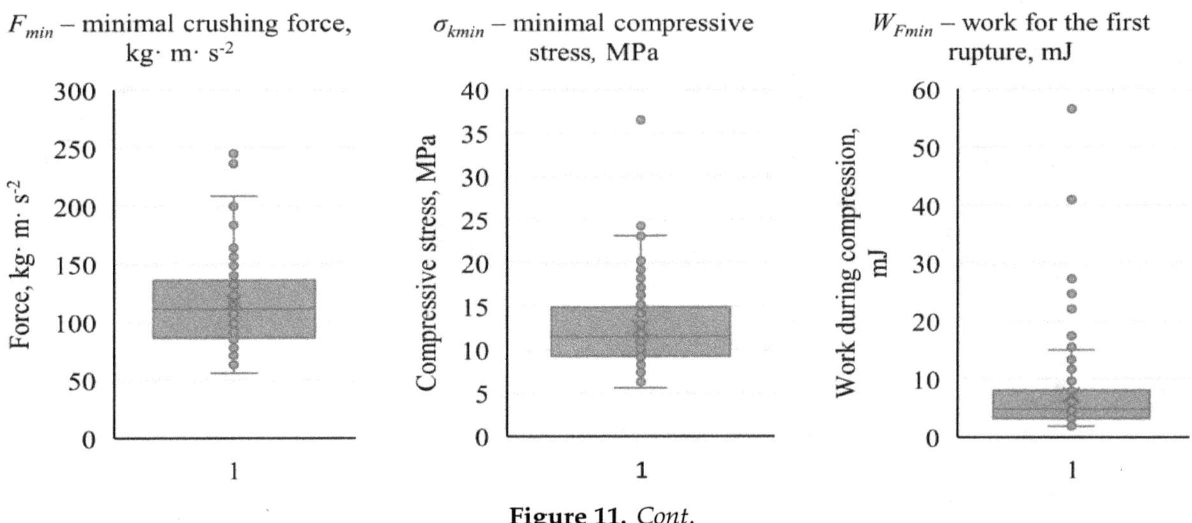

Figure 11. *Cont.*

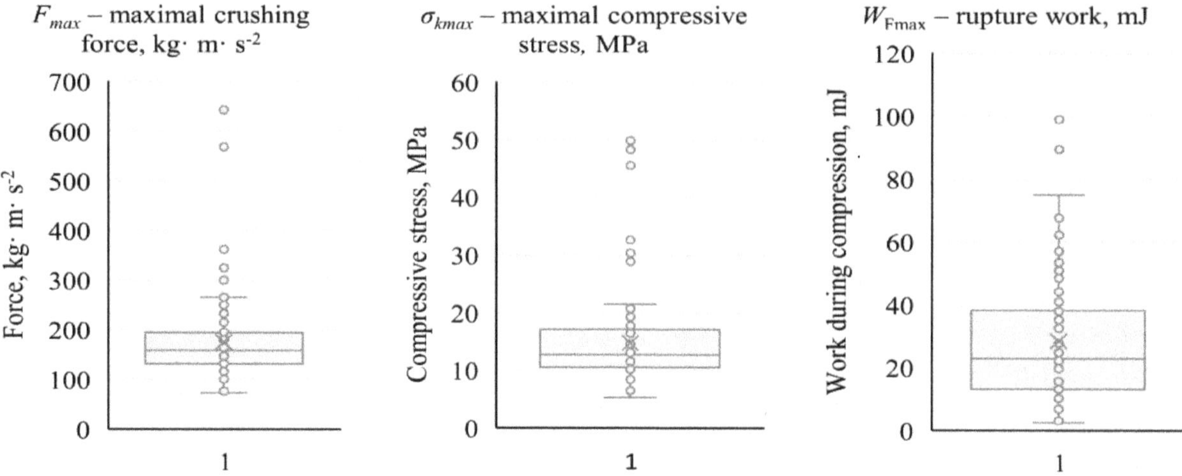

Figure 11. Results of statistical analysis of forces, stresses and energy during compression of rice grains.

Table 1 presents average values and basic results of a statistical analysis of the surveyed quantities. The obtained mean values of forces F_{max} are in accordance with the results presented by Lu and Siebenmorgen [45] obtained in the compression test at the load speed vs. 2 mm·min^{-1} (174.4–188.8 kg·m·s^{-2}), and higher than in the tests carried out by Sadeghi et al. [19] (169.06 kg·m s^{-2} for the Sorkheh type and 125.10 kg·m·s^{-2} for the Sazandegi type with v_s = 1,25 mm·min^{-1}), by Zareiforoush et al. [31] (125.69 kg·m·s^{-2} for the Alikazemi type and 109.96 kg·m·s^{-2} for the Hashemi type with v_s = 5 mm·min^{-1} and 117.38 kg m s^{-2} for the Alikazemi type and 88.33 kg m·s^{-2} for the Hashemi type with v_s = 10 mm·min^{-1}). Differences in the obtained values may result from different grain moistures, loading speeds of the samples and the type of rice used in the tests.

Table 1. Results of statistical analysis of examined mechanical properties of rice grains.

Parameter	Average	Standard Deviation	Median
Minimal crushing force F_{min}, kg m·s^{-2}	117.29	40.71	111.23
Maximal crushing force F_{max}, kg·m·s^{-2}	174.99	80.38	158.12
Minimal compressive stress σ_{kmin}, MPa	12.47	4.83	11.53
Maximal compressive stress σ_{kmax}, MPa	14.71	7.45	12.72
Stiffness C_k, N·mm^{-1}	1275.07	247.72	1322.11
Work for the first rupture W_{Fmin}, mJ	7.26	7.86	4.81
Rupture work W_{Fmax}, mJ	28.03	20.36	22.46

The average value of rupture work W_{Fmax} (28.03 mJ) is similar to the values presented, among others, in the work of Nasirahmadi et al. [26] (26.9 mJ for Fajr and 28.5 mJ for Tarom) and Sadeghi et al. [19] (24.45 mJ for Sazandegi) as well as in the work of Zareiforoush et al. [30,31], where the rupture energy assumed values of about 30 mJ.

Figures 12–18 summarize the results of mechanical properties of rice grains depending on its volume. The graphs show that mechanical properties of rice grains do not depend on its volume.

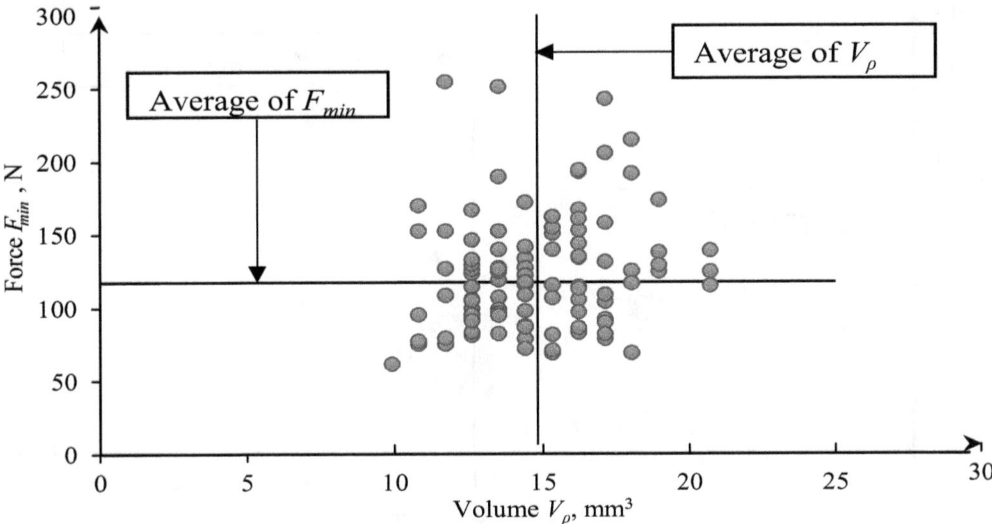

Figure 12. The scatter diagram of forces F_{min} for individual rice grains in relation to volume.

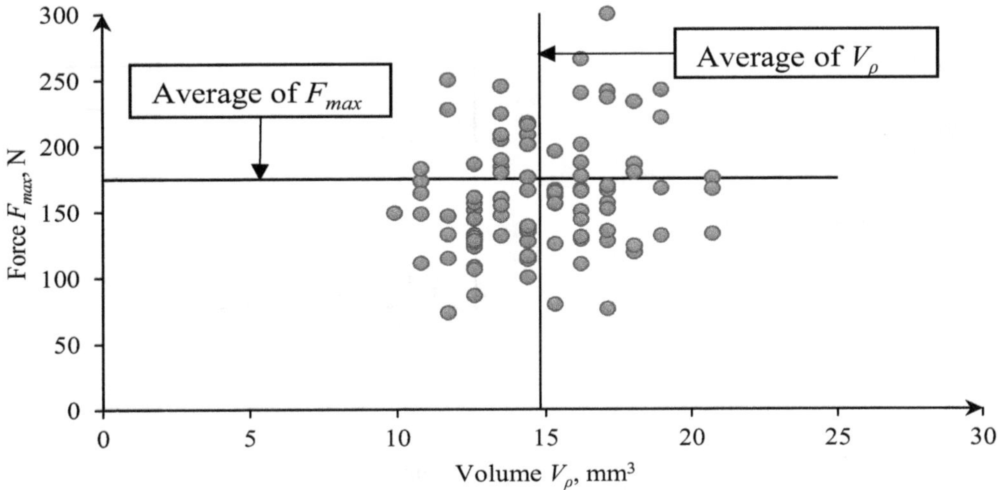

Figure 13. The scatter diagram of forces F_{max} for individual rice grains in relation to volume.

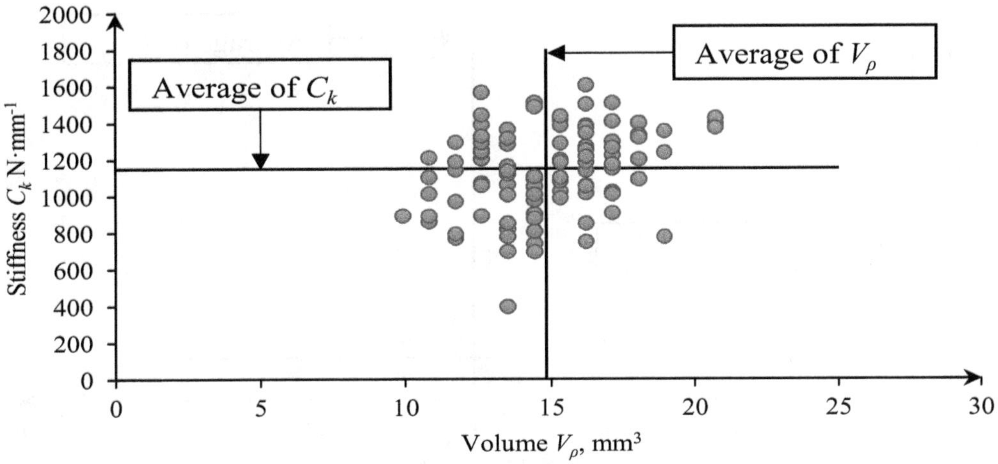

Figure 14. The scatter diagram of stiffnes for individual rice grains in relation to volume.

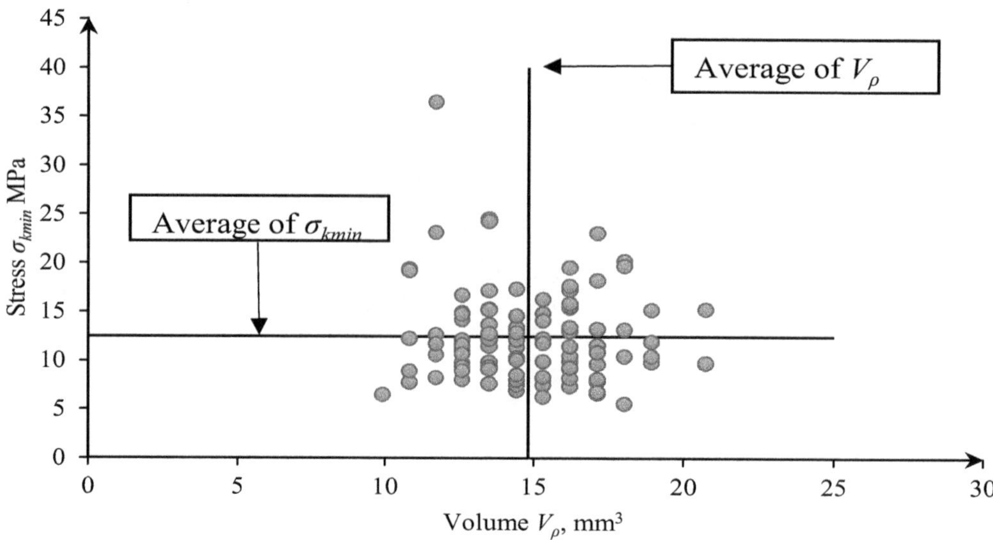

Figure 15. The scatter diagram of stress σ_{kmin} for individual rice grains in relation to volume.

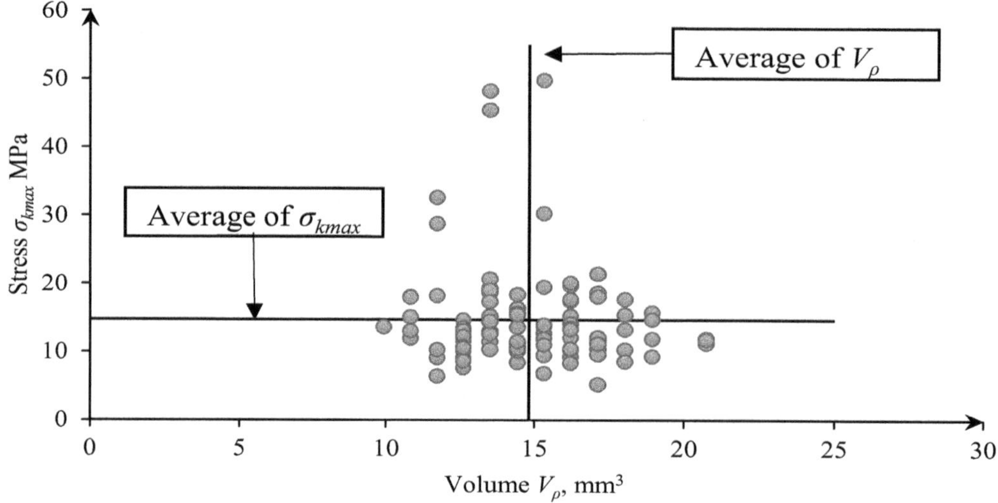

Figure 16. The scatter diagram of stress σ_{kmax} for individual rice grains in relation to volume.

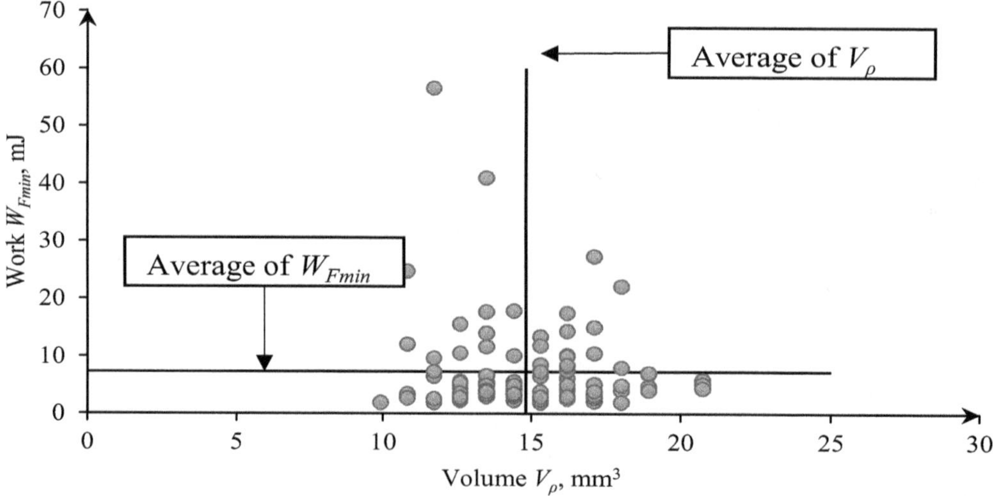

Figure 17. The scatter diagram of stress W_{Fmax} for individual rice grains in relation to volume.

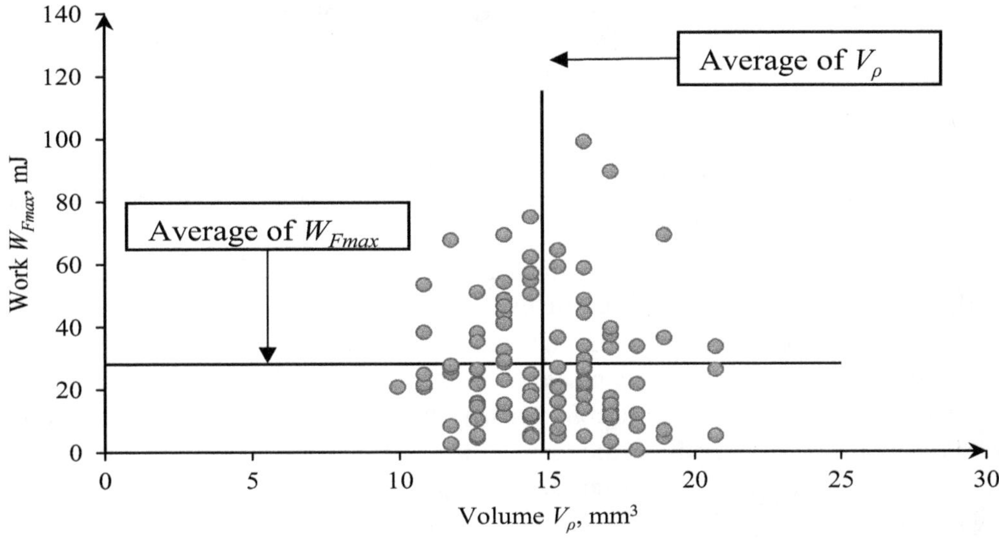

Figure 18. The scatter diagram of stress W_{Fmax} for individual rice grains in relation to volume.

The analysis of the Spearman correlation shows that there are no statistically significant relationships between the volume of the grain V_ρ and the tested strength properties, so there are no interdependencies between the variables (Tab 2). In the case of the grain dimension related volume vs. a low negative correlation between strength σ_{min} and low positive correlation between the force inducing the first crack were found (Table 2). An analysis of the correlation between the grain size related variables, that is, length a_1, width a_2, height a_3, mass m and volumes V_ρ and vs. showed that the volume of grains V_ρ is moderately positively correlated with length a_1 and height a_3 (Table S1). The surface of compression cross-section A_0 was in turn significantly correlated with the grain length a_1 and width a_2 (Table S1). However, no relevant correlations between the grain dimensions, volume and mass and its strength properties and energy needed to crush the grains, were found. Only low correlations occurred, including negative ones between the grain width a_2 and stresses σ_{min} and σ_{max} (Table S1). Contrarily to wheat, a dependence of grinding energy proportionality and its mass was not confirmed [35]. It was not possible to confirm distinct dependencies between the grain thickness (a_3) and the force value, either (for F_{min} positive correlation $R = 0.271$, statistically significant $p < 0.05$, for F_{max} statistically insignificant correlation (Table S1)) and work (W_{Fmin} and W_{Fmax} low correlations), which could be observed for wheat [37]. The volume is related to the grain dimensions, including compression cross-section A_0, whereas the cross section is related to compressive strength (according to dependence (6)). However, it was not possible to show significant dependencies between the grain cross section A_0 and values of compressive strength (low negative correlations between A_0 and σ_{min}, σ_{max} (Table S1)). The obtained results, including the results of dimension and volume scatter, confirm the significant variability and diversity of biological materials within one species. The diversity of the values confirms that each grain is characterized by a different internal structure. Such a diversification can indicate low quality of the grain and poor conditions of cultivation.

Table 2. Results of correlation analysis between the volume of grain and its strength properties.

		F_{min} [1]	F_{max} [2]	C_k [3]	σ_{min} [4]	σ_{max} [5]	W_{Fmin} [6]	W_{Fmax} [7]
V_ρ	rhoSpearman coefficient	0.172	0.176	0.242	−0.018	−0.052	0.071	−0.057
	Significance	0.088	0.08	0.015	0.859	0.604	0.482	0.573
	Number of samples n	100	100	100	100	100	100	100
V_s	rhoSpearman coefficient	0.220	0.062	0.005	−0.280	−0.104	0.185	−0.138
	Significance	0.028	0.543	0.958	0.005	0.305	0.066	0.174
	Number of samples n	100	100	100	100	100	100	100

[1] Minimal crushing force, [2] Maximal crushing force, [3] Minimal compressive stress, [4] Maximal compressive stress, [5] Stiffness, [6] Work for the first rupture, [7] Rupture work.

It is noteworthy that the grains were tested for only one orientation of grains in the strength testing machines. It is expected that the longitudinal and transverse orientation of grains would provide different values of destructive compression, forces and strength [28–30].

4. Conclusions

The main objectives of the study were achieved through the determination of energy (work) and compressive forces in the test of static compression and analysis of the dependence between the grain size and the grain strength parameters and grinding energy.

Based on the analyzes, it was found that the average grain length of rice was 6.38 mm, the average width was 1.91 mm, the average height was 1.51 mm. The average grain volume determined on the basis of the grain weight and density was equal to 14.82 mm^3, while the volume determined on the basis of grain size 18.44 mm^3. The correction factor for given volumes V_ρ and vs. assumed an average value of 0.82. The values of the grain dimension distribution can be a determinant in selection of structural features of the materials used, that is, performance parameters of roller mills: diameters of rollers and the size of the inter-roller gap; selection of the screen eye sieve

The average values of strength properties of rice grains were determined, such as $F_{min} = 117.9$ kg·m s^{-2}, $F_{max} = 174.99$ kg·m·s^{-2}, $\sigma_{min} = 9.80$ MPa, $\sigma_{max} = 14.71$ MPa, $C_k = 1150.26$ N·mm^{-1}, $W_{Fmin} = 7.26$ mJ, $W_{\text{Fmax}} = 28.03$ mJ, which coincided with the results of research carried out by other researchers. The determined ranges of forces, strength and compressive energy (work) are of applicable character and can be used in the design of machines dedicated to process rice. Knowing these values will allow, among others, the estimation of the power of devices, e.g., grinding machines, and roller mills, and in consequence, will minimize energy losses and energy demand for dedicated machines.

The analysis of the Spearman correlation showed that there are no statistically significant relationships between the volume V_ρ of the grain and the tested strength properties, so there are no interdependencies between the variables. In the case of the grain size volume vs. a low negative correlation between strength σ_{min} and low positive correlation between the force inducing the first crack (Table 2) were found. Dependence of grinding energy proportionality and the grain mass as well as clear, distinct dependencies between the grain thickness and the value of force and work (low negative correlation between thickness co a_3 and stresses σ_{min} and work W_{Fmax}, low positive correlation between thickness a_3 and force inducing the first crack F_{min}), that was proven for other biological grainy materials (wheat grains) could not be confirmed either. Based on these results, it was not possible to find significant dependencies between the grain cross-section A_0 and the values of compressive strength (only low negative correlations were found).

Based on the conducted tests, the crack was found to be different for each grain (Figure 9). The results, including the scatter of dimensions and volume, confirm the high variability and diversity of biological materials within one species. Diversification of the obtained values confirms that each grain is characterized by a different internal structure. Such a differentiation can indicate a poor quality of grain and weak cultivation conditions.

An analysis of the test results of rice grain strength properties provides the basis for determining the impact of biomass properties on the grinding process and, in subsequent stages, for the development of procedures for monitoring the grinding process using energy-environmental grinding efficiency models, the original CO_2 emission index for the intelligent monitoring system of usable characteristics of the grinding process.

The results can be used by other researchers to create models of materials (rice grains) for computer simulations of cracking, crushing, mixing using the discrete element method DEM.

Author Contributions: Conceptualization, W.K. and J.F.; methodology, W.K. and R.K.; software, W.K. and P.B.-W.; formal analysis, A.M., A.T., and J.F.; investigation, W.K., K.P. and P.B.-W.; resources, W.K.; data curation, A.M., K.P. and R.K.; writing—original draft preparation, W.K., P.B.-W., and A.T; writing—review and editing, W.K., R.K. and J.F.; visualization, W.K.; supervision, A.T., J.F., A.M. and K.P.; project administration, W.K.; funding acquisition, W.K. All authors have read and agreed to the published version of the manuscript.

List of Symbols

a_1	length of the grain: mm
a_2	width of the grain, mm
a_3	height of the grain, mm
V_s	grain volume calculated based on three dimensions a_1, a_2, a_3, mm^3
V_ρ	grain volume calculated based on the volumetric mass density, mm^3
ρ	volumetric mass density, $kg \cdot m^{-3}$
m	grain weight, g
k_v	correction coefficient of the grain volume
R_c	compressive strength, MPa
F_c	the largest value of the compressive load at which the sample is crushed, $kg \cdot m \cdot s^{-2}$
A_0	the initial cross-section of the sample, m^2
F	compressive force, $kg\ m \cdot s^{-2}$
Δl	sample shortening corresponding to force (F), m
l	the initial length of the sample, m
A_0	area of the initial sample cross-section, m^2
s	displacement, m
W	work, J
dW	elementary work, J
ds	elementary displacement, m
F_{min}	minimal crushing force, $kg \cdot m\ s^{-2}$
F_{max}	maximal crushing force, $kg \cdot m \cdot s^{-2}$
σ_{kmin}	minimal compressive stress, MPa
σ_{kmax}	maximal compressive stress, MPa
C_k	stiffness, $N \cdot mm^{-1}$
W_{Fmin}	work for the first rupture, mJ
W_{Fmax}	rupture work, mJ

References

1. Tomporowski, A.; Flizikowski, J.; Kruszelnicka, W. A new concept of roller-plate mills. *Przem. Chem.* **2017**, *96*, 1750–1755.
2. Tomporowski, A.; Flizikowski, J.; Wełnowski, J.; Najzarek, Z.; Topoliński, T.; Kruszelnicka, W.; Piasecka, I.; Śmigiel, S. Regeneration of rubber waste using an intelligent grinding system. *Przem. Chem.* **2018**, *97*, 1659–1665.
3. Flizikowski, J.B.; Mrozinski, A.; Tomporowski, A. Active monitoring as cognitive control of grinders design. In *AIP Conference Proceedings*; AIP Publishing: Melville, NY, USA, 2017; Volume 1822, p. 020006.
4. Marczuk, A.; Caban, J.; Savinykh, P.; Turubanov, N.; Zyryanov, D. Maintenance research of a horizontal ribbon mixer. *Eksploat. Niezawodn.* **2017**, *19*, 121–125. [CrossRef]
5. Tomporowski, A.; Flizikowski, J. Motion characteristics of a multi-disc grinder of biomass grain. *Przem. Chem.* **2013**, *92*, 498–503.
6. Bochat, A.; Zastempowski, M. Kinematics and dynamics of the movement of the selected constructions of the disc cutting assemblies. In Proceedings of the Engineering Mechanics 2017, Brno University of Technology, Faculty of Mechanical Engineering, Institute of Solid Mechanics, Mechatronics and Biomechanics, Brno-Svratka, Czech Republic, 15–18 May 2017; Volume 23, pp. 170–173.
7. Zastempowski, M.; Bochat, A. Modeling of cutting process by the shear-finger cutting block. *Appl. Eng. Agric.* **2014**, *30*, 347–353.
8. Flizikowski, J.; Macko, M. Method of estimation of efficiency of quasi-cutting of recycled opto-telecommunication pipes. *Polimery* **2001**, *46*, 53–59. [CrossRef]
9. Kaczmarczyk, J.; Grajcar, A. Numerical simulation and experimental investigation of cold-rolled steel cutting. *Materials* **2018**, *11*, 1263. [CrossRef] [PubMed]
10. Kruszelnicka, W.; Shchur, T. Study of rice and maize grains grinding energy. *TEKA Comm. Mot. Energetics Agric.* **2018**, *18*, 71–74.
11. Szyszlak-Barglowicz, J.; Zajac, G. Distribution of heavy metals in waste streams during combustion of *Sida*

hermaphrodita (L.) Rusby biomass. *Przem. Chem.* **2015**, *94*, 1723–1727.

12. Szyszlak-Bargłowicz, J.; Zając, G.; Słowik, T. Hydrocarbon emissions during biomass combustion. *Polish J. Environ. Stud.* **2015**, *24*, 1349–1354. [CrossRef]
13. Kowalczyk-Jusko, A.; Kowalczuk, J.; Szmigielski, M.; Marczuk, A.; Jozwiakowski, K.; Zarajczyk, K.; Maslowski, A.; Slaska-Grzywna, B.; Sagan, A.; Zarajczyk, J. Quality of biomass pellets used as fuel or raw material for syngas production. *Przem. Chem.* **2015**, *94*, 1835–1837.
14. Rudnicki, J.; Zadrag, R. Technical state assessment of charge exchange system of self-ignition engine, based on the exhaust gas composition testing. *Polish Marit. Res.* **2017**, *24*, 203–212. [CrossRef]
15. Nizamuddin, S.; Qureshi, S.S.; Baloch, H.A.; Siddiqui, M.T.H.; Takkalkar, P.; Mubarak, N.M.; Dumbre, D.K.; Griffin, G.J.; Madapusi, S.; Tanksale, A. Microwave hydrothermal carbonization of rice straw: Optimization of process parameters and upgrading of chemical, fuel, structural and thermal properties. *Materials* **2019**, *12*, 403. [CrossRef] [PubMed]
16. Chang, K.-L.; Wang, X.-Q.; Han, Y.-J.; Deng, H.; Liu, J.; Lin, Y.-C. Enhanced enzymatic hydrolysis of rice straw pretreated by oxidants assisted with photocatalysis technology. *Materials* **2018**, *11*, 802. [CrossRef] [PubMed]
17. Mannheim, V. Examination of thermic treatment and biogas processes by Lca. *Ann. Fac. Eng. Hunedoara Int. J. Eng.* **2014**, *12*, 225–234.
18. Zeng, Y.; Jia, F.; Xiao, Y.; Han, Y.; Meng, X. Discrete element method modelling of impact breakage of ellipsoidal agglomerate. *Powder Technol.* **2019**, *346*, 57–69. [CrossRef]
19. Sadeghi, M.; Araghi, H.A.; Hemmat, A. Physico-mechanical properties of rough rice (*Oryza sativa* L.) Grain as affected by variety and moisture content. *Agric. Eng. Int. CIGR J.* **2010**, *12*, 129–136.
20. Cao, W.; Nishiyama, Y.; Koide, S. Physicochemical, mechanical and thermal properties of brown rice grain with various moisture contents. *Int. J. Food Sci. Technol.* **2004**, *39*, 899–906. [CrossRef]
21. Chattopadhyay, P.K.; Hamann, D.D.; Hammerle, J.R. Effect of deformation rate and moisture content on rice grain stiffness1. *J. Food Process Eng.* **1980**, *4*, 117–121. [CrossRef]
22. Buggenhout, J.; Brijs, K.; Celus, I.; Delcour, J.A. The breakage susceptibility of raw and parboiled rice: A review. *J. Food Eng.* **2013**, *117*, 304–315. [CrossRef]
23. Esehaghbeygi, A.; Daeijavad, M.; Afkarisayyah, A.H. Breakage susceptibility of rice grains by impact loading. *Appl. Eng. Agric.* **2009**, *25*, 943–946. [CrossRef]
24. Sarker, M.S.H.; Hasan, S.M.K.; Ibrahim, M.N.; Aziz, N.A.; Punan, M.S. Mechanical property and quality aspects of rice dried in industrial dryers. *J. Food Sci. Technol.* **2017**, *54*, 4129–4134. [CrossRef] [PubMed]
25. Talab, K.T.; Ibrahim, M.N.; Spotar, S.; Talib, R.A.; Muhammad, K. Glass transition temperature, mechanical properties of rice and their relationships with milling quality. *Int. J. Food Eng.* **2012**, *8*.
26. Nasirahmadi, A.; Abbaspour-Fard, M.H.; Emadi, B.; Khazaei, N.B. Modelling and analysis of compressive strength properties of parboiled paddy and milled rice. *Int. Agrophysics* **2014**, *28*, 73–83. [CrossRef]
27. Bonazzi, C.; Courtois, F. Impact of drying on the mechanical properties and crack formation in rice. In *Modern Drying Technology*; John Wiley & Sons, Ltd.: Hoboken, NJ, USA, 2011; pp. 21–49. ISBN 978-3-527-63166-7.
28. Li, Y.-N.; Li, K.; Ding, W.-M.; Chen, K.-J.; Ding, Q. Correlation between head rice yield and specific mechanical property differences between dorsal side and ventral side of rice kernels. *J. Food Eng.* **2014**, *123*, 60–66. [CrossRef]
29. Shu, Y.-J.; Liou, N.-S.; Moonpa, N.; Topaiboul, S. Investigating damage properties of rice grain under compression load. In Proceedings of the International Conference on Experimental Mechanics 2013 and Twelfth Asian Conference on Experimental Mechanics, Bangkok, Thailand, 25–27 November 2013; International Society for Optics and Photonics: Bangkok, Thailand, 2014; Volume 9234, p. 923402.
30. Zareiforoush, H.; Komarizadeh, M.H.; Alizadeh, M.R.; Tavakoli, H.; Masoumi, M. Effects of moisture content, loading rate, and grain orientation on fracture resistance of paddy (*Oryza sativa* L.) grain. *Int. J. Food Prop.* **2012**, *15*, 89–98. [CrossRef]
31. Zareiforoush, H.; Komarizadeh, M.H.; Alizadeh, M.R. Mechanical properties of paddy grains under quasi-static compressive loading. *N. Y. Sci. J.* **2010**, *3*, 40–46.
32. Pandiselvam, R.; Thirupathi, V.; Mohan, S. Engineering properties of rice. *Agric. Eng.* **2015**, *XL*, 69–78.
33. Ligaj, B.; Szala, G. Obliczanie zapotrzebowania energii w procesach rozdrabniania materiałów ziarnistych na przykładzie ziaren zbóż. *Acta Mech. Autom.* **2009**, *3*, 97–99.
34. Tumuluru, J.S.; Tabil, L.G.; Song, Y.; Iroba, K.L.; Meda, V. Grinding energy and physical properties of chopped and hammer-milled barley, wheat, oat, and canola straws. *Biomass Bioenergy* **2014**, *60*, 58–67. [CrossRef]

35. Wiercioch, M.; Niemiec, A.; Roma, L. The impact of wheat seeds size on energy consumption of their grinding process. *Inzynieria Rol.* **2008**, *103*, 367–372.
36. Warechowska, M. Some physical properties of cereal grain and energy consumption of grinding. *Agric. Eng.* **2014**, *1*, 239–249.
37. Dziki, D.; Laskowski, J. Influence of wheat kernel geometrical properties on the mechanical properties and grinding ability. *Acta Agrophysica* **2003**, *2*, 735–742.
38. Warechowska, M.; Warechowski, J.; Skibniewska, K.A.; Siemianowska, E.; Tyburski, J.; Aljewicz, M.A. Environmental factors influence milling and physical properties and flour size distribution of organic spelt wheat. *Tech. Sci.* **2016**, *19*, 387–399.
39. Dziki, D. Ocena energochłonności rozdrabniania ziarna pszenicy poddanego uprzednio zgniataniu. *Inzyneria Rol.* **2007**, *11*, 51–58.
40. Dziki, D.; Cacak-Pietrzak, G.; Miś, A.; Jończyk, K.; Gawlik-Dziki, U. Influence of wheat kernel physical properties on the pulverizing process. *J. Food Sci. Technol.* **2014**, *51*, 2648–2655. [CrossRef] [PubMed]
41. Greffeuille, V.; Mabille, F.; Rousset, M.; Oury, F.-X.; Abecassis, J.; Lullien-Pellerin, V. Mechanical properties of outer layers from near-isogenic lines of common wheat differing in hardness. *J. Cereal Sci.* **2007**, *45*, 227–235. [CrossRef]
42. Greffeuille, V.; Abecassis, J.; Barouh, N.; Villeneuve, P.; Mabille, F.; Bar L'Helgouac'h, C.; Lullien-Pellerin, V. Analysis of the milling reduction of bread wheat farina: Physical and biochemical characterisation. *J. Cereal Sci.* **2007**, *45*, 97–105. [CrossRef]
43. Yenge, G.B.; Kad, V.P.; Nalawade, S.M. Physical properties of maize (*Zea mays* L.) grain. *J. Krishi Vigyan* **2018**, *7*, 125–128. [CrossRef]
44. Korczewski, Z.; Rudnicki, J. An energy approach to the fatigue life of ship propulsion systems marine 2015. In Proceedings of the VI International Conference on Computational Methods in Marine Engineering—The Conference Proceedings, Rome, Italy, 14–17 July 2015; Salvatore, F., Broglia, R., Muscari, R., Eds.; International Center Numerical Methods Engineering: Barcelona, Spain, 2015; pp. 490–501, ISBN 978-84-943928-6-3.
45. Lu, R.; Siebenmorgen, T.J. Correlation of head rice yield to selected physical and mechanical properties of rice kernels. *Trans. ASAE* **1995**, *38*, 889–894. [CrossRef]

14

Risk Assessment for Social Practices in Small Vegetable Farms in Poland as a Tool for the Optimization of Quality Management Systems

Marcin Niemiec [1,*], Monika Komorowska [2], Anna Szeląg-Sikora [3], Jakub Sikora [3], Maciej Kuboń [3], Zofia Gródek-Szostak [4] and Joanna Kapusta-Duch [5]

1 Department of Agricultural and Environmental Chemistry, Faculty of Agriculture and Economics, University of Agriculture in Krakow, Mickiewicz Ave. 21, 31-120 Krakow, Poland
2 Department of Vegetable and Medicinal Plants, Faculty of Biotechnology and Horticulture, University of Agriculture in Krakow, 54 29 Listopada Ave., 31-426 Krakow, Poland
3 Institute of Agricultural Engineering and Informatics, University of Agriculture in Krakow, 116b Balicka St., 30-149 Krakow, Poland
4 Department of Economics an Organization of Enterprises, Cracow University of Economics, 27 Rakowicka St., 31-510 Krakow, Poland
5 Faculty of Food Technology, Department of Human Nutrition, University of Agriculture in Krakow, 122 Balicka St., 30-149 Krakow, Poland
* Correspondence: marcin1niemiec@gmail.com

Abstract: Globalization of the food market is associated with the possibility of selling products into newer markets. However, it is also associated with the necessity to ensure proper quality products. Quality defined by the ISO 9001:2015 standard consists of factors that are part of customers' expectations concerning the safety of products and the technology of their manufacture. Currently, consumers are looking for products with defined and reproducible sensory properties, in which the content of harmful substances is below the critical values specified by legislation. This is observable particularly in developed countries. The second quality factor is the use of a production technology where negative environmental impacts are reduced. Recently, issues associated with protecting workers' rights and social needs have also become very important. In successive versions of quality management systems, such as GLOBAL G.A.P. or SAI Platform, social issues are becoming more and more important. The aim of this study was to assess the role of risk analysis for social practices in small farms in building a quality management system. Surveys were conducted in 2018. The surveys covered 62 vegetables or fruit farms with a cultivated area of up to 20 ha. Their lack of staff was due to the character of production. Where mechanic production is possible in small farms, family members can secure workforce demand. To achieve the research objective, a risk analysis was carried out for the implementation of social practices according to the guidelines of the ISO 31000:2018 standard. The criteria and inventory of identified risks were carried out, based on the guidelines of GLOBAL G.A.P. Risk Assessments on Social Practice (GRASP). Based on the identified risks, the areas relating to social practices, which require improvement in order to satisfy compliance with the GLOBAL G.A.P. standard, were indicated. The results of the conducted research pointed to a high risk of good social practices not being carried out and not meeting compliance with the requirements of the GLOBAL G.A.P. standard. The most important identified problems are associated with the deficiency of competent workers as well as the lack of facilities where workers can rest, eat and drink. A considerable problem is the conformity of employment contracts with local legislation and ensuring that work time and rest time are consistent with the law. In conditions of small farms in Poland, the problem with ensuring compliance with the standard in question is often the small number of workers. Creating an organized quality management system in the area of social practices is difficult in these cases, and sometimes even impossible.

Keywords: GLOBAL G.A.P.; GRASP; quality management systems; certification; primary production; social practice

1. Introduction

The conditions of the modern food market of products are associated with the necessity to increase the effectiveness of the use of the means of production and to reduce the negative impact on the natural environment. The need for rationalization of land use, work, and depletable environmental resources result, on one hand, from the need to lower the costs of food production, and on the other hand, from the needs of the consumer who seeks food with specific quality. Quality is one of the most important factors that influence consumers' choices. It is very often more important than product price, particularly in developed countries. It is one of the most important factors of achieving a competitive advantage. According to the used definition in management systems, quality relates to the degree of customer satisfaction. In the case of agricultural products, the idea of quality has been changing. This has influenced societies and the economic potential of consumers [1].

At the beginning of the first half of the 20th century, quality of agri-food products was identical with their price. In that period, the most important problem in the world was to provide developing societies with an adequate amount of food. Increasing the production efficiency was based mainly on the intensification of fertilization and on increasing the amount of production-boosting chemicals. The consequence of such a strategy of production (both plant and animal) was the emission of a considerable amount of pollutants into the environment, which led to soil and water degradation, as well as air pollution. The second, and very important, effect of agricultural intensification was the deterioration in the quality of products in terms of their chemical composition. The biggest problems included high content of pesticides, nitrates, and heavy metals in plant products, and in the case of animal products, a high content of hormones and pharmaceuticals. Due to the development of agricultural sciences and changes in consumer awareness, the quality of products is more and more associated with the technology of their production. This quality factor has been expressed in the idea of sustainable development of agriculture. Quality is also associated with the way a product is packed and presented [2].

Traditional agriculture has a negative impact on the natural environment—both water, soil, air, and on consumer health [3]. In the 1990s, formalized quality systems in primary production began to develop in developed countries. Those systems took into account the production principles leading to an improvement in food safety at the stage of producing raw materials intended for food or fodder purposes. In that period, the concepts of sustainable production systems for biomass crops began to appear [4]. The fundamental assessment criterion of a given technology in this case is the energy fixation potential of plants, taking into account the expenditure incurred on production. Another important aspect is the issue of the possibility of producing food in areas intended for cultivation of energy crops. According to the adopted ideology, the introduction of specific principles of a system should lead to an improvement in the quality of products, in terms of their safety and chemical composition [5]. Moreover, the principles of modern systems of quality management in agriculture have involved issues associated with rational utilization of soil resources, water resources, and issues connected with decreasing energy consumption in the entire production process [6,7]. The efficient implementation of quality management systems in food production is also associated with creating no-waste or low-waste technologies, as pointed out by Sikora et al. [8]. More and more attention has been paid to the optimization of logistic processes, not only at the production stage, but in the entire supply chain. Optimization of logistic processes is an important part of improving economic efficiency, which was highlighted by Xiao et al. [8], as well as Cupiał et al. [9]. As regards to food safety, many countries have safety assurance systems, created and administered at a national level. They can vary both in scope, as well as in the level of requirements, which result from economic, cultural, climatic, or political conditions [10]. The differences concern mainly environmental and societal issues in production. Therefore, there is a threat that products generated in compliance with local law will not fulfill qualitative criteria requested by consumers in target countries [11]. This applies particularly to the issues connected with environmental and social aspects in developing countries. Food production, compliant with local law of these countries, is very often insufficient to satisfy a conscious consumer. Therefore, one of the main factors of developing agricultural production and the possibility of exporting products, is the implementation of quality management systems in primary production.

One of the most popular quality management standards in primary production in agriculture is the GLOBAL G.A.P. (Good Agriculture Practice) system. It is an independent, optionally-implemented system of assuring product quality and safety in primary agricultural production. The standard was introduced into the market in 1997 under the name EUREPGAP. It was elaborated by members of EUREP (Euro-Retailer Produce Working Group) organization. The purpose of the standard was to develop principles that would be common for the entire primary production, aimed at ensuring compliance with Good Agricultural Practice (GAP) and ensuring food safety. The GLOBAL G.A.P. standard was introduced in exchanged for EUREPGAP on 7 September 2007.

One of the primary objectives of the standard was to minimize the use of fertilizers and plant protection products, so as to limit the negative impact of agriculture on the environment, and also to make sure that good soil culture on areas intended for agricultural production is maintained. One of the most frequently described problems associated with implementation of quality management systems in primary production is a failure to adapt them to small farms [12,13]. Inadequate support from state institutions and non-governmental organizations is a factor that limits the development of quality management systems [14]. To accommodate these problems, the certification system and the manner of implementing the principles of the GLOBAL G.A.P. standard at a farm level were adapted to small farms. Certification is possible within option 1 and 2. In the case of option 1, an individual producer intends to certify their own products in compliance with the requirements of GLOBAL G.A.P., in the scope associated with specific activities. Option 2 relates to the certification of a group of producers (e.g., cooperatives or organizations of producers). In this case, certification concerns agricultural products, in accordance with the requirements of the used range of GLOBAL G.A.P., complemented with requirements pertaining to managing through the implementation of the Quality Management System (QMS). In this option, a certificate is issued for the leading organization, which guarantees that the requirements of the standard are observed by producers that are members of the group. Due to the higher effectiveness of means of production, producer groups have a bigger and bigger share in primary production in Poland. A properly organized quality management system at the level of the producer group allows for the efficient management of quality for all members [5,15]. Within the certification of a producer group, an audit is conducted for the management system in the context of system tools, and the efficiency of communication with particular group members, as well as the effectiveness of identifying non-conformities and monitoring corrective actions. Such a solution makes it possible to minimize the risks associated with the fragmentation of farms and the scattering of production sites [16]. Due to these factors, small producers can be included into the global supply chain, which facilitates the development of small family farms. Including small farms as a part of supply chain, not only on local markets, but also on international scale, is an important part of a sustainable development of rural areas. This is of great importance in developing countries, where plants are cultivated on small areas [17]. However, such an approach associated with certification poses a threat to product quality in the case of an ineffective system of controlling the members of the producer group by head office, which implements the quality management system. GLOBAL G.A.P. is a standard that is based on ethical principles of the producers participating in the system [18]. The aim of this study was to assess the role of risk analysis for social practices in small farms, in building a quality management system.

2. Research Methods

To meet the set objective, surveys were conducted in 2018 in small family farms in Poland, in the following provinces: Świętokrzyskie (11 farms), Mazovia (18 farms), Łódź (19 farms), Lublin (6 farms), and Wielkopolska (8 farms). The research was conducted using the direct interview method. The study involved farms, in which the buyers of products reported the need to implement the GRASP standard. The surveys covered 62 vegetables or farms that grow berry plants, with a cultivated area of up to 20 ha. Among the examined farms, 37 of them grew vegetables, while 25 were growing berry plants. Nearly 30% of the surveyed farms declared that they did not employ workers. These farms were excluded from further analysis, due to the lack of possibility of attaining the GLOBAL G.A.P. GRASP standard. Almost half of the farms employed local workers and family members. Approximately 10% of the farms hired only foreign workers, and 15% hired foreign and local workers Table 1. Other farms employ seasonal workers, mainly citizens of the Ukrainian Republic.

Table 1. Employment structure in the surveyed farms.

Research Group No.	Specification	Number of Farms	Average Farm Area in the Group
1	Lack of workers	17	5.19
2	Local workers, close or extended family	28	9.32
3	Foreign workers	7	14.89
4	Foreign and local workers	10	15.26

Workers were employed on employment contracts only in 10 farms. For separated research groups, a risk analysis was carried out for the implementation of social practices according to the guidelines of the ISO 31000:2018 standard. Inventory of identified risks was carried out based on the guidelines of GLOBAL G.A.P.

In the conducted research, the strategic goal was to adjust the management policy of social practices on the farm to GLOBAL G.A.P. GRASP standards. Based on the conditions in the surveyed farms, (the size and assortment of production, infrastructure equipment, cultural factors, and the mentality of farmers). The matrix used for risk analysis is included in Table 2. According to this table, the level of risk in a three-level scale is defined as small, medium, and large. The identified risks are presented in Table 3. Two values have been identified for each identified risk factor; the probability of risk occurrence and the threat to the strategic goal (compliance with the GLOBAL G.A.P. GRASP standard) in the case of risk.

Table 2. Matrix used for risk assessment.

The Probability of Occurrence of a Risk Factor	The Consequences of Risk		
	Low	Medium	High
Low	L	L	M
Medium	L	M	H
High	M	H	H

H—high risk, M—medium risk, L—low risk.

3. Results and Discussion

The implementation of quality management systems, both at state and private level, is widely considered to be a factor that increases the profitability of farms, as well as a method for reducing the environmental effects associated with agricultural production. Bibliographic data presented by other researchers point to a relationship between the implementation of a quality management system and the amount of income [18,19]. Positive effects are perceived also in relation to the effectiveness of using natural resources in agricultural production. One of the weak points of quality management systems in primary production is associating them with the amount of earnings and workers' lifestyle. Oya et al. [18] draw attention to this problem and point to the need of putting more emphasis on social issues in food production at farm level. Despite the fact that quality systems in primary production, such as Organic Farming, GLOBAL G.A.P., Integrated Production, or sales network quality systems take social aspects into account in their principles. Nevertheless, they can rarely guarantee that farm owners will satisfy workers' social needs. That is the reason why, in many cases, it is essential to certify producers for compliance with the principles of social systems [20]. In recent years, social aspects have permanently entered the range of parameters that are part of the notion of food product quality. The basic components of the quality of agri-food products include:

1. Product safety associated with microbiological, chemical, and physical threats
2. Confirmation that the production principles applied in the used technology are compliant with the principles of sustainable development.
3. Confirmation that, during production, basic principles associated with hygiene and safety of workers have been met.
4. Confirmation that, during production, social practices consistent with principles of good social practices and with principles of international labor conventions have been applied.

All these factors shape the quality of agricultural products and their implementation is connected with incurring costs for technical infrastructure, consultancy, and for administering the system. The surveyed farms are associated in producer groups, and the implementation of the GRASP standard was necessary in order to win a new market in Great Britain. Prior to making a decision about implementing the GRASP standard, a risk analysis was carried out. GRASP is a voluntary, additional module, which not part of the accredited GLOBALG.A.P. certification. It completes the requirements of the standard with respect to good social practices. GRASP certification is only possible when producers are certified for compliance with the GLOBALG.A.P. system or certified in compliance with an equivalent system subjected to benchmarking. The interpretation of GRASP control points depends on the country in which activities are conducted. According to the principles of the standard, the requirements of national legislation replace the GRASP requirements when appropriate legal regulations are more demanding of the GRASP requirements. If there are no legal regulations or if the law is not as demanding, GRASP principles provide minimum compliance criteria.

The foundation of the GRASP standard is a structural organization of a farm that will allow it to eliminate the influence of the management on the workers' representatives. According to the standard, an enterprise must have people acting as the workers' representatives. The workers' representatives are obliged to represent their interests. These representatives should be independent and not work in a position associated with managing the company. The introduced system should guarantee regular meetings between the workers' representatives and the Board, where workers' issues are addressed. Transcripts of these meetings should be made available in company documentation. Workers' representatives should be chosen through voting. They should be chosen by workers and recognized by the Board. A representative can be nominated only in exceptional circumstances. The voting or nomination must take place in the current year or season. Workers' representatives should be aware of their role and rights, and they should be able to have discussions with the Board regarding complaints and suggestions. They should also have knowledge about current legislation and principles of international labor conventions. A quality management system in a farm should be armed with an effective procedure of filing complaints and motions. The manner of filing complaints should be clear and should not generate sanctions for the person who files a complaint or for the workers' representatives who speak for the said person.

Table 3. Risk assessment for social practice in the researched farms.

Specification	Group of Farms			
	1	2	3	4
Selection of the workers' representative	n.a.	H	H	H
Establishing the date of selecting workers	n.a.	H	H	H
Adapting the issue of employee discrimination	n.a.	L	L	M
Adapting the issues relating to contracts with workers	n.a.	H	L	M
Adapting the issue of minimum wages and equal salary	n.a.	H	L	M
Adapting the issues connected with documenting the work time, as well as the level of remuneration connected therewith, overtime hours	n.a.	H	M	M
Adapting the issues relating to employing minors	n.a.	H	L	M
Adapting the issues connected with ensuring proper social conditions for workers living on the farm	n.a.	L	M	H
Total risk	n.a.	H	M	M

H—high risk, M—medium risk, L—low risk, n.a.—not applicable.

Assuring the above identified principles of the standard in the surveyed farms is problematic for objective reasons. In the farms which employ workers, nominating a potential worker's representative eventuated as the biggest problem. The organizational structure of most farms are based on the management and production workers. Among the seasonal workers, it is very difficult to find a person who can assume the responsibility of representing workers' interest and who is knowledgeable of the principles of labor law that are in force in Poland and who knows about international labor conventions. In the majority of the surveyed farms, the only workers who possess the necessary knowledge are those connected with the owner. Establishing the date of selecting the workers' representatives might also be problematic. Therefore, the risk in this respect was assessed as high Table 3. The harvest season is short in the majority of the surveyed farms, and before the season, a limited number of workers are hired. If the election is carried out at the beginning of the vegetation season, then the workers hired in the harvest season will not have the ability to choose their representative. If the election is carried out after hiring pickers, in the period from the beginning of the vegetation season to harvest, the workers' representatives will not operate. In both cases there is a risk that the principles of the standard will not be observed and the certificate will not be granted.

Another principle of the GRASP standard is signing and implementing a declaration to ensure that good social practices and the observance of human rights for all employees. Such a declaration should contain the obligation to comply with the ILO (International Labor Organization) convention with respect to discrimination, legal working age, and the eliminating forced labor. The declaration should have an obligation regarding the acceptance of the rights to freedom of association and to equal remuneration.

In all the studied farms, the risk associated with signing and implementing a declaration of good social practices was assessed as low. Medium risk was determined in farms that employ both foreigners and local population Table 3. An identified risk associated with this field of activities were issues of equal remuneration and minimum wages. To this extent, the functioning systems generally needed improvement, although, to a smaller extent in farms employing foreign workers Table 3.

One of the most significant sources of risk to implementing the GRASP standard was the issue of contracts with workers. According to the principles of the standard, each worker should be employed on a contract consistent with the legislation of the country where activities are carried out. The contract must be signed, should include a full name, nationality, description of work position, date of birth, date of commencing work, working time, remuneration and employment period. For foreign workers—their legal status and work permit. In the case of farms that employ only foreign workers, the risk for the area in question was estimated as low, due to Polish law regarding hiring foreigners. In the case of seasonal local workers, issues associated with employment contracts were assessed at a high level Table 3. During harvest and other activities associated with production, neighbors or both nuclear or extended family members are often hired. Due to the seasonal nature of production and problems with acquiring workers, the turnover of people employed in the farm is very high. In such a situation, the reorganization of the human resources management system would be very difficult and entail incurring additional costs. In addition, adapting employment to legal regulations in many cases may be connected with difficulty in finding employees.

Employing minors on farms is part of a tradition of small farms in many countries [21]. Among the surveyed farms, in group no. 2 that uses local labor force, minors are very frequently hired as help. This particularly applies to producers growing strawberries and raspberries. Because of the harvest period which takes place during school summer holidays, pupils are often hired for the harvest. In these farms, resigning from this source of labor is associated with an increase in costs incurred on production and with the risk of a shortage of workers.

Assurances of adequate social conditions for workers living on the premises of the farm is key to good social practices in agriculture. The parameters of assessment of housing conditions for seasonal workers should be selected by taking into account the workers' nationality and cultural identity that lead to their specific needs [22]. Regardless of the workers background, providing basic social needs is essential to ensuring proper quality of life and conditions for rest. Ensuring adequate housing conditions in farming, particularly in the case of seasonal workers, is a crucial factor associated with social practices on farms, and regardless of the place of agricultural production, which was highlighted by Vallejos et al. [23], Rima et al. [24]. On the other hand, the possibility of living on a farm is a factor which makes it easier to obtain seasonal workers. As a result of the conducted risk analysis, in compliance with the guidelines of the ISO 31000:2018 standard, the risks associated with the scope of these activities was assessed as low in farms using local labor force and medium in the group of farms that offer apartments to their employees Table 3. All the surveyed farms, that offer apartments to their employees, carry out actions for the improvement of housing conditions as this is what employees expect. In the case of farms with low income, adjusting housing conditions to the requirements of the standard can be a considerable strain.

The implementation of quality systems in primary production is associated with the necessity of incurring high costs, consisting of charges incurred on the certification, consultancy, infrastructural changes, as well as changes in production technology, which are frequently related with increasing cost intensity. These factors restrict, and sometimes prevent, the implementation of quality management systems, particularly in countries where small farms with a small production scale are dominant. A consequence of the development of quality management systems in primary production, and of increasing market demand for certified products can be the restructuring of agriculture towards creating large commercial farms, where the implementation of a quality management system is easier. Such changes might be disadvantageous in social and societal terms, particularly in countries where the efficient functioning of small farms is part of local tradition [25,26]. Despite the image benefits and facilitation of selling products from certified farms, economic factors play a key role when deciding to implement a quality management system [27,28]. One of the main problems of the certification of quality management systems is their mal-adjustment to the producer market based on small farms [12,29]. The results of the conducted surveys point to a high risk associated with implementation of the GRASP standard in the group of the smallest farms which use local labor force as well as family. In these cases, the benefits of owning a certificate may not cover the costs arising from implementation of the standard. In bigger farms, which employ foreign workers, the risk associated with the implementation was assessed as medium Table 3. When analyzing the current situation in the market, at a high level of risk, implementation of the GLOBAL G.A.P. GRASP standard is unreasonable. In the case of the medium risk, the decision about implementing the standard should be made basing on a risk analysis carried out for an individual farm. Due to the changing requirements of retail chains, one should expect an increase in farmers interest in the GLOBAL G.A.P. GRASP. Based on the obtained results, it was found that they need to continue in the conditions of small commercial farms.

4. Summary

The results of the conducted surveys point to a substantial level of risk associated with the implementation of the GRASP standard in small vegetable and horticultural farms in Poland. The effective implementation of the standard can be problematic, due to there being no one who fulfils the criteria for workers' representatives. Another problem is establishing a proper date for choosing the workers' representative because of considerable staff rotations during the vegetation season. The obligation to conclude employment contracts or civil law agreements, and to create a system for recording workers' work time is a problem for a lot of farms. Ensuring proper housing conditions is a factor preventing the implementation of the GRASP standard for some farms. In approximately 35% of the surveyed farms, the workers are people from local communities who do not want to work based on civil law agreements. It is the traditional model of purchasing labor in small farms in Poland. Adapting the system to the requirements of the standard would involve the necessity to build, from scratch, a system for managing human resources in farms. This may be problematic, not only in terms of costs, but also because it would require changing the mentality. The results of the conducted surveys indicate a high risk associated with reorganizing the surveyed farms. Based on the conducted risk analysis, it was established that the implementation of the standard in the surveyed farms, which use the local labor force, is currently unreasonable. In the case of bigger farms, which use foreign workers, making a decision about certification should be preceded by an individual risk analysis carried out for a specific farm.

Author Contributions: Conceptualization, M.N., Z.G.-S. and D.K.; methodology, M.K., A.S.-S. and J.K.-D.; resources, M.N., M.K., A.S.-S. and J.S.; formal analysis, M.N., D.K. and M.K.; investigation, D.K., Z.G.-S., M.N. and J.S.; resources, M.N., M.K., M.K. and J.K.-D.; data curation, A.S.-S. and D.K.; writing—M.N., M.K., Anna S.-S. and M.K.; visualization, D.K., J.S. and M.N.; funding acquisition, M.N.

References

1. Wencong, L.; Godwin Seyram Agbemavor Kwasi, H. Transition of small farms in Ghana: Perspectives of farm heritage, employment and networks. *Land Use Policy* **2019**, *81*, 434–452. [CrossRef]
2. Kocira, S.; Kuboń, M.; Sporysz, M. Impact of information on organic product packagings on the consumers decision concerning their purchase. In Proceedings of the 17th International Multidisciplinary Scientific GeoConference SGEM 2017, Sofia, Bulgaria, 27 June–6 July 2017; Volume 17, pp. 499–506. [CrossRef]
3. Chowaniak, M.; Klima, K.; Niemiec, M. Impact of slope gradient, tillage system, and plant cover on soil losses of calcium and magnesium. *J. Elementol.* **2016**, *21*, 361–372. [CrossRef]
4. Ivanyshyn, V.; Nedilska, U.; Khomina, V.; Klymyshena, R.; Hryhoriev, V.; Ovcharuk, O.; Hutsol, T.; Mudryk, K.; Jewiarz, M.; Wróbel, M.; et al. Prospects of growing miscanthus as alternative source of biofuel. *Renew. Energy Sources Eng. Technol. Innov.* **2018**, 801–812. [CrossRef]
5. Szeląg-Sikora, A.; Niemiec, M.; Sikora, J.; Chowaniak, M. Possibilities of Designating swards of grasses and small-seed legumes from selected organic farms in Poland for feed. In Proceedings of the IX International Scientific Symposium "Farm Machinery and Processes Management in Sustainable Agriculture", Lublin, Poland, 22–24 November 2017; pp. 365–370.
6. Niemiec, M.; Komorowska, M.; Szeląg-Sikora, A.; Kuzminova, N. Content of Ba, B, Sr and as in water and fish larvae of the genus Atherinidae L. sampled in three bays in the Sevastopol coastal area. *J. Elem.* **2018**, *23*, 1009–1020. [CrossRef]
7. Niemiec, M.; Sikora, J.; Szeląg-Sikora, A.; Kuboń, M.; Olech, E.; Marczuk, A. Applicability of food industry organic waste for methane fermentation. *Przem. Chem.* **2017**, *69*, 685–688. [CrossRef]
8. Sikora, J.; Niemiec, M.; Szelag-Sikora, A.; Kuboń, M.; Olech, E.; Marczuk, A. Biogasification of wastes from industrial processing of carps. *Przem. Chem.* **2017**, *96*, 2275–2278.

9. Cupiał, M.; Szeląg-Sikora, A.; Niemiec, M. Optimization of the machinery park with the use of OTR-7 software in context of sustainable agriculture. *Agric. Agric. Sci. Proc.* **2015**, *7*, 64–69. [CrossRef]
10. Shukl, S.; Prakash Singh, S.; Shankar, R. Evaluating elements of national food control system: Indian context. *Food Control* **2018**, *90*, 121–130. [CrossRef]
11. Mzoughi, N. Farmers adoption of integrated crop protection and organic farming: Do moral and social concerns matter? *Agric. Econ.* **2011**, *70*, 1536–1545. [CrossRef]
12. Azhar, B.; Prideaux, M.; Razi, N. Sustainability certification of food. *Ref. Modul. Food Sci. Encycl. Food Secur. Sustain.* **2019**, *2*, 538–544. [CrossRef]
13. Tran, D.; Daisaku, D. Impacts of sustainability certification on farm income: Evidence from small-scale specialty green tea farmers in Vietnam. *Food Policy* **2019**, *83*, 70–82. [CrossRef]
14. Gródek-Szostak, Z.; Szeląg-Sikora, A.; Sikora, J.; Korenko, M. Prerequisites for the cooperation between enterprises and business supportinstitutions for technological development. *Bus. Non-Profit Organ. Facing Increased Compet. Grow. Cust. Demand* **2017**, *16*, 427–439.
15. Szeląg-Sikora, A.; Niemiec, M.; Sikora, J. Assessment of the content of magnesium, potassium, phosphorus and calcium in water and algae from the black sea in selected bays near Sevastopol. *J. Elem.* **2016**, *21*, 915–926. [CrossRef]
16. Tey, Y.S.; Rajendran, N.; Brindal, M.; Ahmad Sidique, S.F.; Nasir, M.; Shamsudin, N.M.; Alias, S.; Radam, A.; Hadi, A.H.I. A review of an international sustainability standard (GlobalGAP) and its local replica (MyGAP). *Outlook Agric.* **2016**, *45*, 67–72. [CrossRef]
17. Holzapfel, S.; Wollni, M. Global GAP Certification of small-scale farmers sustainable? Evidence from Thailand. *J. Dev. Stud.* **2014**, *50*, 731–746. [CrossRef]
18. Oya, C.; Schaefera, F.; Skalidou, D. The effectiveness of agricultural certification in developing countries: A systematic review. *World Dev.* **2018**, *112*, 282–312. [CrossRef]
19. Byerlee, D.; Rueda, X. From public to private standards for tropical commodities: A century of global discourse on land governance on the forest frontier. *Forests* **2015**, *6*, 1301–1324. [CrossRef]
20. Raynolds, L.T. Fair-trade labour certification: The contested incorporation of plantations and workers. *Third World Q.* **2017**, *38*, 1473–1492. [CrossRef]
21. Hamenoo, E.S.; Dwomoh, E.A.; Dako-Gyeke, M. Child labor in Ghana: Implications for children'seducation and health. *Child Youth Serv. Rev.* **2018**, *93*, 248–254. [CrossRef]
22. Xiao, Y.; Benoît, C.; Norris, B.N.; Lenzen, M.; Norris, G.; Murray, J. How social footprints of nations can assist in achieving the sustainable development goals. *Ecol. Econ.* **2017**, *135*, 55–65. [CrossRef]
23. Vallejos, Q.M.; Quandt, S.A.; Grzywacz, J.G.; Isom, S.; Chen, H.; Galván, L.; Whalley, L.; Chatterjee, A.B.; Arcury, T.A. Migrant farmworkers' housing conditions across an agricultural season in North Carolina. *Am. J. Ind. Med.* **2011**, *54*, 533–544. [CrossRef] [PubMed]
24. Habib, R.R.; Mikati, D.; Hojeij, S.; El Asmar, K.; Chaaya, M.; Zurayk, R. Associations between poor living conditions and multi-morbidity among Syrian migrant agricultural workers in Lebanon. *Eur. J. Public Health* **2016**, *26*, 1039–1044. [CrossRef] [PubMed]
25. Kibet, N.; Obare, G.A.; Lagat, K.J. Risk attitude effects on Global-GAP certification decisions by smallholder French bean farmers in Kenya. *J. Behav. Exp. Financ.* **2018**, *18*, 18–29. [CrossRef]
26. Partzsch, L.; Kemper, L. Cotton certification in Ethiopia: Can an increasing demand for certified textiles create a "fashion revolution"? *Geoforum* **2019**, *99*, 111–119. [CrossRef]
27. Ibanez, M.; Blackman, A. Is Eco-Certification a win-win for developing country agriculture? Organic coffee certification in Colombia. *World Dev.* **2016**, *82*, 14–27. [CrossRef]
28. Staricco, J.I.; Ponte, S. Quality regimes in agro-food industries: A regulation theory reading of Fair Trade wine in Argentina. *J. Rural. Stud.* **2015**, *38*, 65–76. [CrossRef]
29. Brown, S.; Getz, C. Privatizing farm worker justice: Regulating labor through voluntary certification and labeling. *Geoforum* **2008**, *39*, 1184–1196. [CrossRef]

15

Method for the Reduction of Natural Losses of Potato Tubers during their Long-Term Storage

Tomasz Jakubowski [1] and Jolanta B. Królczyk [2,*]

1 Department of Machinery, University of Agriculture, 30-149 Krakow, Balicka 116B, Poland; tomasz.jakubowski@ur.krakow.pl
2 Faculty of Mechanical Engineering, Opole University of Technology, 5 Mikolajczyka Street, 45-271 Opole, Poland
* Correspondence: j.krolczyk@po.opole.pl

Abstract: The purpose of the study was to establish whether UV-C radiation applied to potato tubers prior to their storage affected their natural losses over a long period of time. A custom-built UV-C radiation stand constructed for the purpose of this experiment was equipped with a UV-C NBV15 radiator generating a 253.7 nm long wave with power density of 80 to 100 $\mu W \cdot cm^{-2}$. Three varieties of edible medium late potatoes, Jelly, Syrena, and Fianna, were the objects of the research. The measurement of tightly controlled storage conditions was carried out over three seasons between 2016/2017 and 2018/2019, in a professional agricultural cold store with automated adjustment of interior microclimate parameters. The obtained data were processed using the variance analysis ($\alpha = 0.05$). There was a statistically significant reduction in transpiration- and respiration-caused losses in the UV-C radiated potato tubers in comparison to those of the control sample. Additionally, the Jelly variety reacted to UV-C radiation demonstrating a reduction in sprout weight.

Keywords: potato; tuber; storage losses; UV-C

---

1. Introduction

The biological effects of UV-C (ultraviolet light) on the preservation of fresh fruits and vegetables is well researched. Treatment with UV-C radiation is one of the methods used to reduce the number of pathogens on the surface of fresh fruits and vegetables [1]. It can be an alternative to other traditional methods, such as disinfectants (chlorine, chlorine dioxide, bromine, iodine, trisodium phosphate, sodium chlorite, sodium hypochlorite, quaternary ammonium compounds, acids, hydrogen peroxide, ozone, permanganate salts) [2–5], modified atmosphere packaging [6–11], low temperature storage [11–13], or the use of edible films [14–18]. The alternative methods mentioned above are selective in reducing the number of pathogens on the surface of fresh fruits and vegetables, whereas the UV-C is a nonselective method. UV-C has a germicidal effect, but it is strongly dependent on the natural resistance of the microorganisms to UV-C [19], and the surface topography on which the microorganisms are attached [20]. By delaying the ripening process, the UV-C treatment extends the shelf life of fruits and vegetables [21–24]. A number of studies showed that UV light can be used to control the fungal decay of citrus fruits [25], kumquats [26], carrots [27,28], apples [29], strawberries [30–33], sweet cherries [30], mandarins [34], bell peppers [35], mangos [36,37], blueberries [38], grapes [39], and persimmon fruits [40]. Recent publications also describe similar effects on potatoes. Pristijono et al. [41] describe their preliminary research on the effect of UV-C irradiation on the sprouting of stored potatoes. Rocha et al. [42] present the use of UV-C radiation and fluorescent light to control postharvest soft rot in potato seed tubers. According to Stevens et al. [43] irradiation of potato tubers (*Ipomea batatas* L.) with UV-C increases their resistance to rot caused by *Fusarium solani* fungi.

Long-term storage of potato tuber crop is associated with quantitative changes involving, inter alia, reduction of the original weight of tubers due to respiration, transpiration, and sprouting. These processes, termed tuber weight losses, are unavoidable, but their extent can be minimized. Respiration, transpiration, and sprouting processes are mainly determined by the temperature and relative humidity of air during storage. Meteorological conditions during the vegetation season, genetic characteristics of varieties, and mechanical damage to tubers caused during harvesting and post-harvest treatment may also be responsible for natural losses of the potato tubers [44–49]. Several researchers [50–52] recommend using physical methods to improve crop condition during the storage alongside the biological and chemical methods of protection. Various systems using the UV-C light exposure have appeared in the literature for crops other than potatoes [53,54]. Zhu et al. [53] evaluated the effect of three UV-C wavelengths (222, 254, and 282 nm) on degradation of the mycotoxin patulin introduced into apple juice and apple cider. Falguera et al. [54] investigated the influence of ultraviolet irradiation (UV) on some quality attributes (color, pH, soluble solids content, formol index, total phenolics, sugars, and vitamin C) and enzymatic activities (polyphenol oxidase, peroxidase, and pectinolytic enzymes) of fresh apple juice. Another pilot experiment carried out by Jakubowski [55] analyzed the effect of UV-C radiation on the possible reduction of storage losses caused by transpiration and respiration of potato tubers of the following varieties: Lord, Vineta, Owacja, Ditta, Finezja, and Tajfun. It was assumed that ultraviolet radiation in the C band (253.7 nm) applied to potato tubers before storage would cause a reduction of pests present on the potato periderm. This would indirectly cut down the overall storage losses. It was also assumed that following the exposure of potato tubers to UV-C radiation, the population of pathogens typically present during storage processes would also be limited. This, in turn, could lead to a more effective repair of the periderm damage caused by mechanical harvesting, transport, and initial crop storage. In order to verify the research hypothesis, the tubers of selected varieties were initially mechanically damaged under laboratory conditions, so that the continuity of periderm was broken. An MTS Insight 2 strength testing machine (a device allowing for damage of the potato tuber pulp so that the shape and size of the damage are identical in the tested material regardless of the size, weight, or mechanical properties of the tuber) was used to damage the tubers. The results of the pilot research proved a reduction in the natural loss of potato compared to the control sample across all the used varieties. For Vineta and Ditta, these differences were statistically significant ($\alpha = 0.05$). The experiment described above was conducted over a 5-month storage period, and the tuber weight loss (represented by very early, early, and medium-early varieties) was analyzed only before and after the storage time. It is therefore reasonable to measure the reaction of stored potato tubers to UV-C radiation in subsequent stages during their storage, focusing on the losses caused by respiration and transpiration (particularly after a period of physiological dormancy when the sprouting process begins, which is accompanied by an increased transpiration process).

The purpose of this project was to determine the impact of pre-storage UV-C irradiation of potato tubers on potato tuber natural losses measured over a long period of storage.

2. Materials and Methods

2.1. Potato Tubers

The medium-late varieties of edible potatoes used in the research were Jelly, Syrena, and Fianna. Each tuber went through an identical annual agrotechnical procedure. The tubers for analysis, under the stage of full technical maturity and mechanically undamaged, were randomly selected from the crop commercial fraction ($\Phi = 35$–55 mm). The selection of tubers size was necessary to provide uniform UV-C exposure doses in the research procedure. When determining the minimum sample size, the *t*-test was applied for a single sample (at population mean and standard deviation values from pilot studies), the test target power was assumed to be equal 0.9, and the probability of a type I error $\alpha = 0.05$. Each experiment combination involved three replications (with 30 pcs for single replication).

The scope of the research encompassed the measurements of tuber weight losses arising from the processes of respiration, transpiration, and sprouting. The tuber weight was determined before storage (M0), during the initial storage period corresponding to the cooling stage (M1), during the proper storage period (M2), during the first signs of sprouting (M3), and immediately after the storage process (M4). The tuber weight losses were calculated as the difference between M0 and Mn, n = 1–4. Immediately after the storage process, the potato tuber sprout weight and number (Mk, Lk) were also determined.

2.2. *Period of Trials and Conditions*

Accurate storage experiments were carried out over three seasons between 2016/2017 and 2018/2019 in a professional agricultural cold store with automated adjustment of interior microclimate parameters. After the crop was harvested in the second to third decades of September, the selected tubers underwent weight examination and ultraviolet irradiation in the C band. Then, they were placed in a cold store for initial storage. To standardize the conditions of heat and mass exchange with the environment while storing and to minimize the impact of possible temperature and humidity differentiation, the tubers were stored in wooden cases in single layers and the free spaces between the cases were of similar size. Initial storage lasted for approximately 10 days at a temperature of 15 °C and a relative humidity of 90%–95%. After this period, the storage temperature was gradually reduced to 7 °C. This process took approximately 14 days. The air relative humidity during this period amounted to 92%–95%. From the second to third decades of October to the first decade of March, tubers were stored at a temperature of approximately 7 °C and a relative humidity of 92%–95%. At 10 days before the expected end of the storage period, mid-March, the temperature in the cold chamber was increased to 10 °C.

2.3. *Equipment*

Potato tuber weight and sprout weight were determined using the AS310.R2 analytical scale (d = 0.1 mg) with the RS232 interface. Potato tubers underwent ultraviolet irradiation (Figure 1) in the C band for 900 s at a constant height of the UV-C radiator (0.7 m) above the surface of the rollers rotating at a constant speed of 25 rpm (exposure time and working parameters of the station were selected based on the results of pilot studies [55]). In order to expose potato tubers to UV-C, the custom-built stand illustrated in Figure 1 was used. A UV-C NBV15 radiator was used (light wave length 253.7 nm, power 15 W, and power density from 80 to 100 $\mu W \cdot cm^{-2}$), equipped with a precise timer (AURATON 100). The lifetime of the NBV15 radiator applied in the research, guaranteeing stability of its operational parameters (UV-C radiation intensity of 0.9 $W \cdot m^{-2}$ at a distance of 1 m from the radiator), is 8000 h. The radiator is equipped with a reflector made of high quality aluminum with a high reflection coefficient (similar to the coefficient of a mirror). The potato tuber radiation stand is equipped with a system of exchangeable, parallel, and sliding rollers acting as the bottom of the chamber. The rollers, with a diameter of 45 to 55 mm, are installed on a rail, and the distances between them range from 15 to 25 mm. They are driven electrically, with rotational speed control ranging from 20 to 35 rpm. The speed range was selected so that the potato tubers placed on the rollers were set in rotation, but at the same time they were not displaced along the chamber, which allowed for equal irradiation of the entire surface of the potato tubers. The presented test stand, together with the technology allowing for limitation of storage losses of potato tubers, was submitted as an invention to the Patent Office of the Republic of Poland (P.419392, P.425887: More details about patents are mentioned below in the Patents section).

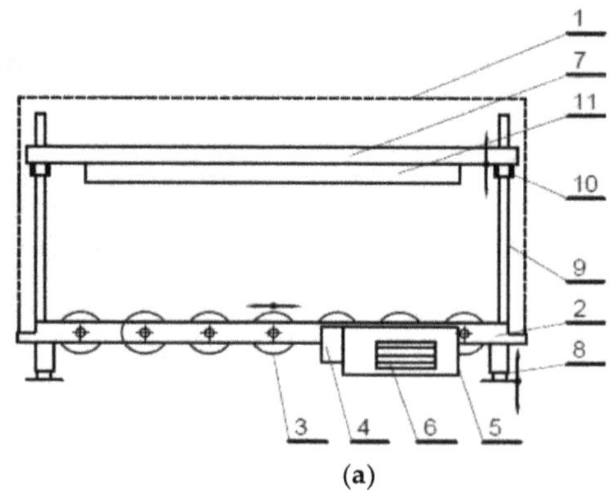

(a)

(b)

Figure 1. Stand for potato tubers UV-C radiation: (**a**) stand layout, (**b**) view of rotating rollers. L: 1: chamber housing, 2: device frame, 3: rotating rollers, 4: mechanism for gradeless regulation of roller rotation speed with the device switch, 5: engine, 6: engine cooling system, 7: radiator frame, 8: height-adjustable foot (screw mechanism for leveling the device), 9: radiator frame guide, 10: gradeless UV-C radiator control above the bottom of the chamber (rollers), 11: UV-C radiators.

2.4. Statistical Analysis

The analysis of variance, preceded by the test regarding the distribution normality in the samples (Kolmogorov–Smirnov test) and the variance homogeneity test (Levene's test), was carried out. The zero hypothesis was verified based on the F-Snedecor test. When examining tuber weight data, the variance was analyzed for arrangements with repeated measurements (with Mauchley sphericity test), and for sprouting data, the variance was analyzed for main effects. Due to the biological nature of the examined object, a possibility of a lack of parity in the samples was assumed a priori. Differences between statistically significant averages were examined using the Spjotvoll–Stoline multiple comparison test (generalization of the Tukey procedure for samples with different N). Groups of homogeneous variables were set. In the analysis of data presented in graphical form (Figures 2–4), the Student's t-test was used for related variables. The obtained results were analyzed at the significance level $\alpha = 0.05$ using the STATISTICA 13.3 package.

3. Results

The results of the experiment are presented in Tables 1–3 and in Figures 2–4. Test results from Kolmogorov–Smirnov, Levene, and Mauchley tests allowed for the variance analysis presented in the methodology. The variance analysis proved a significant effect of potato tuber UV-C irradiation before storage on weight loss after storage (predictor: "UV-C exposition {3}", received values: $F = 6.868$; $p = 0.0089$) (Table 1). The variance analysis for arrangements with repeated measurements proved a significant impact of repeated measurement (time: measurement in subsequent storage stages; predictor: "Time {4}", received values: $F = 4498.591$; $p = 0.0000$) and its interaction with other qualitative predictors (predictor: "Time {4} x Variety {2} x UV-C exposition {3}"; received values: $F = 2.329$; $p = 0.0301$) (Table 1). In all the experiment variants, smaller natural losses of potato tuber weight (caused by transpiration and respiration) were observed for the UV-C radiated tubers in comparison to the control sample (Figure 2). Multiple comparisons of average pairs and homogeneous-variables groups determined on this basis (Table 2) showed significant differences in the size of potato tuber natural weight loss (caused by transpiration and respiration) between subsequent storage stages. However, in stages I-III no differences resulting from the variety nor UV-C radiation were shown (letters "a", "b", "c": no statistical differences) (Figure 2). Significant differences in the size of potato tuber natural weight loss (caused by transpiration and respiration) were observed in stage IV for the Jelly variety (3.787 $g \cdot g^{-1}$ for sample Jelly+UV-C and 4.308 $g \cdot g^{-1}$ for control sample; letters "d" and "e") (Table 2). The variance analysis for the impact of the year of the experiment, variety, and UV-C radiation on the weight and number of potato tuber sprouts after the storage period proved a significant impact (Table 3) on the last two quality predictors and their interactions on dependent variables (predictors: Variety {2} x UV-C exposition {3}; $F = 5.20$; $p = 0.0004$).

Table 1. Results of the analysis of variance for repeated measurements: Impact of the year of testing, variety, and UV-C irradiation on potato tuber weight loss in individual storage periods.

Qualitive Predictor and Interaction	Sum of Square	Degrees of Freedom	Mean Square	Value of F-Snedecor test	Probability Test
Free Word	38815.53	1	38815.53	5391.355	0.0000
Year {1}	4.98	2	2.49	0.346	0.7078
Variety {2}	22.46	2	11.23	1.560	0.2104
UV-C exposition {3}	49.45	1	49.45	6.868	0.0089
{1}x{2}	55.41	4	13.85	1.924	0.1039
{1}x{3}	11.46	2	5.73	0.796	0.4513
{2}x{3}	12.18	2	6.09	0.846	0.4293
{1}x{2}x{3}	37.48	4	9.37	1.301	0.2673
Error	11533.74	1602	7.20		
Time {4}	8273.54	3	2757.85	4498.591	0.0000
{4}x{1}	0.80	6	0.13	0.217	0.9717
{4}x{2}	6.28	6	1.05	1.706	0.1152
{4}x{3}	23.42	3	7.81	12.732	0.0000
{4}x{1}x{2}	12.45	12	1.04	1.692	0.0618
{4}x{1}x{2}x{3}	1.83	6	0.31	0.498	0.8102
{4}x{2}x{3}	8.57	6	1.43	2.329	0.0301
{4}x{1}x{2}x{3}	13.60	12	1.13	1.849	0.0357
Error	2946.30	4806	0.61		

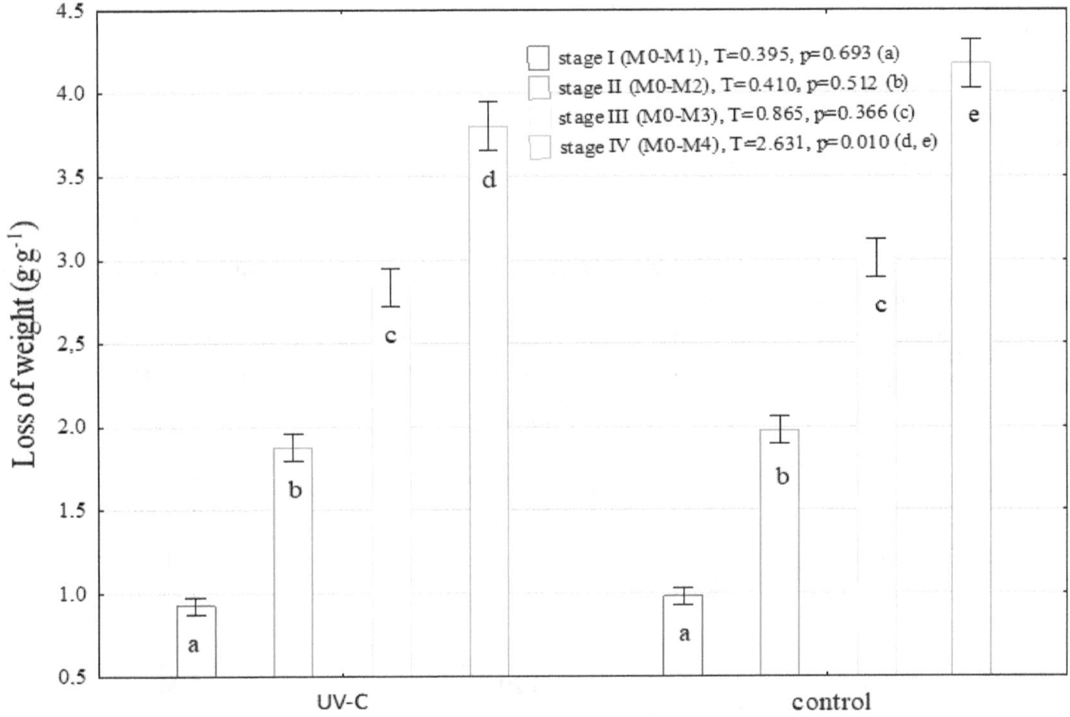

Figure 2. Effect of UV-C irradiation on weight loss (resulting from transpiration and respiration) of potato tubers in individual storage periods (error = mean value +/− 95% of confidence interval).

Table 2. Effect of variety and UV-C irradiation on weight loss (resulting from transpiration and respiration) of potato tubers in individual storage periods.

Variety	Exposition	Time	Loses ($g \cdot g^{-1}$)	Homogeneous Groups					
				1	2	3	4	5	6
Fianna	UV-C	I	0.887	****					
Jelly	UV-C	I	0.927	****					
Fianna	control	I	0.948	****					
Syrena	UV-C	I	0.961	****					
Jelly	control	I	0.977	****					
Syrena	control	I	1.012	****					
Fianna	UV-C	II	1.785		****				
Jelly	UV-C	II	1.849		****				
Fianna	control	II	1.964		****				
Jelly	control	II	1.974		****				
Syrena	UV-C	II	1.991		****				
Syrena	control	II	1.992		****				
Fianna	UV-C	III	2.741			****			
Jelly	UV-C	III	2.804			****			
Fianna	control	III	2.863			****			
Syrena	UV-C	III	2.968			****			
Syrena	control	III	2.998			****			
Jelly	control	III	3.172			****			
Fianna	UV-C	IV	3.649				****		
Jelly	UV-C	IV	3.787				****	****	
Syrena	UV-C	IV	3.971				****	****	****
Fianna	control	IV	4.088				****	****	****
Syrena	control	IV	4.120					****	****
Jelly	control	IV	4.308						****

(****) Arrangement of homogeneous groups (Spjotvoll–Stoline test).

Table 3. Impact of the year of experiment, variety, and UV-C irradiation on weight and number of potato tuber sprouts after storage.

Qualitive Predictor and Interaction	Value of F-Snedecor Test	Probability Test
Free Word	10787.16	0.0000
Year {1}	0.22	0.9276
Variety {2}	12.58	0.0000
UV-C exposition {3}	10.11	0.0001
{1}x{2}	1.43	0.1767
{1}x{3}	1.72	0.1420
{2}x{3}	5.20	0.0004
{1}x{2}x{3}	1.82	0.0680

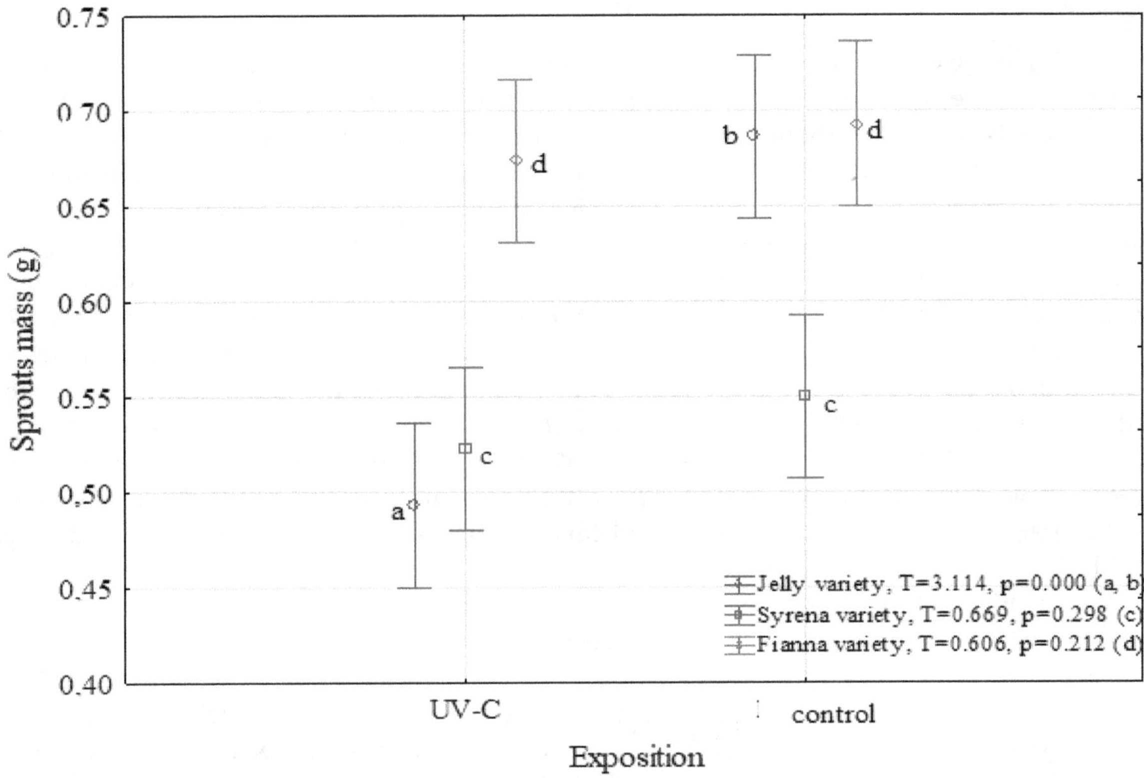

Figure 3. Effect of potato tuber UV-C irradiation on sprout weight after storage (statistically significant difference for the Jelly variety), (error = mean value +/– 95% of confidence interval).

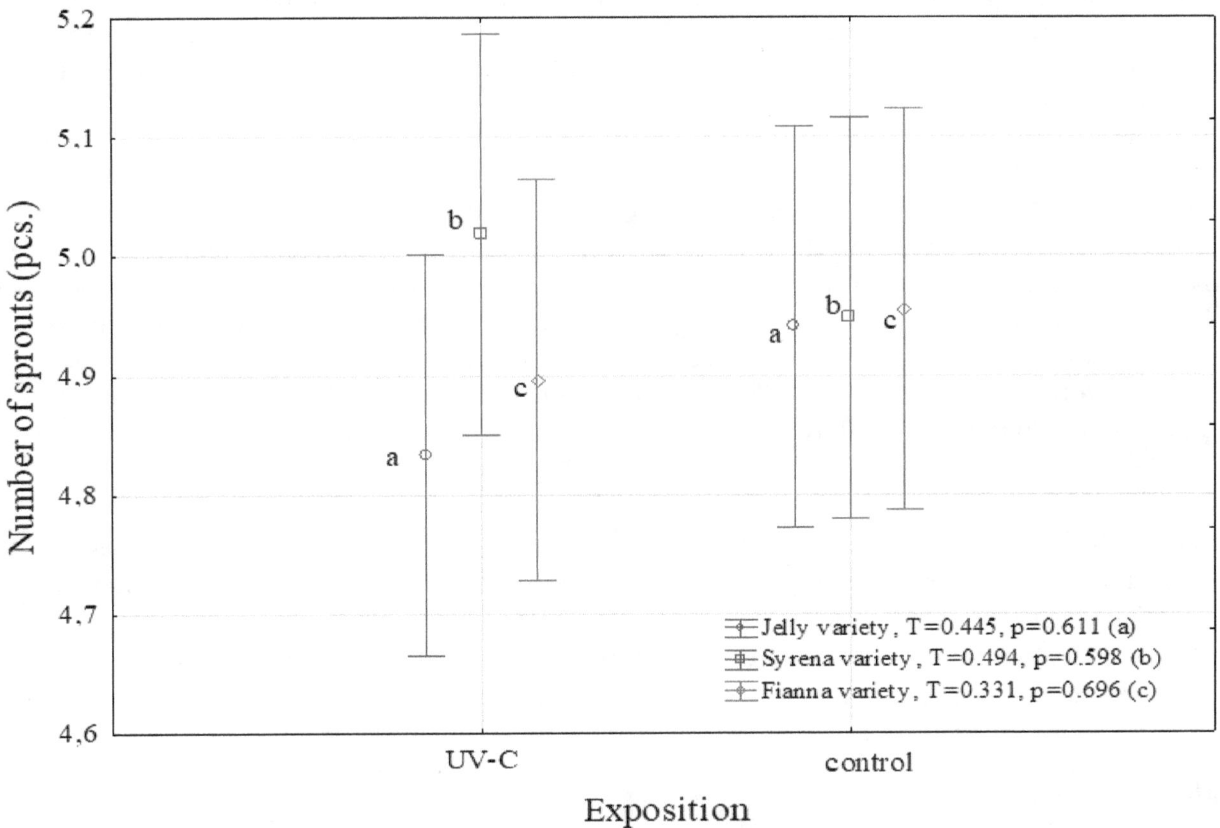

Figure 4. Effect of potato tuber UV-C irradiation on number of sprouts after storage (statistically insignificant differences), (error = mean value +/– 95% of confidence interval).

4. Discussion

The results of the conducted experiment, in which the reduction of natural losses measured after 6.5 months of storage is achieved by means of the UV-C irradiation of potato tubers before storage, were expected to be similar to the results obtained for early and medium-early varieties (Vineta and Ditta) and described by Jakubowski [55]. The tests carried out in this research and the results obtained allowed for a description of the phenomenon causing reduction of natural losses and a determination of the stage of potato storage in which the physical factor, in the form of UV-C radiation, significantly affects the tubers under exposure. The results of the experiment prove that the effects of UV-C on stored potato tubers occur in the final stage (IV) of their storage (Table 2), which corresponds to the phase of tubers awakening and beginning their sprouting. In stage IV of storage, all the tested varieties reacted to UV-C, demonstrating the reduction in natural losses in comparison to the control, and for the Jelly variety, these differences were statistically significant ($F = 2.329$; $p = 0.0301$). The reduction in the weight loss of the pre-treated tubers during storage could occur through changes in the epi- and cuticular wax morphology. Some information on this aspect of treatment are presented in other research, such as Charles et al. [56,57]. Sprouting of the tubers indicates the break of dormancy. Oxidative atmospheres, such as chlorine atmospheres, are known to control sprouting [58].

The analyzed potato tubers reacted to UV-C radiation demonstrating the reduction in sprout weight by 80 $g \cdot g^{-1}$, on average, for all the varieties, which amounts to approximately 14.2% compared to the control. For Jelly tubers, these differences were statistically significant (difference of 194 $g \cdot g^{-1}$, which is approximately 39% compared to the control) (Figure 3). The tubers of Jelly and Fiana varieties subjected to UV-C irradiation were also characterized by a lower number of sprouts compared to the tubers not being under exposure, by 0.11 and 0.05 pcs, respectively. A higher number of sprouts resulting from UV-C irradiation was noted for the tubers of Syrena variety, by 0.07 pcs compared to the control (Figure 4). The results of the experiment suggest that UV-C, in addition to neutralizing pests in areas of damaged potato periderm [55], may act as a sprouting inhibitor, which reduces respiration and, more particularly, transpiration. The obtained results also allow for the conclusion that UV-C does not interfere with the process of tuber transpiration and respiration at earlier stages (I-III) of storage. The inhibitory effect of ultraviolet in the C band may result from the fact that the effect of 253.7 nm wave on a biological object may lead to damaging its DNA chains, and at the same time UV is absorbed by DNA, RNA [59–63], protein, free purine, and pyrimidine bases, acting mutagenetically and inhibiting cell division in the irradiated organism [64–66]. The preliminary tests carried out by Pristijono et al. [41] on freshly harvested potatoes (Solanum tuberosum 'Innovator') that were exposed to UV-C light revealed that UV-C irradiation significantly affected the number of sprouts. UV-C irradiation also affected the sprout length since irradiated potatoes had significantly shorter sprouts than those of untreated potatoes. Despite the fact that storage conditions were different in this experiment (storage in air 20°C), the authors [41] conclude that these results indicate promise for UV-C as a potential postharvest treatment to reduce the incidence of sprouting in potato tubers.

5. Conclusions

1. A significant impact of potato tuber UV-C irradiation on the size of natural losses was observed.
2. A reduction in potato tuber weight loss caused by transpiration and respiration was shown in comparison to the control sample.
3. Jelly variety reacted to UV-C radiation, demonstrating the reduction in the sprout weight.
4. The result of the experiment indicates that the proposed physical UV-C method can be applied in practice and can be used as a way of reducing the natural defects of stored potato tubers.

6. Patents

Jakubowski T. Patent: The method and device for increasing the storage life of potato tubers with the participation of radiation UV-C (in Polish; Sposób i urządzenie do zwiększania trwałości przechowalniczej bulw ziemniaczanych przy udziale promieniowania UV-C: P.419392, data zgłoszenia 07-11-2016).

Jakubowski T., Sobol Z. Patent: The method for modifying the color of potato products and a device to modify the color of potato products (in Polish; Sposób modyfikowania barwy wyrobów z ziemniaków i urządzenie do modyfikowania barwy wyrobów z ziemniaków: P.425887, data zgłoszenia 11-06-2018).

Author Contributions: Conceptualization, T.J. and J.B.K.; methodology, T.J.; validation and formal analysis, T.J. and J.B.K.; investigation, resources, and data curation, T.J. and J.B.K.; writing—original draft preparation, writing—review and editing, and visualization, J.B.K. All authors have read and agreed to the published version of the manuscript.

References

1. Turtoi, M. Ultraviolet light treatment of fresh fruits and vegetables surface. *J. Agroaliment. Process. Technol.* **2013**, *19*, 325–337.
2. Beuchat, L.R. *Surface Decontamination of Fruits and Vegetables Eaten Raw: A Review*; World Health Organization: Geneva, Switzerland, 1998.
3. Yaun, B.; Sumner, S.; Eifert, J.; Marcy, J. Inhibition of pathogens on fresh produce by ultraviolet energy. *Int. J. Food Microbiol.* **2004**, *90*, 1–8. [CrossRef]
4. Allende, A.; McEvoy, J.; Tao, Y.; Luo, Y. Antimicrobial effect of acidified sodium chlorite, sodium chlorite, sodium hypochlorite, and citric acid on *Escherichia coli* O157:H7 and natural microflora of fresh-cut cilantro. *Food Control* **2009**, *20*, 230–234. [CrossRef]
5. Oms-Oliu, G.; Rojas-Graü, A.; Gonzáles, L.A.; Varela, P.; Soliva-Fortuny, R.; Hernando Hernando, I.; Pérez Munuera, I.; Fiszman, S.; Martín-Belloso, O. Recent approaches using chemical treatments to preserve quality of freshcut fruit: A review. *Postharvest Biol. Technol.* **2010**, *57*, 139–148. [CrossRef]
6. Kader, A.A.; Zagory, D.; Kerbel, E.L. Modified atmosphere packaging of fruits and vegetables. *Crit. Rev. Food Sci. Nutr.* **1989**, *28*, 1–30. [CrossRef] [PubMed]
7. Soliva-Fortuny, R.C.; Elez-Martínez, P.; Martín-Belloso, O. Microbiological and biochemical stability of fresh-cut apples preserved by modified atmosphere packaging. *Innov. Food Sci. Emerg. Technol.* **2004**, *5*, 215–224. [CrossRef]
8. Saxena, A.; Singh Bawa, A.; Srinivas Raju, P. Use of modified atmosphere packaging to extentd shelflife of minimally processed jackfruit (*Artocarpus heterophyllus* L.) bulbs. *J. Food Eng.* **2008**, *87*, 455–466. [CrossRef]
9. Oliveira, M.; Usall, J.; Solsona, C.; Alegre, I.; Vinas, I.; Abadias, M. Effects of packaging type and storage temperature on the growth of foodborne pathogens on shreddred 'Romaine' lettuce. *Food Microbiol.* **2010**, *27*, 455–466. [CrossRef]
10. Sandhya. Modified atmosphere packaging of fresh produce: Current status and future needs. *LWT—Food Sci. Technol.* **2010**, *43*, 381–392. [CrossRef]
11. Abadias, M.; Alegre, I.; Oliveira, M.; Altisent, R.; Vinas, I. Growth potential of *Escherichia coli* O157:H7 on fresh-cut fruits (melon and pineapple) and vegetables (carrot and escarole) stored under different conditions. *Food Control* **2012**, *27*, 37–44. [CrossRef]
12. Harvey, J.M. Optimum environments for the transport of fresh fruits and vegetables. *Int. J. Refrig.* **1981**, *4*, 293–298. [CrossRef]
13. Tano, K.; Oule, M.K.; Doyon, G.; Lencki, R.W.; Arul, J. Comparative evaluation on the effect of storage temperature fluctuation on modified atmosphere packages of selected fruit and vegetables. *Postharvest Biol. Technol.* **2007**, *46*, 212–221. [CrossRef]
14. Zhang, R.; Beuchat, L.R.; Chinnan, M.S.; Shewflet, R.L.; Haung, Y.W. Inactivation of Salmonella Montevideo on tomatoes by applying cellulose-based edible Films. *J. Food Prot.* **1996**, *59*, 808–812. [CrossRef] [PubMed]
15. Vina, S.Z.; Mugridge, A.; Garcia, M.A.; Ferreyra, R.M.; Martino, M.N.; Chaves, A.R.; Zaritzky, N.E. Effects of polyvinylchloride films and edible starch coatings on quality aspects of refrigerated Brussels sprouts. *Food Chem.* **2007**, *103*, 701–709. [CrossRef]
16. Raybaudi-Massilia, R.M.; Mosqueda-Melgar, J.; Martin-Belloso, O. Edible alginate-based coating as carrier of antimicrobials to improve shelf-life and safety of fresh-cut melon. *Int. J. Food Microbiol.* **2008**, *121*, 313–327. [CrossRef]

17. Falguera, V.; Quintero, H.P.; Jimeez, A.; Munoz, J.A.; Ibarz, A. Edible films and coatings: Structures, active functions and trends in their use. *Trends Food Sci. Technol.* **2011**, *22*, 292–303. [CrossRef]
18. Gol, N.B.; Patel, P.R.; Ramana Rao, T.V. Improvement of quality and shelf-life of strawberries with edible coatings enriched with chitosan. *Postharvest Biol. Technol.* **2013**, *85*, 185–195. [CrossRef]
19. Shama, G. Ultraviolet light. In *Handbook of Food Science, Technology, and Engineering*; Hui, Y.H., Ed.; CRC Press: Boca Raton, FL, USA, 2005; pp. 122-1–122-14.
20. Gardner, D.W.; Shama, G. Modeling UV-induced inactivation of microorganisms on surfaces. *J. Food Prot.* **2000**, *63*, 63–70. [CrossRef]
21. Lu, J.Y.; Stevens, C.; Khan, V.A.; Kabwe, M.; Wilson, C.L. The effect of ultraviolet irradiation on shelf-life and ripening of peaches and apples. *J. Food Qual.* **1991**, *14*, 299–305. [CrossRef]
22. D'hallewin, G.; Schirra, M.; Manueddu, E.; Piga, A.; Ben-Yehoshua, S. Scoparone and scopoletin accumulation and ultraviolet-C induced resistance to postharvest decay in oranges as influenced by harvest date. *J. Am. Soc. Hortic. Sci.* **1999**, *124*, 702–707. [CrossRef]
23. Lamikanra, O.; Kueneman, D.; Ukuku, D.; Bett-Garber, K.L. Effect of Processing Under Ultraviolet Light on the Shelf Life of Fresh-Cut Cantaloupe Melon. *J. Food Sci.* **2005**, *70*, C534–C539. [CrossRef]
24. Darvishi, S.; Fatemi, A.; Davari, K. Keeping quality of use of fresh 'Kurdistan' strawberry by UV-C radiation. *World Appl. Sci. J.* **2012**, *17*, 826–831.
25. Ben-Yehoshua, S.; Rodov, V.; Kim, J.J.; Carmeli, S. Preformed and induced antifungal materials of citrus fruits in relation to the enhancement of decay resistance by heat and ultraviolet treatments. *J. Agric. Food Chem.* **1992**, *40*, 1217–1221. [CrossRef]
26. Rodov, V.; Ben-Yehoshua, S.; Kim, J.J.; Shapiro, B.; Ittah, Y. Ultraviolet illumination induces scoparone production in kumquat and orange fruit and improves decay resistance. *J. Am. Soc. Hortic. Sci.* **1992**, *117*, 788–792. [CrossRef]
27. Mercier, J.; Arul, J.; Julien, C. Effect of UV-C on phytoalexin accumulation and resistance to Botrytis cinerea in stored carrots. *Phytopathology* **1993**, *139*, 17–25. [CrossRef]
28. Mercier, J.; Roussel, D.; Charles, M.T.; Arul, J. Systemic and local responses associated with UV and pathogen-induced resistance to Botrytis cinereal in stored carrot. *Phytopathology* **2000**, *90*, 981–986. [CrossRef]
29. De Capdeville, G.; Wilson, C.L.; Beer, S.V.; Aist, J.R. Alternative disease control agents induce resistance to blue mold in harvested 'Red Delicious'apple fruit. *Phytopathology* **2002**, *92*, 900–908. [CrossRef]
30. Marquenie, D.; Michiels, C.; Geeraerd, A.; Schenk, A.; Soontjens, C.; Van Impe, J.; Nicolai, B. Using survival analysis to investigate the effect of UV–C and heat treatment on storage rot of strawberry and sweet cherry. *Int. J. Food Microbiol.* **2002**, *73*, 187–196. [CrossRef]
31. Marquenie, D.; Geeraerd, A.H.; Lammertyn, J.; Soontjens, C.; Van Impe, J.F.; Michiels, C.W.; Nicolai, B.M. Combinations of pulsed white light and UV-C or mild heat treatment to inactivate conidia of Botrytis cinerea and *Monilia fructigena*. *Int. J. Food Microbiol.* **2003**, *85*, 185–196. [CrossRef]
32. Marquenie, D.; Michiels, C.W.; Van Impe, J.F.; Schrevens, E.; Nicolai, B.M. Pulsed white light in combination with UV-C and heat to reduce storage rot of strawberry. *Postharvest Biol. Technol.* **2003**, *28*, 455–461. [CrossRef]
33. Lammertyn, J.; De Ketelaere, B.; Marquenie, D.; Molenberghs, G.; Nicolai, B.M. Mixed models for multicategorical repeated response: Modeling the time effect of physical treatments on strawberry sepal quality. *Postharvest Biol. Technol.* **2003**, *30*, 195–207. [CrossRef]
34. Kinay, P.; Yildiz, F.; Sen, F.; Yildiz, M.; Karacali, I. Integration of pre- and postharvest treatments to minimize Penicillium decay of *Satsuma mandarins*. *Postharvest Biol. Technol.* **2005**, *37*, 31–36. [CrossRef]
35. Artes, F.; Conesa, A.; Lopez-Rubira, V.; Artes-Hernandez, F. UV–C treatments for improving microbial quality in whole and minimally processed bell peppers. In *The Use of UV as a Postharvest Treatment: Status and Prospects, Proceedings of the International Conference on Quality Management of Fresh Cut Produce, Bangkok, Thailand, 6–8 August 2007*; Ben-Yehoshua, S., D'Hallewin, G., Erkan, M., Rodov, V., Lagunas, M., Eds.; ISHSS: Leuven, Belgium, 2006; pp. 12–17.
36. Gonzalez-Aguilar, G.A.; Wang, C.Y.; Buta, J.G.; Krizek, D.T. Use of UV-C irradiation to prevent decay and maintain postharvest quality of ripe 'Tommy Atkins' mangoes. *Int. J. Food Sci. Technol.* **2001**, *36*, 767–773. [CrossRef]
37. Gonzalez-Aguilar, G.A.; Zavaleta-Gatica, R.; Tiznado-Hernandez, M.E. Improving postharvest quality of mango 'Haden' by UV-C treatment. *Postharvest Biol. Technol.* **2007**, *45*, 108–116. [CrossRef]

38. Perkins-Veazie, P.; Collins, J.K.; Howard, L. Blueberry fruit response to postharvest application of ultraviolet radiation. *Postharvest Biol. Technol.* **2008**, *47*, 280–285. [CrossRef]
39. Romanazzi, G.; Mlikota Gabler, F.; Smilanick, J.L. Preharvest chitosan and postharvest UV irradiation treatments suppress gray mold of table grapes. *Plant Dis.* **2006**, *90*, 445–450. [CrossRef]
40. Khademi, O.; Zamani, Z.; Poor Ahmadi, E.; Kalantari, S. Effect of UV-C radiation on postharvest physiology of persimmon fruit (*Diospyros kaki* Thunb.) cv. 'Karaj' during storage at cold temperature. *Int. Food Res. J.* **2013**, *20*, 247–253.
41. Pristijono, P.; Bowyer, M.C.; Scarlett, C.J.; Vuong, Q.V.; Stathopoulos, C.E.; Golding, J.B. Effect of UV-C irradiation on sprouting of potatoes in storage. In Proceedings of the VIII International Postharvest Symposium: Enhancing Supply Chain and Consumer Benefits-Ethical and Technological Issues, Cartagena, Spain, 21–24 June 2016; pp. 475–478.
42. Rocha, A.B.; Honório, S.L.; Messias, C.L.; Otón, M.; Gómez, P.A. Effect of UV-C radiation and fluorescent light to control postharvest soft rot in potato seed tubers. *Sci. Hortic.* **2015**, *181*, 174–181. [CrossRef]
43. Stevens, C.; Khan, V.A.; Lu, J.Y.; Wilson, C.L.; Chalutz, E.; Droby, S.; Kabwe, M.K.; Haung, Z.; Adeyeye, O.; Pusey, L.P.; et al. Induced resistance of sweet potato to Fusarium root rot by UV-C hormesis. *Crop Prot.* **1999**, *18*, 463–470. [CrossRef]
44. Clasen, B.; Stoddard, T.; Luo, S.; Demorest, Z.; Li, J.; Cedrone, F. Improving cold storage and processing traits in potato through targeted gene knockout. *Plant. Biotechnol. J.* **2016**, *14*, 169–176. [CrossRef]
45. Elmore, E.; Briddon, A.; Dodson, T.; Muttucumaru, N.; Halford, G.; Mottram, S. Acrylamide in potato crisps prepared from 20 UK-grown varieties: Effects of variety and tuber storage time. *Food Chem.* **2016**, *182*, 1–8. [CrossRef] [PubMed]
46. Hardigan, D.; Hirsch, N.; Manrique-Carpintero, A. The contribution of the Solanaceae coordinated agricultural project to potato breeding. *Potato Res.* **2014**, *57*, 215–224. [CrossRef]
47. El-Awady Aml, A.; Moghazy, M.; Gouda, A.; Elshatoury, A. Inhibition of sprout growth and increase storability of processing potato by antisprouting agent. *Trends Hortic. Res.* **2014**, *4*, 31–40. [CrossRef]
48. Castronuovo, D.; Tataranni, G.; Lovelli, S.; Candido, V.; Sofo, A.; Scopa, A. UV-C irradiation effects on young tomato plants: Preliminary results. *Pak. J. Bot.* **2014**, *46*, 945–949.
49. Katerova, Z.; Ivanov, S.; Prinsen, E.; Van Onckelen, H.; Alexieva, V.; Azmi, A. Low doses of ultraviolet-B or ultraviolet-C radiation affect ACC, ABA and IAA levels in young pea plants. *Biol. Plant.* **2009**, *53*, 365–368. [CrossRef]
50. Hassan, H.; Abd El-Rahman, A.; Liela, A. Sprouting suppression and quality attributes of potato tubers as affected by post-harvest UV-C treatment under cold storage. *Int. J. Adv. Res.* **2016**, *4*, 241–253. [CrossRef]
51. Pietruszewski, S.; Martínez, E. Magnetic field as a method of improving the quality of sowing material. *Int. Agrophysics* **2015**, *29*, 377–389. [CrossRef]
52. Kasyanov, G.; Syazin, I.; Grachev, A.; Davidenko, T.; Vazhenin, E. Features of usage of electromagnetic field of extremely low frequency for the storage of agricultural products. *J. Electromagn. Anal. Appl.* **2013**, *5*, 236–241. [CrossRef]
53. Zhu, Y.; Koutchma, T.; Warriner, K.; Zhou, T. Reduction of Patulin in Apple Juice Products by UV Light of Different Wavelengths in the UVC Range. *J. Food Prot.* **2014**, *77*, 963–967. [CrossRef]
54. Falguera, V.; Pagán, J.; Ibarz, A. Effect of UV irradiation on enzymatic activities and physicochemical properties of apple juices from different varieties. *Food Sci. Technol.* **2011**, *44*, 115–119. [CrossRef]
55. Jakubowski, T. Use of UV-C radiation for reducing storage losses of potato tubers. *Bangladesh J. Bot.* **2018**, *47*, 533–537. [CrossRef]
56. Charles, M.T.; Mercier, J.; Makhlouf, J.; Arul, J. Physiological basis of UV-C-induced resistance to Botrytis cinerea in tomato fruit: I. Role of pre-and post-challenge accumulation of the phytoalexin-rishitin. *Postharvest Biol. Technol.* **2008**, *47*, 10–20. [CrossRef]
57. Charles, M.T.; Makhlouf, J.; Arul, J. Physiological basis of UV-C induced resistance to Botrytis cinerea in tomato fruit: II. Modification of fruit surface and changes in fungal colonization. *Postharvest Biol. Technol.* **2008**, *47*, 21–26. [CrossRef]
58. Tweddell, R.J.; Boulanger, R.; Arul, J. Effect of chlorine atmospheres on sprouting and development of dry rot, soft rot and silver scurf on potato tubers. *Postharvest Biol. Technol.* **2003**, *28*, 445–454. [CrossRef]

59. Onik, J.; Xie, Y.; Duan, Y.; Hu, X.; Wang, Z.; Lin, Q. UV-C treatment promotes quality of early ripening apple fruit by regulating malate metabolizing genes during postharvest storage. *PLoS ONE* **2019**, *14*, e0215472. [CrossRef]
60. Wu, X.; Guan, W.; Yan, R.; Lei, J.; Xu, L.; Wang, Z. Effects of UV-C on antioxidant activity, total phenolics and main phenolic compounds of the melanin biosynthesis pathway in different tissues of button mushroom. *Postharvest Biol. Technol.* **2016**, *118*, 51–58. [CrossRef]
61. Najeeb, U.; Xu, Z.; Ahmed, M.; Rasheed, G.; Jilani, M.; Naeem, W.; Shen, W. Ultraviolet-C mediated physiological and ultrastructural alterations in *Juncus effuses* L. shoots. *Acta Physiol. Plant.* **2011**, *33*, 481–488. [CrossRef]
62. Cools, K.; Alamar, M.; Terry, L. Controlling sprouting in potato tubers using ultraviolet-C irradiance. *Postharvest Biol. Technol.* **2014**, *98*, 106–114. [CrossRef]
63. Kowalski, W. *Ultraviolet Germicidal Radiation Handbook*; Springer: Berlin/Heidelberg, Germany, 2009; pp. 38–90. ISBN 978-3-642-01998-2.
64. Bhattacharjee, C.; Sharan, R. UV-C radiation induced conformational relaxation of pMTa4 DNA in *Escherichia coli* may be the cause of single strand breaks. *Int. J. Radiat. Biol.* **2005**, *81*, 919–927. [CrossRef]
65. Schreier, W.; Schrader, T.; Koller, F.; Gilch, P.; Crespo-Hernandez, C.; Swaminathan, V.; Carell, T.; Zinth, W.; Kohler, B. Thymine dimerization in DNA is an ultrafast photoreaction. *Science* **2007**, *315*, 625–629. [CrossRef]
66. Quek, P.; Hu, J. Indicators for photoreactivation and dark repair studies following ultraviolet disinfection. *J. Ind. Microbiol. Biotechnol.* **2008**, *35*, 533–541. [CrossRef] [PubMed]

16

Evaluation of Dust Concentration during Grinding Grain in Sustainable Agriculture

Paweł Sobczak [1], Jacek Mazur [1,*], Kazimierz Zawiślak [1], Marian Panasiewicz [1], Wioletta Żukiewicz-Sobczak [2], Jolanta Królczyk [3] and Jerzy Lechowski [4]

1 Department of Food Engineering and Machines, University of Life Sciences in Lublin, 20-612 Lublin, Poland
2 Department of Public Health, Pope John Paul II State School of Higher Education in Biala Podlaska, 21-500 Biala Podlaska, Poland
3 Faculty of Mechanical Engineering, Opole University of Technology, 45-271 Opole, Poland
4 Department of Biochemistry and Toxicology, University of Life Sciences, 20-950 Lublin, Poland
* Correspondence: jacek.mazur@up.lublin.pl

Abstract: This work analyses the organic dust concentration during a wheat grinding process which was carried out using two types of grinders: A hammer mill and a roller mill. DustTrak II aerosol monitor was used to measure the concentration of the dust PM10 (particles with the size smaller than 10 μm), PM4.0, and PM1.0. An increase of the grain moisture to 14% resulted in the reduction in PM10 when grinding grain using the hammer mill. An inverse relationship was obtained when grain was ground using the roller mill. A smaller amount of the fraction below 0.1 mm was observed for larger diameter of the holes in the screen and smaller size of the working gap in the roller mill. For both mills, the obtained concentration of the PM10 fraction dust exceeded the acceptable level. To protect farmers health, it is necessary to use dust protection equipment or to modify the grinding technology by changing the grain moisture content and/or the grinding parameters.

Keywords: grinding; organic dust; sustainable agriculture

---

1. Introduction

Management and processing of agricultural goods in a sustainable agriculture farm is of particular importance in terms of further use of these products. In agricultural holdings, straight after harvest, cereals usually have a higher moisture content than the one necessary for successful storage. This is the result of varied harvest periods. Installation of a specialized drying facility is one of the ways to lower the moisture content, however, this generates additional costs and entails higher energy consumption [1,2]. There are approaches to reduce costs of drying by using the heat produced by combustion of pellets derived from by-products of agricultural processing, e.g., Straw, rapeseed oil cakes, soybean hulls [3,4]. Moreover, technologies are also developed for processing agricultural goods, including cereals, directly after harvesting without drying [5]. In most cases such raw materials are used directly as livestock fodder. Grinding cereals and other seeds is the key production step in animal fodder production. Several factors, including moisture content and the type of raw material, its hardness and brittleness, as well as chemical composition influence the course, efficiency and effectiveness of grinding. In addition to physical properties, the milling process is determined by both structural and operating characteristics of a grinder. One of the consequences of using this group of machines is their wear, which also affects the degree of fragmentation of raw material. Among the elements of grinding machines that are affected by use wear; one should mention hammers, screens, and rollers. Irrespective of the milling method, organic dust is produced when cereal raw materials are ground. For health reasons, of particular importance is the continuous monitoring of aerosol concentration in the air and the use of equipment protecting against dust. Dust is one of the main harmful factors occurring in the work environment of farmers. Harmful influence of dust on the human body may cause many diseases, including pneumoconiosis and cancer. According to the data from the literature of the subject [6]; a human inhales daily approximately 12.3 m^3 of air, which is a mixture of gas and particles of liquids as well as solids. Dust suspended in the air is a mixture with varied chemical composition and physical characteristics. Organic dust present in the air with a particle diameter greater than 10 μm quickly settles on surface and is called deposited dust. At the same time, smaller fractions are suspended in the air. PM10 fraction refers to particles with the size smaller than 10 μm, while PM1.0 to the particles with the diameter smaller than 1 μm. Dust with dimensions smaller than 10 μm (PM10) enters the respiratory system and those with particle size smaller than 1 μm, may penetrate alveoli and thus enter bloodstream and all other systems [7–9]. As evidenced by studies, dangerous mycotoxins enter the human body together with inhaled organic dust [10,11]. The presence of dust during the grinding process is very common. Primarily, particles of a greater size are present (PM10), but there are also those with smaller particle size (PM2.5). As the result of their further spread, and frequently mutual collision, their additional fragmentation takes place, which increases the amount of fine fraction PM2.5 and very fine fraction PM1.0. Therefore, a PM4.0 fraction concentration might be a good determinant of dust contamination in its initial phase.

The European Commission takes into account mainly dust particles of PM10 and PM2.5 fraction [12]. According to the data from 2018, it is estimated that in Poland's agriculture and waste disposal there is approx. 16% with PM10 particulate size and approx. 15.2% with PM2.5 particulate size in the overall amount of generated dusts [13]. According to the data of the World Health Organization [14]; for a PM10 acceptable level of 24 h continuous concentration of dust is 50 $\mu g \cdot m^{-3}$, while in the case of PM2.5 the acceptable level is 25 $\mu g \cdot m^{-3}$. According to epidemiological research [12], it is difficult to identify the threshold values of dust concentration below which no adverse effects on human organism are observed. Many research works were performed on the harmfulness of suspended particulate matter on the human body [15–18]. In the research by Zwoździak et al. [19] concentration of PM10, PM2.5, and PM1 at school was examined by comparing it with the results of the children examined using a spirometer. The average concentration of PM10 was equal to 115.2 $\mu g \cdot m^{-3}$, while that of PM2.5 was 46.4 $\mu g \cdot m^{-3}$. A short-term cause and effect relationship was documented between the concentration of PM2.5 fraction inside the school and the change of values of the parameters measured in spirometry.

Health statistics show that most of occupational diseases reported in Polish farmers is caused by pathogens present in organic dusts. In Poland, lung diseases are more common in farmers than in the rest of the population, just as in other countries. Therefore, the problem is serious, and there is a need to take appropriate preventive measures [10,20–22].

The studies carried out were intended to assess concentration of organic dust produced during grinding grain in a hammer mill and roller mill, i.e., the most popular grinding devices in agricultural holdings. The test presented in the paper were carried out for the grain with three different levels of moisture content in order to assess the impact of moisture content on the level of dust contamination.

2. Materials and Methods

The test material was wheat of Zyta variety. The wheat grain is the typical kind of grain used in the farms. The initial moisture content of the wheat, the so-called storage humidity, was 9% [23]. The studies were carried out also on wheat with the moisture content of 14% and 18%. For this purpose, the test material was moistened to the desired moisture content by adding the appropriate amount of water and storing in cold storage conditions for 24 h in order to balance the moisture content. The quantity of water needed for moistening was calculated using the following formula:

$$M_W = \frac{x_2 - x_1}{100 - x_2} \cdot M_N$$

where:

M_w—amount of water necessary for moistening,

M_N—moistened grain mass,
X_1—initial moisture content,
X_2—desired moisture content.

The first moisture content (9%) it is mainly moisture of grain after drying or storage, moisture 14% this is a maximum safe moisture content for storage and 18% and more this is a moisture content of grain after harvesting. Some farmers processed the grain immediately after harvesting. The wheat grain is the typical kind of grain used in the farms.

Grinding was performed using two different types of mills, i.e., A hammer mill with interchangeable screens and a roller mill with the diameter of rollers equal to 200 mm. In the hammer mill three different screens with mesh with the diameter of openings respectively: $\varphi = 3$, $\varphi = 5$ and $\varphi = 8$ mm were used, while in the case of the roller mill three different working gaps were set, i.e., 0.4, 0.7, and 1.0 mm. The operating parameters of machines are different according to the destination of the ground grain. The disintegration at holes at screen size of 3 mm is typical feed for pigs, 5 mm is typical for poultry and 8 mm is designed for hens in the hammer mill. The size of gap was set according to the same destiny of feed in roller mill.

Concentration of organic dust, as measured using a DustTrak II monitor, with which three interchangeable size-selective inlets were used, allowing the measuring of the concentration of dust with particulate matter size: PM10, PM4.0 and PM1.0. The duration of the measurement was set to 1 min. During this time, 30 measurements were made, i.e., one every 2 s. The device used to measure the dust concentration was installed at a distance of 1 m from the grinding mill, at an employee head height, i.e., 1.5 m above the ground. Each time the device was set in the same place. The measurements were taken at the same distance from the milling device, thus simulating the typical position of personnel handling the grinder (Figure 1). Additionally, a control test was also made by measuring dust concentration in the room when grinding process was not taking place. In the discussed study, the PM4.0 was used instead of the typical value of PM2.5, taking into account a phenomenon typically occurring during the process of dynamic milling that is mutual collision of particles and resulting from it gradual formation of a certain quantity of finer particles [24–26].

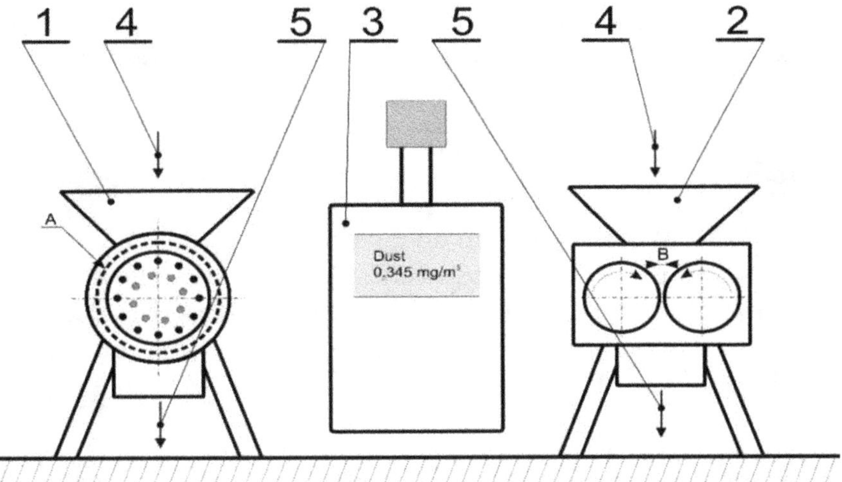

Figure 1. The layout of the dust measuring station: 1—hammer mill, 2—roller mill, 3—DustTrak II dust monitor, 4—inlet of the raw material prior to grinding, 5—outlet of the raw material after grinding, A—screen with interchangeable mesh with circular openings, B—adjustable working gap between the rollers.

The obtained wheat middlings were analyzed on sieves in order to determine the finest fraction, i.e., with the size smaller than 0.1 mm, as well as the average particle size. The study was performed on a Retsch sieve separator in accordance with PN-R-64798: 2009 standard [27].

Designations presented in Table 1 were used for descriptive purposes in this work.

Table 1. Description of the designations used in this work.

Designation	Description	Designation	Description
W9-h3	Grain moisture content of 9%, the diameter of openings in the screen of the mill equal to 3 mm	W9-h0.4	Grain moisture content of 9%, the working gap between rollers of the mill equal to 0.4 mm
W9-h5	Grain moisture content of 9%, the diameter of openings in the screen of the mill equal to 5 mm	W9-h0.7	Grain moisture content of 9%, the working gap between rollers of the mill equal to 0.7 mm
W9-h8	Grain moisture content of 9%, the diameter of openings in the screen of the mill equal to 8 mm	W9-h1	Grain moisture content of 9%, the working gap between rollers of the mill equal to 1 mm
W14-h3	Grain moisture content of 14%, the diameter of openings in the screen of the mill equal to 3 mm	W14-h0.4	Grain moisture content of 14%, the working gap between rollers of the mill equal to 0.4 mm
W14-h5	Grain moisture content of 14%, the diameter of openings in the screen of the mill equal to 5 mm	W14-h0.7	Grain moisture content of 14%, the working gap between rollers of the mill equal to 0.7 mm
W14-h8	Grain moisture content of 14%, the diameter of openings in the screen of the mill equal to 8 mm	W14-h1	Grain moisture content of 14%, the working gap between rollers of the mill equal to 1 mm
W18-h3	Grain moisture content of 18%, the diameter of openings in the screen of the mill equal to 3 mm	W18-h0.4	Grain moisture content of 18%, the working gap between rollers of the mill equal to 0.4 mm
W18-h5	Grain moisture content of 18%, the diameter of openings in the screen of the mill equal to 5 mm	W18-h0.7	Grain moisture content of 18%, the working gap between rollers of the mill equal to 0.7 mm
W18-h8	Grain moisture content of 18%, the diameter of openings in the screen of the mill equal to 8 mm	W18-h1	Grain moisture content of 18%, the working gap between rollers of the mill equal to 1 mm

3. Results and Discussion

Prior to the commencement of the research on the grinding process the average concentration of organic dust in the room was determined. The concentrations were as follows: For PM1.0—26 $\mu g \cdot m^{-3}$, for PM4.0—35 $\mu g \cdot m^{-3}$, and for PM10—53 $\mu g \cdot m^{-3}$. The same analysis was also performed 5 min after completion of the grinding process and turning off the machinery. The concentration of dust was as follows: For PM1.0—42 $\mu g \cdot m^{-3}$, for PM4.0—328 $\mu g \cdot m^{-3}$, and for PM10—101 $\mu g \cdot m^{-3}$.

The average values of organic dust concentration in the course of grinding with the hammer mill are shown in Figure 2, while Figure 3 presents the corresponding values in the case of the roller mill. In addition, in Tables 2–4 respectively, the maximum and minimum concentration of dust particulates during the grnding process is shown.

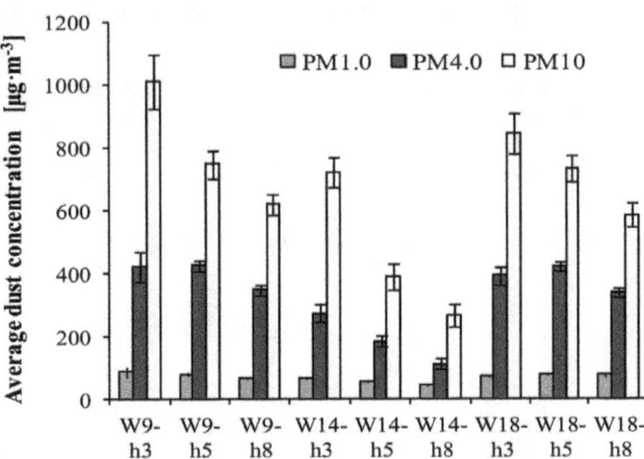

Figure 2. The distribution of organic dust concentration during grinding using the hammer mill in relation to the moisture content and the size of the openings in the screen.

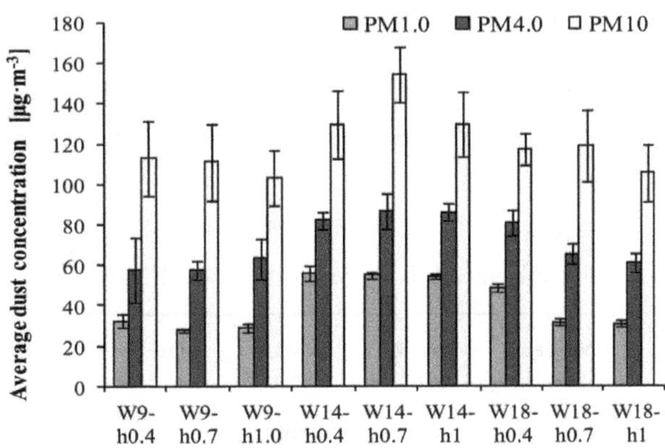

Figure 3. Distribution of organic dust concentration during grinding using the roller mill depending on the moisture content and the size of the working gap between the rollers.

In the case of the hammer mill, at a constant moisture content, a significant decrease of concentration of the organic particulate matter of the PM10 fraction was observed for increased diameter of screen openings. In the case of the roller mill, a similar trend can be observed for increased working gap, however, only for the raw material with the moisture content of 9%. It should be noted that in this case the trend is noticeable, though not confirmed statistically. Additionally, in the case of the hammer mill, for the moisture content of 9% and 14%, a tendency can be observed of lowering the concentration of PM1.0 fraction with the increase of the diameter of screen openings, although statistical analysis does not confirm the importance of these differences at 0.05 significance level. In the case of the roller mill, a similar relationship was noted for the raw material with the moisture content of 18% and increased size of the working gap between the rollers. For a particulate matter of the PM4.0 fraction produced in the course of grinding using the hammer mill, a decrease of dust concentration was observed for larger mesh openings in the case of moisture content of 14%. A similar tendency is noticeable for the roller mill for the increased size of the working gap in the case of the raw material with the moisture content of 18%.

Table 2. Statistical analysis of dust concentration PM1.0 during grinding.

Mill	Moisture Content of the Raw Material [%]	Screen [mm]	Max. [$\mu g \cdot m^{-3}$]	Min. [$\mu g \cdot m^{-3}$]	Dusts Average [$\mu g \cdot m^{-3}$]	Dusts—S.D.	Homogeneous Group
Hammer mill	-	Control	50	33	38.7	4.4	
	9	3	121	69	84	14.3	
		5	82	69	77.5	3.7	d
		8	77	59	65.3	4	c
	14	3	71	57	63.6	4.4	c
		5	66	51	55.9	2.9	b
		8	49	39	42.2	2	b
	18	3	77	63	69.4	3.6	
		5	83	72	77.7	2.6	d
		8	92	66	74	5.5	d
Roller mill	-	Control	39	25	27.1	3	a
	9	0.4	39	28	32.3	3.2	
		0.7	29	26	27.7	0.9	a
		1	36	26	28.6	2	a
	14	0.4	68	51	56.2	3.8	
		0.7	58	52	54.9	1.6	b
		1	56	52	54.2	1	b
	18	0.4	52	43	48.6	1.9	
		0.7	38	29	31.6	1.9	
		1	34	28	30.7	1.6	a

Table 3. Statistical analysis of dust concentration PM4.0 during grinding.

Mill	Moisture Content of the Raw Material [%]	Screen [mm]	Max. [$\mu g \cdot m^{-3}$]	Min. [$\mu g \cdot m^{-3}$]	Dusts Average [$\mu g \cdot m^{-3}$]	Dusts—S.D.	Homogeneous Group
Hammer mill	-	Control	140	106	120.4	7.5	b
	9	3	530	344	421.6	46.9	d
		5	463	404	423.6	15.5	d
		8	375	307	345.8	17.1	c
	14	3	354	234	272.2	26.7	
		5	213	156	183.1	15.6	
		8	131	81	110.7	15	b
	18	3	445	342	391.5	28.9	
		5	451	391	421	15	d
		8	359	296	335	13.6	c
Roller mill	-	Control	55	37	41.9	3.7	
	9	0.4	105	42	57.7	15.8	
		0.7	65	49	57.6	4.5	
		1	84	48	63.1	9.5	a
	14	0.4	90	75	82.3	4.1	
		0.7	104	74	86.8	8.7	
		1	94	78	86.2	3.9	
	18	0.4	94	68	81	6.1	
		0.7	75	56	65.4	5	
		1	69	53	61.3	4.6	a

Table 4. Statistical analysis of dust concentration PM10 during grinding.

Mill	Moisture Content of the Raw Material [%]	Screen [mm]	Max. [mg·m^{-3}]	Min. [mg·m^{-3}]	Dusts Average [mg·m^{-3}]	Dusts—S.D.	Homogeneous Group
Hammer mill	-	Control	294	158	200.3	30.4	
	9	3	1200	832	1011.9	87.7	
		5	837	670	746.6	43.8	c
		8	698	552	620.3	32.9	
	14	3	795	647	720.9	48.1	c
		5	460	297	387.2	40	
		8	314	184	262.8	35.7	
	18	3	1020	702	842.3	63.5	
		5	811	637	730.9	41	c
		8	667	525	583.2	39.4	
Roller mill	-	Control	86	54	68.8	8.8	a
	9	0.4	160	78	113.1	18.7	
		0.7	157	80	111	18.7	
		1	150	90	102.8	13.6	a,b
	14	0.4	165	106	129.3	16.9	b
		0.7	178	134	154.5	13.5	
		1	161	100	129.6	16.1	
	18	0.4	129	104	117.4	8	
		0.7	144	86	119	18	
		1	128	72	105.3	14	b

When grinding in the hammer mill, for each of the tested sizes of mesh openings in the screen, the highest value of organic dust concentration of PM10 fraction was measured for the grain with initial moisture content of 9%, while the lowest concentration characterized the grain with an initial moisture content of 14%. The highest value of concentration of the PM10 fraction was recorded for the moisture content of 9% and the screen openings size equal to 3 mm (1102 µg·m^{-3}), while the lowest concentration was measured for the moisture content of 14% and the screen openings size of 8 mm (263 µg·m^{-3}). In the case of the PM4.0 fraction, the highest concentrations of organic dust was observed for the raw material with the initial moisture content of 9% and 18%. It reached a similar value both for the openings with the diameter of 3 as well as 5 mm (approx. 420 µg·m^{-3}). For all tested initial moisture contents, the use of screens with the openings size of 8 mm resulted in the decrease of concentration of the organic dust by approx. 24% (to the level of approx. 340 µg·m^{-3}). In the case of the very fine particulate matter PM1.0, the highest value of concentration was recorded for the input material with moisture content of 9% and the screen openings size of 3 mm (84 µg·m^{-3}). For the entire range of tested moisture content of the input material, the lowest value of the PM1.0 fraction concentration was observed for grain with the moisture content of 14% (42 µg·m^{-3}). Using the roller mill reduced the amount of produced PM10, PM4.0, and PM1.0 dust as compared to the amount obtained during grinding of the raw material in the hammer mill. In this case, for all examined mill gaps between the rollers, the highest value of the PM10 fraction concentration was measured for grain with the input moisture content of 14%. The highest concentration was equal to 155 µg·m^{-3} for the gap of 0.7 mm, followed by the gaps of 0.4 and 1 mm (approximately 129 µg·m^{-3}). The lowest value of dust concentration of the PM10 fraction was recorded during grinding grain with the initial moisture content of 9% for the mill gap of 1 mm; this value was 103 µg·m^{-3}. When grinding grain using the roller mill the concentration of dust of the PM4.0 fraction was also the highest for the input material with the moisture content of 14% (87 µg·m^{-3} for the mill gap of 0.7 mm). In contrast, the lowest concentration of organic dust of the PM4.0 fraction was recorded during grinding grain with the moisture content of 9% (the lowest concentration of 57 µg·m^{-3} was measured for the mill gap of 0.7 mm). For the roller mill the highest concentration of the finest particulate matter of the PM1.0 fraction was recorded for the initial humidity of 14%, and mill gap of 0.4 mm (52 µg·m^{-3}). As in the above described variants, considering the analyzed mill gaps, also in the case of the finest particulate matter the lowest values of organic dust concentration were observed for grain with the initial moisture content of 9% (the lowest concentration equal to 28 $\mu g \cdot m^{-3}$ was measured for the gap of 0.7 mm). In the research by Dacarro [28], the average concentration of organic dust of 790 $\mu g \cdot m^{-3}$ was obtained during the grinding stage of grain processing in mills, while the highest concentration of inhaled dust was obtained at the cleaning stage, when it was equal to 2763 $\mu g \cdot m^{-3}$, with the concentration of dust of the smallest particulate size (respirable dust) was, respectively: 321 and 1400 $\mu g \cdot m^{-3}$.

The use of the roller mill, instead of the hammer mill in the process of grinding grain with moisture content of 9%, resulted in an average reduction of the PM10 fraction by approx. 724%, the PM4.0 fraction by approx. 671%, and the PM1.0 fraction by 256%, and these were the largest decreases registered among the analyzed moisture content levels. For grain with the moisture content of 14% which is characterized by the lowest dynamics of the decrease these values were, respectively: PM10—337%, PM4.0—223%, and PM1.0—98%. Moreover, in the case of the last one even an approx. 22% increase in the amount of PM1.0 fraction was recorded for the roller mill with the mill gap of 1 mm in comparison to the amount of dust produced by the hammer mill with the screen openings with the diameter of 8 mm.

Concentration of organic dust originating from the production of industrial fodder was determined in the research by Sobczak [29], when concentration of this type of dust was measured at various stages of production. The research showed that the highest average concentration was recorded during grinding grain, and for the PM10 fracture it was equal to 1250 $\mu g \cdot m^{-3}$, however, neither the machinery nor the grinding parameters were provided. In the study presented here, the maximum instantaneous concentration of the organic dust for the PM10 fraction was equal to 1200 $\mu g \cdot m^{-3}$, and it was recorded while grinding grain using the hammer mill with the screen openings with the diameter of 3 mm and input material moisture content of 9%. In addition, the research done by Karpaciński [30] confirms the high concentration of organic dust in grain processing plants. This study was performed in Canadian mills, where the determined concentration of dust was higher than 10,000 $\mu g \cdot m^{-3}$, which is significantly above the acceptable level. Concentration of organic dust in grain milling industry and herbal plants during the packing stage is much lower, and amounts approximately to 100 $\mu g \cdot m^{-3}$ for dust of the PM1.0 fraction [31,32].

Table 5 shows the average size of particles after the grinding process, and the share of the finest fraction (below 0.1 mm). The average size of particles depended on grain moisture content and the choice of grinding method. The highest amount of the fine fraction was obtained during grinding grain using the hammer mill with the screen of 3 mm mesh size and the material moisture content of 9%. In this case, the share of the finest fraction was equal to 3.98%. Analogously, when using the roller mill with the smallest tested mill gap, i.e., 0.4 mm, the amount of the finest dust fraction was equal to 2.78%. When grinding grain with higher moisture content the amount of the finest fraction decreased. The average size of particles obtained during grinding grew with the increase of grain moisture content. The largest average size of particles was obtained after grinding grain in the roller mill with the mill gap between the rollers of 1 mm and grain moisture content of 18%. For the hammer mill, the largest average size of particles was obtained when grinding grain with the moisture content of 18% and using the screen with the mesh having the diameter of the openings equal to 8 mm. A research was conducted analyzing the average particle size and production of fine fraction below 0.2 mm when using a roller mill for grinding hard and soft grain into flour. Sprouted grain with balanced moisture content of 12% was used as the soft grain [21]. The study showed that grinding sprouted grain produced greater amount of fine fraction, which might result from the bonding strength of starch molecules present in germinated material. In the study presented here, the change of moisture content, and therefore the change of grain hardness, had an opposite effect, which was caused by lack of enzymatic reactions that take place during germination.

Table 5. The average size of particles and the share of the finest fraction in ground middlings from hammer and roller mill.

Hammer mill	W9-h3	W9-h5	W9-h8	W14-h3	W14-h5	W14-h8	W18-h3	W18-h5	W18-h8
The share particles smaller than 0.1 mm [%]	3.98	1.08	0.6	1.99	0.51	0.31	1.78	0.34	0.06
The average size of particles [mm]	0.563	0.975	1.094	0.651	1.067	1.309	0.609	1.106	1.693
Roller mill	W9-h0.4	W9-h0.7	W9-h1	W14-h0.4	W14-h0.7	W14-h1	W18-h0.4	W18-h0.7	W18-h1
The share particles smaller than 0.1 mm [%]	2.78	1.57	1.7	1.26	0.46	0.24	0.38	0.22	0.14
The average size of particles [mm]	1.003	1.334	1.942	1.928	2.521	2.877	2.775	2.955	2.997

4. Conclusions

The average concentration of organic dust during wheat grinding depends on the type of the milling machinery and the moisture content of the input material. The highest concentration of organic dust was recorded when grinding the raw material with the lowest tested moisture content, i.e., 9%, which can be related to the hardness and at the same time brittleness of grain that during the grinding process breaks more easily into small fractions. An increase of grain moisture to 14% resulted in the reduction in particulate matter concentration (PM10) inhaled by employees when grinding grain using the hammer mill. An inverse relationship was obtained when grain was ground using the roller mill, which may be the result of different construction of working parts. For both the mills, the obtained concentration of the PM10 fraction dust was high, and in every case, it exceeded the acceptable level, which for most European countries is 100 $\mu g \cdot m^{-3}$ [33]. However, when comparing both types of the mills, statistically smaller concentration of organic dust was produced when grinding using the roller mill. The sieve analysis revealed that the amount of the finest fraction, i.e., below 0.1 mm, was smaller when grinding using the roller mill as compared with the hammer mill. The amount of the finest fraction depends on the size of the openings in the screen mesh, i.e., the larger the diameter of the holes in the screen, the smaller the amount of the fraction below 0.1 mm. Similar is the relationship with the width of the mill gap in the roller mill, i.e., the smaller the gap, the higher the share of the finest fraction. In some cases, instantaneous maximum concentration of organic dust during the process of grinding grain exceeded even ten times the accepted limit values. Having in mind the health safety of farmers, it is necessary to use dust protection equipment or to modify the grinding technology by changing the grain moisture content or choosing appropriate grinding parameters.

Author Contributions: All authors contributed equally to this paper.

References

1. Amantea, R.P.; Fortes, M.; Ferreira, W.R.; Santos, G.T. Energy and exergy efficiencies as design criteria for grain dryers. *Dry. Technol.* **2018**, *36*, 491–507. [CrossRef]
2. Lutfy, O.F.; Noor, S.B.M.; Abbas, K.A.; Marhaban, M.H. Some control strategies in agricultural grain driers: A review. *J. Food Agric. Environ.* **2008**, *6*, 74–85.
3. Niedziółka, I.; Szpryngiel, M.; Kachel-Jakubowska, M.; Kraszkiewicz, A.; Zawiślak, K.; Sobczak, P.; Nadulski, R. Assessment of the energetic and mechanical properties of pellets produced from agricultural biomass. *Renew. Energy* **2015**, *76*, 312–317. [CrossRef]
4. Kraszkiewicz, A.; Kachel-Jakubowska, M.; Niedziółka, I.; Zaklika, B.; Zawiślak, K.; Nadulski, R.; Sobczak, P.; Wojdalski, J.; Mruk, R. Impact of various kinds of straw and other raw materials on physical characteristics of pellets. *Rocz. Ochr. Środowiska* **2017**, *19*, 270–287.
5. Zawiślak, K. Przetwarzanie Ziarna Kukurydzy na cele Paszowe. In *Rozprawy Naukowe*; Akademia Rolnicza w

Lublinie: Lublin, Poland, 2006; pp. 1–95. Available online: http://yadda.icm.edu.pl/yadda/element/bwmeta1.element.agro-article-2e48e26a-bef0-41d8-b73f-ecd603e23ba6.

6. Mahajan, S.P. *Air Pollution Control Commonwealth of Learning*; Ramachandra, T.V., Ed.; Capital Publishing Company: New Delhi, India, 2006.
7. Kuskowska, K.; Dmochowski, D. Analiza rozkładu stężeń pyłu zawieszonego frakcji PM_{10}, $PM_{2.5}$ i $PM_{1.0}$ na różnych wysokościach Mostu Gdańskiego. *Zesz. Nauk. SGSP* **2016**, *53*, 101–119.
8. Frąk, M.; Majewski, G.; Zawistowska, K. Analysis of the quantity of microorganisms adsorbed on particulate matter PM_{10}. *Sci. Rev.–Eng. Environ. Sci.* **2014**, *64*, 140–149.
9. Warren, K.J.; Wyatt, T.A.; Romberger, D.J.; Ailts, I.; West, W.W.; Nelson, A.J.; Nordgren, T.M.; Staab, E.; Heires, A.J.; Poole, J.A. Post-Injury and Resolution Response to Repetitive Inhalation Exposure to Agricultural Organic Dust in Mice. *Safety* **2017**, *3*, 10. [CrossRef] [PubMed]
10. Żukiewicz-Sobczak, W.; Cholewa, G.; Krasowska, E.; Chmielewska-Badora, J.; Zwoliński, J.; Sobczak, P. Grain dust originating from organic and conventional farming as a potential source of biological agents causing respiratory diseases in farmers. *Postepy Dermatol. I Alergol.* **2013**, *6*, 358–364. [CrossRef]
11. Niculita-Hirzel, H.; Hantier, G.; Storti, F.; Plateel, G.; Roger, T. Frequent Occupational Exposure to Fusarium Mycotoxins of Workers in the Swiss Grain Industry. *Toxins* **2016**, *8*, 370. [CrossRef]
12. European Environment Agency. *EEA 2017. Air Quality of Europe*; EEA: København, Denmark, 2017; ISSN 1977-8449. Available online: https://www.eea.europa.eu/publications/air-quality-in-europe-2017. (accessed on 26 February 2019).
13. Klimat dla Polski – Polska dla klimatu: 1988-2018-2050. Available online: www.kobize.pl/pl/fileCategory/id/1/opracowania (accessed on 26 February 2019).
14. WHO. *WHO Air Quality Guidelines for Particulate Matter, Ozone, Nitrogen Dioxide and Sulfur Dioxide*; Global Update 2005; WHO: Geneva, Switzerland, 2006; Available online: http://whqlibdoc.who.int/hq/2006/WHO_SDE_PHE_OEH_06.02_eng.pdf (accessed on 26 February 2019).
15. Ashmore, M.R.; Dimitroulopoulou, C. Personal exposure of children to air pollution. *Atmos. Environ.* **2009**, *43*, 128–141. [CrossRef]
16. Chen, C.; Zhao, B. Review of relationship between indoor and outdoor particles: I/O ratio, infiltration factor and penetration factor. *Atmos. Environ.* **2011**, *45*, 275–288. [CrossRef]
17. Grahame, T.J.; Schlesinger, R.B. Evaluating the health risk from secondary sulfates in Eastern North American regional ambient air particulate matter. *Toxicology* **2005**, *17*, 15–27. [CrossRef] [PubMed]
18. Pawłowski, L. How heavy metals affect sustainable development. *Rocz. Ochr. Środowiska (Annu. Set Environ. Prot.)* **2011**, *13*, 51–64.
19. Zwoździak, A.; Sówka, I.; Fortuna, M.; Balińska-Miśkiewicz, W.; Willak-Janc, E.; Zwoździak, J. Wpływ stężeń pyłów (PM_1, $PM_{2.5}$, PM_{10}) w środowisku wewnątrz szkoły na wartości wskaźników spirometrycznych u dzieci. *Rocz. Ochr. Środowiska (Annu. Set Environ. Prot.)* **2013**, *15*, 2022–2038.
20. Żukiewicz-Sobczak, W.; Cholewa, G.; Krasowska, E.; Chmielewska-Badora, J.; Zwoliński, J.; Sobczak, P. Rye grains and the soil derived from under the organic and conventional rye crops as a potential source of biological agents causing respiratory diseases in farmers. *Postępy Dermatol. I Alergol.* **2013**, *6*, 373–380. [CrossRef] [PubMed]
21. Dziki, D.; Laskowski, J. Study to analyze the influence of sprouting of the wheat grain on the grinding process. *J. Food Eng.* **2010**, *96*, 562–567. [CrossRef]
22. Hameed Hassoon, W.; Dziki, D. The Study of Multistage Grinding of Rye. In Proceedings of the IX International Scientific Symposium; Farm Machinery and Processes Management in Sustainable Agriculture, Lublin, Poland, 22–24 November 2017. [CrossRef]
23. Polski Komitet Normalizacyjny. *PN-EN ISO 712:2009. Ziarno Zbóż i Przetwory Zbożowe. Oznaczanie Wilgotności*; Polski Komitet Normalizacyjny: Warszawa, Poland, 2009.
24. Shieh, J.Y.; Ku, C.H.; Christiani, D.C. Respiratory effects of the respirable dust ($PM_{4.0}$). *Epidemiology* **2004**, *15*, 4–166. [CrossRef]
25. De Lima Gondim, F.; Lima, Y.C.; Melo, P.O.; dos Santos, G.R.; Serra, D.S.; Araújo, R.S.; de Oliveira, M.L.M.; Lima, C.C.; Cavalcante, F.S.A. Exposure to $PM_{4.0}$ from the combustion of cashew nuts shell in the respiratory system of mice previously exposed to cigarette smoke. *Int. J. Recent Sci. Res.* **2017**, *8*, 16762–16769. [CrossRef]
26. Josino, J.B.; Serra, D.S.; Gomes, M.D.M.; Araújo, R.S.; de Oliveira, M.L.M.; Cavalcante, F.S.Á. Changes of

respiratory system in mice exposed to $PM_{4.0}$ or TSP from exhaust gases of combustion of cashew nut shell. *Environ. Toxicol. Pharmacol.* **2017**, *56*, 1–9. [CrossRef]

27. Polski Komitet Normalizacyjny. *PN-R-64798:2009. Pasze–Oznaczanie Rozdrobnienia*; Polski Komitet Normalizacyjny: Warszawa, Poland, 2009.
28. Dacarro, C.; Grisoli, P.; Del Frate, G.; Villani, S.; Grignani, E.; Cottica, D. Micro-organisms and dust exposure in an Italian grain mill. *J. Appl. Microbiol.* **2004**, *98*, 163–171. [CrossRef]
29. Sobczak, P.; Zawiślak, K.; Żukiewicz-Sobczak, W.; Wróblewska, P.; Adamczuk, P.; Mazur, J.; Kozak, M. Organic dust in feed industry. *Pol. J. Environ. Stud.* **2015**, *24*, 5–2177. [CrossRef]
30. Karpaciński, E.A. Exposure to inhalable flour dust in Canadian flour mills. *Appl. Occup. Environ. Hyg.* **2003**, *18*, 1022–1030. [CrossRef] [PubMed]
31. Buczaj, A. Poziom zapylenia w wybranych zakładach przemysłu zbożowego w województwie lubelskim. *Inżynieria Rol.* **2011**, *1*, 7–13.
32. Zawiślak, K.; Sobczak, P.; Kozak, M.; Mazur, J.; Panasiewicz, M.; Żukiewicz-Sobczak, W.; Wojdalski, J.; Mieszkalski, L. Microbiological analysis and concentration of organic dust in an herb processing plant. *Pol. J. Environ. Study* **2019**, *28*, 1–7. [CrossRef]
33. Directive 2008/50/EC of the European Parliament and of the Council of 21 May 2008 on ambient air quality and cleaner air for Europe OJ L. *Off. J. Eur. Union* **2008**, *152*, 1–44.

17

Studies of a Rotary–Centrifugal Grain Grinder using a Multifactorial Experimental Design Method

Andrzej Marczuk [1], Agata Blicharz-Kania [2,*], Petr A. Savinykh [3], Alexey Y. Isupov [3], Andrey V. Palichyn [4] and Ilya I. Ivanov [4]

1 Department of Agricultural, Forestry and Transport Machines, University of Life Sciences in Lublin, Głęboka 28, 20-612 Lublin, Poland; andrzej.marczuk@up.lublin.pl
2 Department of Biological Bases of Food and Feed Technologies, University of Life Sciences in Lublin, Głęboka 28, 20-612 Lublin, Poland
3 Federal Agricultural Research Centre of the North-East Named after N.V. Rudnitskiy, Kirov 610007, Russia; peter.savinykh@mail.ru (P.A.S.); isupoff.aleks@yandex.ru (A.Y.I.)
4 FSBEI HE Vologda State Dairy Farming Academy (DSFA) named after N.V. Vereshchagin, Vologda—Molochnoye 160555, Russia; zmij.hh@yandex.ru (A.V.P.); kadyichna@mail.ru (I.I.I.)
* Correspondence: agata.kania@up.lublin.pl

Abstract: A scientific and technical literature review on machines designed to grind fodder grain revealed that the existing designs of grinding machines—those based on destruction by impact, cutting, or chipping—have various drawbacks. Some disadvantages include high metal and energy intensity, an uneven particle size distribution of the ground (crushed) product, a high percentage of dust fraction, the rapid wear of work tools (units), and heating of the product. To eliminate most of the identified shortcomings, the design of a rotary–centrifugal grain grinder is proposed in this paper. The optimization of the grinder's working process was carried out using experimental design methodology. The following factors were studied: the grain material feed, rotor speed (rpm), opening of the separating surface, number of knives (blades) on the inner and outer rings, technical conditions of the knives (sharpened or unsharpened), and the presence of a special insert that is installed in the radial grooves of the distribution bowl. The optimization criteria were based on the amount of electricity consumed by and the performance of the rotary–centrifugal grain grinder. The quality of performance was quantified by the finished product, based on the percentage of particles larger than 3 mm in size. An analysis of the results of the multifactorial experiment allowed us to establish a relationship (interaction) between the factors and their influence on the optimization criteria, as well as to determine the most significant factors and to define further directions for the research of a centrifugal–rotary grain grinder. From our experimental results, we found that the grinder is underutilized in the selected range of factor variation. Furthermore, the number of knives installed at the second stage of the grinder, the gap (clearance) of the separating surface, and the technical condition of the knives are among the most important factors influencing the power consumption and the quality of the resulting product. A reduction in the number of knives at the first stage has a positive effect on all the selected optimization criteria; and by varying the factors in the selected range, it is possible to obtain a product corresponding to medium and coarse grinding.

Keywords: grain grinding; rotary–centrifugal grinder; construction optimization criteria

1. Introduction

In the cost structure of feed production, grinding represents a very labor- and energy-intensive process, constituting, according to various datasets, at least half of the total costs associated with the conservation, storage, and preparation of feed mixtures. Hence, implementing a sustainable design of tools and processes is becoming increasingly important for manufacturers [1]. There is an increasing use of concentrated feed in the composition of mixed fodder [2–4]. An important factor for improving the digestibility and ensuring the most complete extraction of the potential energy of the feed is the method of its grinding. To date, the most common method in agricultural production is grinding with hammer mills (disintegrators) [5–7]. However, modern requirements for the quality of the feed obtained in the grinding process require the minimization of metal and energy intensity.

The main approach for solving the problems discussed above is to improve existing machines and devices (units) by optimizing structural elements and the subsequent control of technological operating modes. Furthermore, one can develop and implement technical solutions based on new physical principles or by combining several previously known approaches. Thus, recently, the design of grinders has become of great practical interest in the field of grinding grain. It is possible to significantly reduce the specific energy consumption for the production of concentrated feed and obtain a product with a high uniformity of particle size by combining the grinding process with shearing and chipping [5,8–15].

Pushkarev A.S. and Fomin V.V. [3,16], in their research on the parameters of the work tools (units) of a centrifugal–rotary grain grinder, found that by modifying the curvilinear shape of the cutting pair, the specific energy intensity of the grinding process of grain is reduced; increasing the grinder's performance without changing its energy consumption. Further they found, the moisture content of the material being crushed has a significant effect on the specific energy intensity of the grinding process; an increase in the moisture content of grain, increases the energy consumption. However, the trend towards a decrease in energy intensity remains when curvilinear work tools are used. Finally, by taking into account the change in the optimal cutting angle of the material being processed as it moves in the channel of the work tool of a small-size centrifugal–rotary grinder, the grain size distribution (granulometric composition) is equalized, the dust fraction decreases, and there are no whole grains in the finished product [3,16].

Based on the analysis of Druzhynin R.A. and Ivanov V.V. [2,4] on the research of theorists and practitioners, it is recommended to use two-stage grinding for the manufacture of coarse and medium grinding products. Additionally, multi-stage crushers and grinders are more effective to obtain a fine grinding product. At the same time, the disadvantages of knife-type grinders compared to impact crushers are also noted [17]: intensive wear of the working parts (knives), a sharp decline in the quality of grinding as a result of wear of the working bodies (knives), and a higher specific energy consumption than crushers, with greater productivity.

Considering the existing industrial designs of grinders that have knives (blades) as work tools, it can be seen that they are used mainly in personal subsidiary farms, where high performance and drive power are not required, but a low price and simplicity of design are needed. Such crushers are widely represented on the market by the following manufacturers: Electromash, Greentechs, Bison, Kolos, Yarmash, Cyclone, Niva, Fermer, ZDN, etc. [17].The analysis of existing models of grinders on the market and scientific research on grinding and crushing suggests that the search continues for a method of grinding feed which improves the economic efficiency of the process, while simultaneously ensuring a high quality of the resulting product of concentrated feed.

The purpose of this study is to assess the structural elements and various operating modes of a rotary–centrifugal grinder according to key performance indicators. The task of the study is to determine the significance of factors (x_1–x_7) according to certain optimization criteria (y_1–y_3), allowing for more specific implementation in further research of rotary–centrifugal grain grinders.

2. Materials and Methods

2.1. *Design of a Rotary–Centrifugal Device for Grinding Grain*

Based on the analysis of the designs of crushers and grinders, we have developed and proposed the design of a rotary–centrifugal device for grinding grain [18]. The proposed device (Figure 1) consists of a stationary case or housing (1) with loading (input, 2) and output (3) nozzles. Two adjacent disks are coaxially mounted inside the housing (1): the upper (4) (Figure 2a) and the lower movable disks (5) (Figure 2b). On the work surface of the lower disk (5) there are annular protrusions (6), and on the work surface of the upper disk (4) there are installed knives (7), which are diamond-shaped with small cutting angles with respect to the large diagonals. The outer row of knives (8) forms a separating surface; thus, changing the angle of the knives makes it possible to continuously adjust the degree of grinding for the material. The lower disk (5) has grooves (9) in the radial direction which are made opposite in terms of angle to the direction of rotation of this disk. The lower disk (5) is mounted on the flange of the drive shaft (10) and is rotated by means of a pulley (11) mounted on it. The upper disk (4) is rigidly fixed to the stationary case (1). In the upper part of the stationary case, a receiving chamber (12) is installed, which is formed by vertical walls. The receiving chamber (12) communicates in its upper part with the loading nozzle (2) and is connected with the working chamber (14), which is the space between disks (4) and (5), through the radial windows (13).

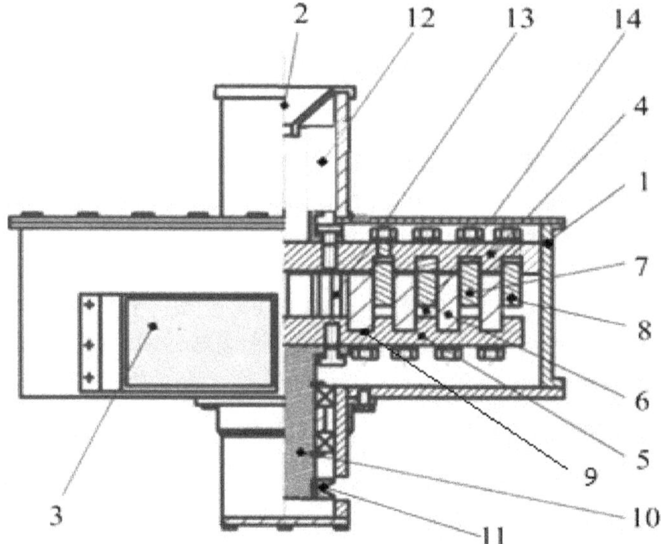

Figure 1. Design overview of a rotary–centrifugal device for grinding bulk materials. Parts include: 1—case; 2—loading (input); 3—output nozzles; 4—upper disk; 5—lower movable disk; 6—annular protrusions; 7—knives; 8—outer row of knives; 9—groove; 10—drive shaft; 11—pulley; 12—receiving chamber; 13—radial windows; and 14—working chamber.

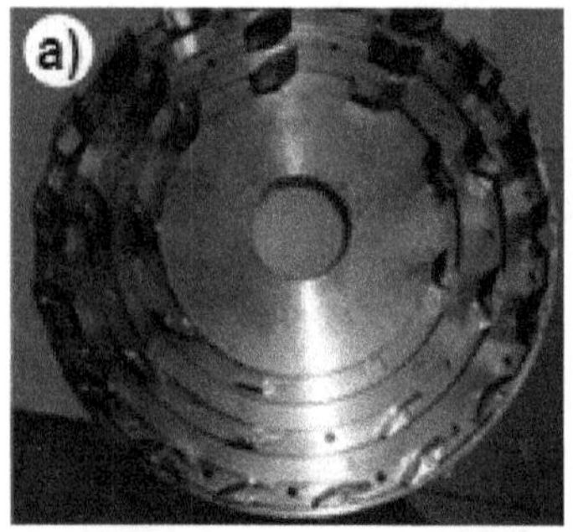

Figure 2. Photographs of the (**a**) upper disk and (**b**) lower disk.

The working process of grinding grain in the rotary–centrifugal device is carried out as follows (Figure 1). The incoming grain is subjected to the mechanical action of the first cutting pair; then, under the action of centrifugal forces, the pre-ground material moves along the grooves to the next pair. Then, the ground grains (groats or stock feed), having reached the outer row of the knives (8) which form the separating surface, pass into the gap between the knives (8) and, under the influence of the air flow created by the rotating lower disk (5), leave the housing (1) through the outlet nozzle (3).

The studies were carried out on an experimental installation of a rotary–centrifugal grinder (Figure 3) made on-site at the Federal State Budgetary Educational Institution of Higher Professional Education Vologda State Dairy Farming Academy (DSFA) named after N.V. Vereshchagin.

Figure 3. Overview of the experimental installation.

2.2. Description of a Multifactorial Experimental Design

To achieve the objectives of the study, the method of a multifactorial experimental design was applied. The analysis of the design of the grinder and a review of scientific and technical literature on grain grinding made it possible to identify a number of factors (Table 1) which allow us to most fully describe the process of grinding in the proposed centrifugal–rotary grinder. Below is a brief description of the selected factors (x_1–x_7) and how they vary:

- Grain was fed to the cumulative bunker of the experimental rotary–centrifugal installation with an auger conveyor and a frequency-controlled drive x_1, with a power inverter-controlled electric motor, Hyundai N700-220HF (Seul, South Korea);
- The rotation frequency of the lower disk (Figure 2b) x_2 was varied using a frequency-controlled drive with an electric motor controlled by an another Hyundai N700-220HF power inverter;
- The opening of the separating surface x_3 was adjusted by setting an appropriate size between the parallel planes of two adjacent knives, h_1 (Figure 5) on the outer row of knives (Figure 2a);
- The knives at the first x_4 and second x_5 stages of the upper disk (Figure 2a) were installed evenly, depending on the required quantity;
- To assess the impact of knife sharpening loss during operation, a factor of the technical condition of the knives x_6 was introduced as an experimental variable; i.e., "new" knives with a given angle of sharpening $\chi = 24°$ and "old" knives that have a much larger angle of sharpening (unsharpened knives); i.e., imitating their bluntness;
- The presence of inserts x_7 (Figure 4), installed in the slot of the distribution bowl of the lower disk (Figure 2b), was also considered. Eight inserts were installed in order to change the trajectory and rate of the material feed to the cutting pairs.

Table 1. Grinder experimental factors and the levels of their variations.

Factors	x_1 (kg·s^{-1})	x_2 (min^{-1})	x_3 (mm)	x_4 (pcs)	x_5 (pcs)	x_6	x_7
−1	0.023	800	2.5	9	18	"Old" ones	Present
+1	0.038	1200	3.2	3	9	"New" ones	Absent

Figure 4. Geometry of the inserts with appropriate dimensions labeled.

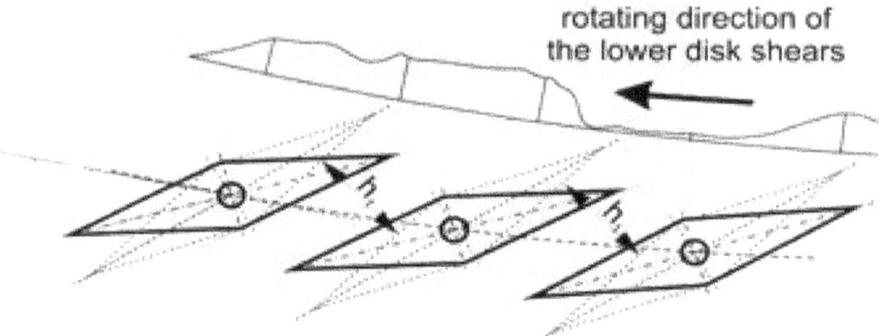

Figure 5. Geometry of the opening of the separating surface. h_1—size between the parallel planes of two adjacent knives.

Barley with 14% moisture content was used as the ground material. To reduce the amount of research, a type 2^{7-2} matrix of the fractional factor experiment was used.

The optimization criteria included the power consumption y_1 (kW) and the grinder performance (capacity) y_2 (kg/s). The measurement of power consumption was carried out using a Mercury 221 electrical energy meter connected to a PC via USB in CAN/RS-232/RS485 (Figure 6). The indicators were monitored instantly using a K-505 measurement kit.

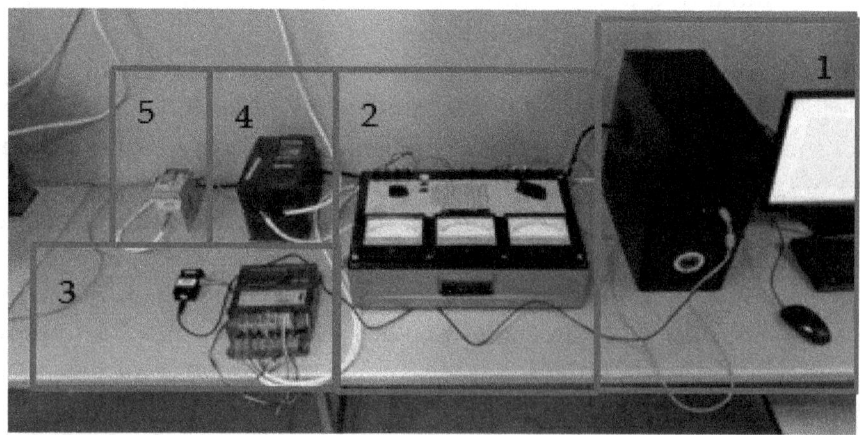

Figure 6. A set of measuring equipment used for the research. 1—personal computer; 2—K-505; 3—Mercury 221 electric energy meter with USB-CAN/RS-232/RS485 adapter; 4—frequency converter Hyundai N700-220HF; 5—circuit breaker

The performance (capacity) of the experimental installation y_2 was determined by monitoring the mass of the ground material per unit of time under a steady operating mode.

Zoo-technical requirements for mixed fodder concentrate according to GOST 9268-2015, GOST R 51550-2000, and other guidelines highlight a number of criteria for assessing the quality of the resulting product. However, the criteria directly dependent on the design and operating mode of the grain grinder. Some criteria are the percentage of particles greater than 3 mm y_3 after grinding, the grinding coarseness y_4, and the presence of whole grains in the grinding results y_5.

3. Results and Discussion

The results of the sieve analysis reveal that the most critical parameter for the compliance of the finished product with the zoo-technical requirements is the percentage of particles greater than 3 mm, y_3, which in our results is up to 60%. Screen sizing also shows the complete absence of whole grains y_5 in all samples. Calculations of the grain size (grinding coarseness) y_4 shows that grains are within the range of 1.5–3.2 mm. However, the use of this indicator as an optimization criterion at this stage is not informative, since the grain size depends largely on the percentage of particles greater than 3 mm. Thus, in this study, the most significant criterion for assessing the quality of the product is the percentage of particles greater than 3 mm, y_3. The critical particle size in different experiments depends on many factors (type of material, intended use, and subsequent processing) [19,20].

The processing of the results of the multifactorial experiment was carried out using the StatGraphics software package. A multivariate analysis of variance ANOVA (Table 2) and the regression results in Equations (1)–(3) show that the models obtained are statistically significant and describe the current processes with a reliability of at least 95%.

$$y_1 = 3.89 + 0.12x_1 + 0.2x_2 - 0.15x_3 + 0.3x_5 - 1.3x_6 + 0.28x_1x_3 + 0.12x_1x_5 - 0.28x_2x_4 + \\ +0.14x_3x_6 - 0.18x_3x_7 + 0.12x_4x_7 \quad (1)$$

$$y_2 = 0.021 + 0.004x_1 + 0.003x_2 - 0.001x_3 + 0.002x_6 + 0.002x_1x_2 - 0.001x_1x_3 + \\ +0.001x_2x_3 + 0.001x_2x_4 \quad (2)$$

$$y_3 = 13.1 + 1.71x_1 - 3.38x_2 - 5.6x_3 + 2.4x_4 + 5.75x_5 + 8.88x_6 - 2.6x_7 + 3.01x_1x_4 - \\ -1.68x_2x_6 - 8.16x_3x_6 + 3.2x_3x_7 \quad (3)$$

Table 2. Results of the model's analysis of variance (ANOVA).

	y_1	y_2	y_3
p-value	<0.0001	<0.0001	<0.0001
Error d.f.	20	18	20
Standard Error	0.464976	0.001784	4.21375
R-squared	0.9426	0.9564	0.9589

The search for a compromise solution aimed at reducing the amount of energy consumed y_1 and the percentage of particles exceeding 3 mm y_3, while increasing the performance (capacity) y_2 of the centrifugal–rotary grain grinder, was carried out using the StatGraphics software package (Table 3). The optimal values of the factors, obtained for the compromise solution in the selected range of factor variation, are presented in Table 4.

Table 3. Optimization criteria for the three factors, calculated using the StatGraphics software package.

Optimization Factor	Goal	Sensitivity	Lower Level	Upper Level	Average Predicted Value	Lower 95.0% Limit	Upper 95.0% Limit	Goal Achieved
y_1 (kW)	Minimize	Medium	-	-	2.59	2.00	3.18	0.76
y_2 (kg·s^{-1})	Maximize	Medium	-	-	0.032	0.0303	0.034	0.86
y_3 (%)	Minimize	Medium	0.0	10.0	−0.000022	−4.91	4.91	1.0

Table 4. Optimal values of the factors obtained using the StatGraphics software package.

Factor	x_1	x_2	x_3	x_4	x_5	x_6	x_7
Specified value	1	1	1	0.057	−1	1	1
Actual value	0.038 kg·s^{-1}	1200 min^{-1}	3.2 mm	6 pcs	9 pcs	"new" knives	no insert

The conducted analysis of the influence of factors on electricity consumption (1), and the performance of the centrifugal–rotary grinder (2), as well as the content of particles larger than 3 mm after grinding (3) reveals the following:

One of the most significant factors influencing the performance of the grinder y_2 is the feed of grain x_1 (Figure 7d). At the same time, an increase in the grain feed x_1 leads to an increase in all indicators y_1, y_2, and y_3. This is a consequence of an increased volume of material being transported by the lower disk of the grinder and an increased speed of transportation through the centrifugal–rotary grinder.

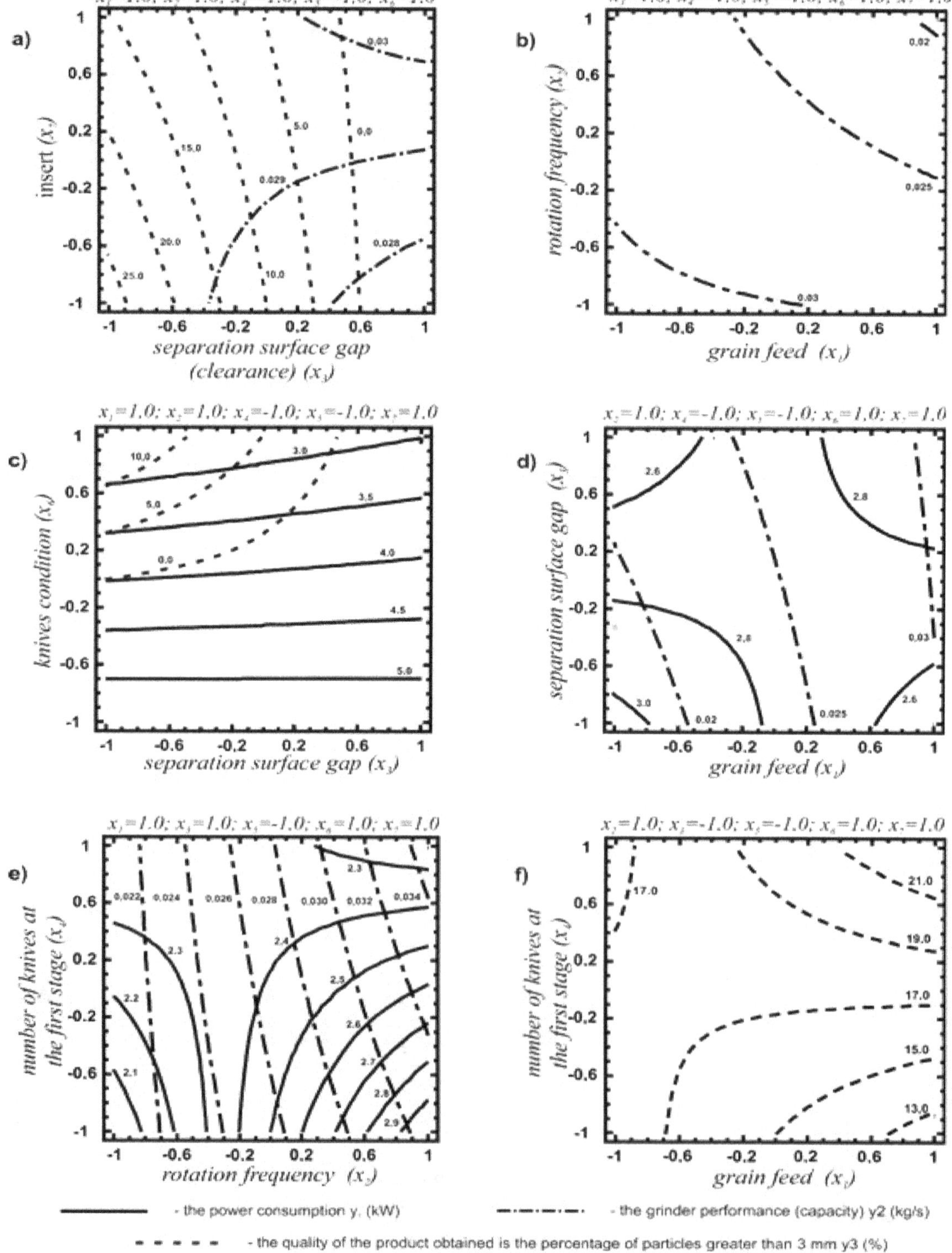

Figure 7. Two-dimensional sections of the response surface.

The increased rotation frequency of the lower disk x_2 leads to a directly proportional increase in the linear velocity of the work units, $v = 2\pi nR$ (n—rotation frequency, R—disc radius), at all stages, and an increase in the centrifugal inertial force. $F = 8m\pi^2n^2R$ (m—mass, n—rotation frequency, R—disc radius). As a result, the number of contacts of the grain material with the work units of the grinder is increased, while the time of its exposure in the work area is decreased. Thus, with increased values of the factor x_2, a decrease in the percentage of particles larger than 3 mm is observed with a simultaneous increase in the performance of the centrifugal–rotary grinder y_2. However, this leads to increased work for grinding and transporting the material, which is reflected in the total power consumption y_1. Bitra [21] notes that increasing speed affects the effective specific energy of hammers in different ways, depending on the type of raw material. In our experiments, the effective specific energy increases at certain rate and then decreases. Similar dependencies were shown by Moiceanu et al. [6].

Reducing the gap (clearance) of the separating surface x_3 naturally reduces the grinder performance y_2 and increases the power consumption y_1. As a result, there is an increase in the time required for the removal of the ground material from the work unit, as well as an increase in the mass of the transported ground material along the separating surface of the grinder. At the same time, reducing the gap of the separating surface x_3 does not lead to a logical decrease in the percentage of particles larger than 3 mm y_3. This phenomenon is explained by the fact that when there is a small gap of the separating surface, the particles of the material are mostly reflected from this "smooth" surface and then move along it until they pass through it. Likewise, a larger gap of the separating surface h (Figure 5) has an effect on the third grinding stage, since the knives of the separating surface have a larger approach angle and, as a result, have a lower reflectivity.

Changing the number of knives at the first grinding stage x_4 does not have a significant effect on the power consumption y_1 and grinder capacity y_2. An analysis of Equation (3), characterizing the quality of the product obtained, shows that reducing the number of knives from 9 to 3 at the first stage x_4 allows for a reduction of the percentage of particles over 3 mm in the finished product y_3 to 4.8%.

Reducing the number of knives from 18 to 9 at the second stage x_5 contributes to a reduction of particles over 3 mm y_3 in grinding by more than 11% and a reduction of the power consumption y_1 by 0.6 kW. This is understood as a result of an increase in the speed of transportation of the grain material through the grinder and a decrease in the mass accelerated by the lower disk.

Replacing the "old" knives x_6 with "new" ones is the main factor affecting energy consumption, and saves at least 2.6 kW of electricity. This can be explained by the fact that with the use of "old" knives, grinding takes place mainly by impact rather than cutting, resulting in a decrease in the number of particles larger than 3 mm y_3. At the same time, the performance (capacity) y_2 is reduced and the power consumption of the bulk grinder y_1 is increased. Kováč et al. [22] noticed that the basic factors affecting the properties of the ground material and the unit energy consumption in milling process are, among others, the knife angle and number of knives.

The presence of a special insert x_7 installed in the distribution bowls of the lower disk of the centrifugal–rotary grinder does not have any significant effect on the power consumption y_1 and the performance y_2. However, its installation increases the content of particles larger than 3 mm y_3 by 5.2%.

An analysis of the interactions of factors in the regressions, Equations (1)–(3), was carried out using (among others) two-dimensional sections (Figure 7), and shows the following results:

In general, installing a special insert x_7 into the distribution bowl of the grinder, in conjunction with the opening of the separating surface x_3, has negative effects on the percentage of particles larger than 3 mm y_3, and on the performance of the centrifugal–rotary grinder y_2. Thus, at a minimum gap (clearance) of the separating surface $x_3 = -1$, when the insert is installed, the percentage of particles larger than 3 mm is more than 25% and without the insert this value is 15%. However, when the gap is $x_3 = 0.6$ (h = 30.5 mm), particles larger than 3 mm are not observed in any case. Also, an increase in the gap (clearance) of the separating surface x_3 with the insert x_7 installed in the distribution bowl leads to a drop in the performance of the centrifugal–rotary grinder by 0.025 kg·s^{-1}, whereas without this insert the capacity increases by 0.008 kg^{-1} (Figure 7a);

The interaction of the factors x_1 and x_2 is most significant for regulating the performance of the grinder (y_2); reducing the feed and rotation frequency of the lower disk of the grinder leads to increased performance (Figure 7b);

The interaction of the factors of the condition of knives x_6 and the gap (clearance) of the separating surface x_3 is most significant for the percentage of particles larger than 3 mm (y_3). At the same time, regardless of the angle of sharpening of the knives, the number of particles larger than 3 mm y_3 decreases with an increasing gap of the separating surface. The deterioration of the sharpening of the knives leads to a twofold increase in power consumption y_1, which is more than 5 kW (Figure 7c). Branco et al. [23] suggested that knives should be replaced or sharpened periodically to ensure high efficiency in grinding.

Some of the determining parameters that affects the power consumption y_1 during grinding are the interactions of factors x_1 and x_3, as well as x_2 and x_4. When solving a compromise problem of increasing the grinder performance y_2 and reducing its power consumption y_1, it is necessary to increase the values of the factors x_3, x_2, and x_4, while decreasing the value of x_1 (Figure 7d,e).

With an increase in the feed x_1 and an increase in the number of knives at the first grinding stage x_4, the percentage of particles larger than 3 mm y_3 increases. As a result, the number of knives should be minimized (Figure 7f).

The search for a compromise solution was made using the StatGraphics software package, with equal significance of optimization criteria with the desired result shown in Table 4. The results of the optimization are shown in Table 4. These results suggest the compromised optimum is achieved by selecting the maximum values of the grain feed x_1 (in the selected range of factor variation, Table 1), the rotation frequency x_2, and opening of the separating surface x_3. Furthermore, "new" knives x_6 should be used without inserts in the distribution bowl x_7, where nine knives at the second stage x_5, and six at the first stage x_4 (Table 4) should be employed. Under these optimum conditions, it is possible to achieve a power consumption y_1 of 2.59 kW, a grinder capacity y_2 of 0.032 kg^{-1}, and the complete absence of particles larger than 3 mm y_3 after grinding. At the same time, the obtained values (Table 4) correspond in general to only 87.8% of the desired results, which indicates the insufficiency of the selected ranges for the factors to achieve the most optimal values of the power consumption y_1, performance (capacity) y_2, and quality of the resulting product, in terms of the content of particles exceeding 3 mm in size y_3.

4. Conclusions

The analysis of the results of this study (using the method of a multifactorial experimental design) into the operation of a centrifugal–rotary grinder suggests that the grinder is underutilized in the selected range of factor variation. The installation of special inserts in the distribution bowl of the lower disk (x_7) generally has a negative impact on the quality of the resulting product in terms of the content of particles larger than 3 mm. The number of knives installed at the second stage of the grinder (x_5), the gap (clearance) of the separating surface (x_3), and the technical condition of the knives (x_6) are among the most important factors influencing the power consumption and the quality of the resulting product. A reduction in the number of knives at the first stage (x_4) has a positive effect on all the selected optimization criteria. Finally, by varying the factors in the selected range, it is possible to obtain a product corresponding to medium and coarse grinding.

Summarizing the results of the study, we can conclude that, in further research the material feed (x_1) and rotor speed (x_2) should be increased, and the range of variation of the opening of the separating surface (x_3) should be extended. Further, the installation of a special insert (x_7) in the distribution bowl of the lower disk should be abandoned, or additional research related to changes in its shape and size should be carried out. Finally, the number of knives at the first stage (x_4) and at the second stage (x_5) should be reduced, and "new" knives (x_6) should be used in all cases.

Author Contributions: Conceptualization, A.M.; Formal analysis, A.M. and A.B.-K.; Funding acquisition, A.M.; Investigation, P.A.S., A.Y.I., A.V.P. and I.I.I.; Methodology, P.A.S., A.Y.I., A.V.P. and I.I.I.; Writing—original draft, P.A.S., A.Y.I., A.V.P. and I.I.I.; Writing—review & editing, A.M. and A.B.-K.

Nomenclature

h_1—size between the parallel planes of two adjacent knives (mm); x_1—grain feed (kg·s^{-1}); x_2—rotation frequency of the lower disk (min^{-1}); x_3—separation surface gap (mm); x_4—number of knives at the first stages (pcs); x_5—number of knives at the second stages (pcs); x_6—knives condition ("Old" ones/"New" ones); x_7—presence of inserts (present/absent); y_1—power consumption (kW); y_2—grinder performance (capacity) (kg·s^{-1}); y_3—particles greater than 3 mm (%); v—linear velocity (m·s^{-1}); n—rotation frequency (s^{-1}); R—disc radius (m); F—centrifugal inertial force (N); m—mass (kg).

References

1. Linke, B.S.; Dornfeld, D.A. Application of axiomatic design principles to identify more sustainable strategies for grinding. *J. Manuf. Syst.* **2012**, *31*, 412–419. [CrossRef]
2. Druzhynin, R.A. Improving the Working Process of an Impact Centrifugal Grinding Machine. Ph.D. Thesis in Engineering Science, Voronezh State University, Voronezh, Russia, 2014; p. 169.
3. Fomin, V.V. Reducing the Energy Consumption and Improving the Uniformity of Grain Grinding in A Small-Size Centrifugal-Rotary Grinder. Abstract of a Ph.D. Dissertation, National Agricultural University FSEI HPE, Novosibirsk, Russia, 2010; p. 23.
4. Ivanov, V.V. Improving the Operating Modes of a Disk Grinder of Feed Grains. Abstract of a Ph.D. Dissertation, FSBEI HPE "Don State University", Rostov-on-Don, Russia, 2014; p. 132.
5. Ghorbani, Z.; Masoumi, A.; Hemmat, A. Specific energy consumption for reducing the size of alfalfa chops using a hammer mill. *Biosyst. Eng.* **2010**, *105*, 34–40. [CrossRef]
6. Moiceanu, G.; Paraschiv, G.; Voicu, G.; Dinca, M.; Negoita, O.; Chitoiu, M.; Tudor, P. Energy Consumption at Size Reduction of Lignocellulose Biomass for Bioenergy. *Sustainability* **2019**, *11*, 2477. [CrossRef]
7. Svihus, B.; Kløvstad, K.H.; Perez, V.; Zimonja, O.; Sahlström, S.; Schüller, R.B.; Prestløkken, E. Physical and nutritional effects of pelleting of broiler chicken diets made from wheat ground to different coarsenesses by the use of roller mill and hammer mill. *Anim. Feed Sci. Technol.* **2004**, *117*, 281–293. [CrossRef]
8. Ahmad, F.; Weimin, D.; Qishou, D.; Rehim, A.; Jabran, K. Comparative Performance of Various Disc-Type Furrow Openers in No-Till Paddy Field Conditions. *Sustainability* **2017**, *9*, 1143. [CrossRef]
9. Bulatov, S.Y.; Nechayev, V.N.; Savinykh, P.A. The development of a grain crusher for farm households, and the results of research on the optimization of its structural and technological pa-rameters. In *Theory, Development, Methods, Experiment, Analysis, Monograph*; Nizhniy Novgorod State Engineering and Economic Institute: Knyaginino, Russia, 2014; p. 156.
10. Marczuk, A.; Misztal, W.; Savinykh, P.; Turbanov, N.; Isupov, A.; Zyryanov, D. Improving efficiency of horizontal ribbon mixer by optimizing its constructional and operational parameters. *Ekspolatacja I Niezawodn. Maint. Reliab.* **2019**, *21*, 220–225. [CrossRef]
11. Sukhlyayev, V.A.; Molin, A.A.; Mezlyakov, I.N. *Device for Grinding Bulk Materials*; No 146644 RF 20-10-2014. IPC B02C 13/00; Bulletin No 29; Russian Federation: Moscow, Russia, 2014.
12. Sysuev, W.A.; Aleškin, A.V.; Savinyh, P.A.; Marczuk, A.; Wrotkowski, K.; Misztal, W. *Badania Rozwojowo-optymalizacyjne Urządzeń Do Obróbki Ziarna Zbóż I Pasz Objętościowych O Podwyższonej Wilgotności*; Towarzystwo Wydawnictw Naukowych Libropolis: Lublin, Poland, 2017; p. 156.
13. Sysuev, V.A.; Aleškin, A.V.; Savinyh, P.A.; Marczuk, A.; Wrotkowski, K.; Misztal, W. *Badanie Mobilnych Rozdrabniaczy Oraz Rozdrabniaczy-Mieszarek Pasz*; Towarzystwo Wydawnictw Naukowych Libropolis: Lublin, Poland, 2016; p. 103.
14. Sysuyev, V.A.; Aleshkin, A.V.; Savinykh, P.A. Feed-processing machines. In *Theory, Development, Experiment*; North-Eastern Zonal Agricultural Research & Development Institute: Kirov, Russia, 2008; Volume I, p. 640.
15. Yancey, N.; Wright, C.T.; Westover, T.L. Optimizing hammer mill performance through screen selection and hammer design. *Biofuels* **2013**, *4*, 85–94. [CrossRef]
16. Pushkarev, A.S. Using Work Units with Curvilinear-Shaped Cutting Elements. Abstract of a Ph.D. Dissertation, FSBEI HE Altai State Technical University named after I.I. Polzunova, Barnaul, Altai Krai, Russia, 2018; p. 22.

17. Mironov, K.Y. Improving the Efficiency of Grain Grinding Process with the Justification of the Parameters of Work Units of an Impeller Impact Crusher. Abstract of a Ph.D. Dissertation, GBOU VO "Nizhny Novgorod State Engineering and Economic University", Nizhny Novgorod, Russia, 2018; p. 142.
18. Savinykh, P.A. Device for grinding bulk materials. No 2656619 Russia, 06-06-2018. Solntsev, R.V. Centrifugal grain grinder. *Bull. Altai Agrar. Uni-Versity* **2010**, *4*, 76–80.
19. Murphy, A.; Collins, C.; Phillpotts, A.; Bunyan, A.; Henman, D. Influence of hammermill screen size and grain source (wheat or sorghum) on the growth performance of male grower pigs. In *Report Prepared for the Co-Operative Research Centre for An Internationally Competitive Pork Industry*; Project IB-107; Pork CRC: Willaston, SA, Australia, 2009.
20. Vidal, B.C.; Dien, B.S.; Ting, K.C.; Singh, V. Influence of Feedstock Particle Size on Lignocellulose Conversion—A Review. *Appl. Biochem. Biotechnol.* **2011**, *164*, 1405–1421. [CrossRef] [PubMed]
21. Bitra, V.S.; Womac, A.R.; Chevanan, N.; Miu, P.I.; Igathinathane, C.; Sokhansanj, S.; Smith, D.R.; Cannayen, I. Direct mechanical energy measures of hammer mill comminution of switchgrass, wheat straw, and corn stover and analysis of their particle size distributions. *Powder Technol.* **2009**, *193*, 32–45. [CrossRef]
22. Kovač, J.; Krilek, J.; Mikleš, M. Energy consumption of chipper coupled to a universal wheel skidder in the process of chipping wood. *J. For. Sci.* **2011**, *57*, 34–40. [CrossRef]
23. Branco, F.P.; Naka, M.H.; Cereda, M.P. Granulometry and Energy consumption as indicators of disintegration efficiency in a hammer mill adapted to extracting arrowroot starch (*Maranta arundinacea*) in comparison to starch extraction from cassava. *Eng. Agrícola* **2019**, *39*, 341–349. [CrossRef]

PERMISSIONS

All chapters in this book were first published in MDPI; hereby published with permission under the Creative Commons Attribution License or equivalent. Every chapter published in this book has been scrutinized by our experts. Their significance has been extensively debated. The topics covered herein carry significant findings which will fuel the growth of the discipline. They may even be implemented as practical applications or may be referred to as a beginning point for another development.

The contributors of this book come from diverse backgrounds, making this book a truly international effort. This book will bring forth new frontiers with its revolutionizing research information and detailed analysis of the nascent developments around the world.

We would like to thank all the contributing authors for lending their expertise to make the book truly unique. They have played a crucial role in the development of this book. Without their invaluable contributions this book wouldn't have been possible. They have made vital efforts to compile up to date information on the varied aspects of this subject to make this book a valuable addition to the collection of many professionals and students.

This book was conceptualized with the vision of imparting up-to-date information and advanced data in this field. To ensure the same, a matchless editorial board was set up. Every individual on the board went through rigorous rounds of assessment to prove their worth. After which they invested a large part of their time researching and compiling the most relevant data for our readers.

The editorial board has been involved in producing this book since its inception. They have spent rigorous hours researching and exploring the diverse topics which have resulted in the successful publishing of this book. They have passed on their knowledge of decades through this book. To expedite this challenging task, the publisher supported the team at every step. A small team of assistant editors was also appointed to further simplify the editing procedure and attain best results for the readers.

Apart from the editorial board, the designing team has also invested a significant amount of their time in understanding the subject and creating the most relevant covers. They scrutinized every image to scout for the most suitable representation of the subject and create an appropriate cover for the book.

The publishing team has been an ardent support to the editorial, designing and production team. Their endless efforts to recruit the best for this project, has resulted in the accomplishment of this book. They are a veteran in the field of academics and their pool of knowledge is as vast as their experience in printing. Their expertise and guidance has proved useful at every step. Their uncompromising quality standards have made this book an exceptional effort. Their encouragement from time to time has been an inspiration for everyone.

The publisher and the editorial board hope that this book will prove to be a valuable piece of knowledge for researchers, students, practitioners and scholars across the globe.

LIST OF CONTRIBUTORS

Elías Arilla, Marta Igual, Javier Martínez-Monzó and Purificación García-Segovia
Food Investigation and Innovation Group, Food Technology Department, Universitat Politècnica de València, Camino de Vera s/n, 46022 Valencia, Spain

Pilar Codoñer-Franch
Department of Pediatrics Obstetrics and Gynecology, University of Valencia, Avenida de Blasco Ibáñez, No. 15, 46010 Valencia, Spain
Department of Pediatrics, University Hospital Dr. Peset, Foundation for the Promotion of Health and Biomedical Research in the Valencian Region (FISABIO), Avenida Gaspar Aguilar, No. 90, 46017 Valencia, Spain

Ellis Skinner, Marin Tuleu and Adam Hawkes
Department of Chemical Engineering, Imperial College London, London SW7 2AZ, UK

Iván García Kerdan
Department of Chemical Engineering, Imperial College London, London SW7 2AZ, UK
Department of the Built Environment, School of Design, University of Greenwich, London SE10 9LS, UK
Instituto de Ingeniería, Universidad Nacional Autónoma de México, Mexico City 04510, Mexico

Sara Giarola
Department of Earth Science & Engineering, Imperial College London, London SW7 2AZ, UK

Paweł Satora, Magdalena Skotniczny and Szymon Strnad
Department of Fermentation Technology and Microbiology, Faculty of Food Technology, University of Agriculture in Krakow, Balicka 122, 30-149 Krakow, Poland

Katarína Ženišová
Department of Microbiology, Molecular Biology and Biotechnology, National Agricultural and Food Centre, Food Research Institute, Priemyselna 4, Bratislava, Slovakia

Gürel Soyer
11th Regional Directorate, General Directorate of State Hydraulic Works, Ministry of Agriculture and Forestry, 22100 Edirne, Turkey

Ersel Yilmaz
Department of Biosystems Engineering, Aydın Adnan Menderes University, 09020 Aydın, Turkey

Katarzyna Kozłowicz, Sybilla Nazarewicz, Dariusz Góral and Marek Domin
Department of Biological Bases of Food and Feed Technologies, University of Life Sciences in Lublin, Gł,eboka 28, 20-612 Lublin, Poland

Anna Krawczuk
Department of Machinery Exploitation and Management of Production Processes, University of Life Sciences in Lublin, Akademicka 13, 20-950 Lublin, Poland

Gabriela Malik
Higher School of Economics and Computer Science in Krakow, Kraków 31-510, Poland

Danuta Kajrunajtys
Department of International Management, Cracow University of Economics, Krakow 31-510, Poland

Agnieszka Szparaga and Bartosz Płóciennik
Department of Agrobiotechnology, Koszalin University of Technology, Racławicka 15–17, 75-620 Koszalin, Poland

Sylwester Tabor
Department of Production Engineering, Logistics and Applied Computer Science, Agricultural University Krakow, Balicka 116B, 30-149 Krakow, Poland

Maciej Kuboń
Department of Production Engineering, Logistics and Applied Computer Science, Agricultural University Krakow, Balicka 116B, 30-149 Krakow, Poland
Institute of Agricultural Engineering and Informatics, University of Agriculture in Krakow, 116b Balicka St., 30-149 Krakow, Poland

Sławomir Kocira
Department of Machinery Exploitation and Management of Production Processes, University of Life Sciences in Lublin, Akademicka 13, 20-950 Lublin, Poland

Ewa Czerwińska
Department of Biomedical Engineering, Koszalin University of Technology, ´Sniadeckich 2, 75-453 Koszalin, Poland

Pavol Findura
Department of Machines and Production Biosystems, Slovak University of Agriculture in Nitra, Tr. A. Hlinku 2, 949 76 Nitra, Slovakia

Andrzej Anders, Dariusz Choszcz, Piotr Markowski, Adam Józef Lipiński, Zdzisław Kaliniewicz and Elwira Ślesicka
Department of Heavy Duty Machines and Research Methodology, University of Warmia and Mazury in Olsztyn, Olsztyn 10-957, Poland

Paweł A. Kluza and Izabela Kuna-Broniowska
Department of Applied Mathematics and Computer Science, University of Life Sciences in Lublin, 20-612 Lublin, Poland

Stanisław Parafiniuk
Department of Machinery Exploitation and Management of Production Processes, University of Life Sciences in Lublin, 20-612 Lublin, Poland

Andrzej Marczuk and Wojciech Misztal
Department of Agricultural, Forestry and Transport Machines, Faculty of Production Engineering, University of Life Sciences in Lublin, 28 Gleboka Street, 20-612 Lublin, Poland

Sergey Bulatov
Department of Technical Service, SBEI HE Nizhniy Novgorod State Engineering and Economic University, 22a Oktyabrskaya Street, Knyaginino 606340, Russia

Vladimir Nechayev
SBEI HE Nizhniy Novgorod State Engineering and Economic University, 22a Oktyabrskaya Street, Knyaginino 606340, Russia

Petr Savinykh
Federal State Budget Scientific Institution. Federal Agricultural Research Centre of the North-East named after Rudnitskiy N.V. 166a Lenin Street, Kirov 610007, Russia

Dorota Plewik and Małgorzata Tokarska-Rodak
Pope John Paul II State School of Higher Education in Biala Podlaska, Sidorska 95/97, 21-500 Biala Podlaska, Poland

Agnieszka Latawiec and Bernardo Strassburg
International Institute for Sustainability, Estrada Dona Castorina 124, Horto, Rio de Janeiro 22460-320, Brazil
Department of Geography and the Environment, Rio Conservation and Sustainability Science Centre, Pontifical Catholic University of Rio de Janeiro, Rio de Janeiro 22451-900, Brazil
Department of Production Engineering, Logistics and Applied Computer Science, Faculty of Production and Power Engineering, University of Agriculture in Kraków, Balicka 116B, 30-149 Kraków, Poland

Paweł Sobczak
Department of Food Engineering and Machines, University of Life Sciences in Lublin, Akademicka 13, 20-950 Lublin, Poland

Dariusz Andrejko, Franciszek Kluza and Leszek Rydzak
Department of Biological Bases of Food and Feed Technologies, University of Life Sciences in Lublin, 20-612 Lublin, Poland

Zbigniew Kobus
Department of Technology Fundamentals, University of Life Sciences in Lublin, 20-612 Lublin, Poland

Weronika Kruszelnicka, Robert Kasner, Patrycja Bałdowska-Witos, Józef Flizikowski and Andrzej Tomporowski
Department of Manufacturing Techniques, Faculty of Mechanical Engineering, University of Science and Technology in Bydgoszcz, 85-796 Bydgoszcz, Poland

Katarzyna Piotrowska
Faculty of Mechanical Engineering, Lublin University of Technology; 20-618 Lublin, Poland

Marcin Niemiec
Department of Agricultural and Environmental Chemistry, Faculty of Agriculture and Economics, University of Agriculture in Krakow, Mickiewicz Ave. 21, 31-120 Krakow, Poland

Monika Komorowska
Department of Vegetable and Medicinal Plants, Faculty of Biotechnology and Horticulture, University of Agriculture in Krakow, 54 29 Listopada Ave., 31-426 Krakow, Poland

Anna Szeląg-Sikora and Jakub Sikora
Institute of Agricultural Engineering and Informatics, University of Agriculture in Krakow, 116b Balicka St., 30-149 Krakow, Poland

Zofia Gródek-Szostak
Department of Economics an Organization of Enterprises, Cracow University of Economics, 27 Rakowicka St., 31-510 Krakow, Poland

Joanna Kapusta-Duch
Faculty of Food Technology, Department of Human Nutrition, University of Agriculture in Krakow, 122 Balicka St., 30-149 Krakow, Poland

Tomasz Jakubowski
Department of Machinery, University of Agriculture, 30-149 Krakow, Balicka 116B, Poland

Jolanta B. Królczyk
Faculty of Mechanical Engineering, Opole University of Technology, 5 Mikolajczyka Street, 45-271 Opole, Poland

Jacek Mazur, Kazimierz Zawiślak and Marian Panasiewicz
Department of Food Engineering and Machines, University of Life Sciences in Lublin, 20-612 Lublin, Poland

Wioletta Żukiewicz-Sobczak
Department of Public Health, Pope John Paul II State School of Higher Education in Biala Podlaska, 21-500 Biala Podlaska, Poland

Jolanta Królczyk
Faculty of Mechanical Engineering, Opole University of Technology, 45-271 Opole, Poland

Jerzy Lechowski
Department of Biochemistry and Toxicology, University of Life Sciences, 20-950 Lublin, Poland

Agata Blicharz-Kania
Department of Biological Bases of Food and Feed Technologies, University of Life Sciences in Lublin, Głęboka 28, 20-612 Lublin, Poland

Petr A. Savinykh and Alexey Y. Isupov
Federal Agricultural Research Centre of the North-East Named after N.V. Rudnitskiy, Kirov 610007, Russia

Andrey V. Palichyn and Ilya I. Ivanov
FSBEI HE Vologda State Dairy Farming Academy (DSFA) named after N.V. Vereshchagin, Vologda—Molochnoye 160555, Russia

Index

Printed in the USA
CPSIA information can be obtained
at www.ICGtesting.com
JSHW050845251023
50683JS00018B/72

9 781647 40418